Perspectives on Global Biodiversity Scenarios and Environmental Services in the 21st Century

Naveen Kumar Chourasia
Government College, Bichhua, India

Kavita Chahal
Government College, Bichhua, India

A volume in the Advances in Environmental Engineering and Green Technologies (AEEGT) Book Series

IGI Global
PUBLISHER of TIMELY KNOWLEDGE

Published in the United States of America by
IGI Global
Engineering Science Reference (an imprint of IGI Global)
701 E. Chocolate Avenue
Hershey PA, USA 17033
Tel: 717-533-8845
Fax: 717-533-8661
E-mail: cust@igi-global.com
Web site: http://www.igi-global.com

Library of Congress Cataloging-in-Publication Data

Names: Chourasia, Naveen Kumar, 1979- editor. | Chahal, Kavita, 1979- editor.
Title: Perspectives on global biodiversity scenarios and environmental services in the 21st century / edited by Naveen Kumar Chourasia, Kavita Chahal.
Description: Hershey, PA : Engineering Science Reference, [2023] | Includes bibliographical references and index. | Summary: "Perspectives on Global Biodiversity Scenarios and Environmental Services in the 21st Century makes biodiversity scenarios understandable, relevant, and valuable to stakeholders by using effective language and focused communication techniques. Instead of merely showing the potential effects of global change on biodiversity, scenarios should consider the feedback connecting environmental forces. Biodiversity provides numerous essential environmental services crucial to human well-being both now and in the future. The climate is a critical component of ecosystem functioning and directly and indirectly affects human health. The target audience includes biology and environmental science students and faculty, scientists, social workers who generate and collaborate on biodiversity scenarios, policymakers, and corporations with a basic science understanding"-- Provided by publisher.
Identifiers: LCCN 2023019971 (print) | LCCN 2023019972 (ebook) | ISBN 9781668490341 (hardcover) | ISBN 9781668490358 (paperback) | ISBN 9781668490365 (ebook)
Subjects: LCSH: Biodiversity--Monitoring. | Biodiversity--Climatic factors. | Biodiversity--Forecasting.
Classification: LCC QH541.15.B56 P 2023 (print) | LCC QH541.15.B56 (ebook) | DDC 333.95--dc23/eng/20230614
LC record available at https://lccn.loc.gov/2023019971
LC ebook record available at https://lccn.loc.gov/2023019972

This book is published in the IGI Global book series Advances in Environmental Engineering and Green Technologies (AEEGT) (ISSN: 2326-9162; eISSN: 2326-9170)

British Cataloguing in Publication Data
A Cataloguing in Publication record for this book is available from the British Library.

Advances in Environmental Engineering and Green Technologies (AEEGT) Book Series

ISSN:2326-9162
EISSN:2326-9170

Editor-in-Chief: Sang-Bing Tsai, Zhongshan Institute, University of Electronic Science and Technology of China, China & Wuyi University, China, Ming-Lang Tseng, Lunghwa University of Science and Technology, Taiwan, Yuchi Wang, University of Electronic Science and Technology of China Zhongshan Institute, China

MISSION

Growing awareness and an increased focus on environmental issues such as climate change, energy use, and loss of non-renewable resources have brought about a greater need for research that provides potential solutions to these problems. Research in environmental science and engineering continues to play a vital role in uncovering new opportunities for a "green" future.

The **Advances in Environmental Engineering and Green Technologies (AEEGT)** book series is a mouthpiece for research in all aspects of environmental science, earth science, and green initiatives. This series supports the ongoing research in this field through publishing books that discuss topics within environmental engineering or that deal with the interdisciplinary field of green technologies.

COVERAGE

- Contaminated Site Remediation
- Green Transportation
- Radioactive Waste Treatment
- Renewable Energy
- Alternative Power Sources
- Water Supply and Treatment
- Electric Vehicles
- Pollution Management
- Policies Involving Green Technologies and Environmental Engineering
- Air Quality

IGI Global is currently accepting manuscripts for publication within this series. To submit a proposal for a volume in this series, please contact our Acquisition Editors at Acquisitions@igi-global.com or visit: http://www.igi-global.com/publish/.

The Advances in Environmental Engineering and Green Technologies (AEEGT) Book Series (ISSN 2326-9162) is published by IGI Global, 701 E. Chocolate Avenue, Hershey, PA 17033-1240, USA, www.igi-global.com. This series is composed of titles available for purchase individually; each title is edited to be contextually exclusive from any other title within the series. For pricing and ordering information please visit http://www.igi-global.com/book-series/advances-environmental-engineering-green-technologies/73679. Postmaster: Send all address changes to above address.

Titles in this Series

For a list of additional titles in this series, please visit:
http://www.igi-global.com/book-series/advances-environmental-engineering-green-technologies/73679

Palynology and Human Ecology of Africa
Emuobosa Akpor Orijemie (Department of Archaeology and Anthropology, University of Ibadan, Nigeria) and Sylvester Onoriode Obigba (Palynology Laboratory, Department of Archaeology and Anthropology, University of Ibadan, Nigeria)
Engineering Science Reference • © 2024 • 300pp • H/C (ISBN: 9781668478011) • US $265.00

Nematode-Plant Interactions and Controlling Infection
Waleed Fouad Abobatta (Horticulture Research Institute, Agriculture Research Center, Egypt) and Rehab Yasin Ghareeb (Plant Protection and Biomolecular Diagnosis Department, Arid Lands Cultivation Research Institute, City of Scientific Research and Technological Applications, Egypt)
Engineering Science Reference • © 2023 • 289pp • H/C (ISBN: 9781668480830) • US $240.00

Promoting Sustainable Management Through Technological Innovation
Supriya Lamba Sahdev (ISBR Business School, India) Chitra Krishnan (Amity University, Noida, India) and Ahdi Hassan (Global Institute for Research Education and Scholarship, India)
Engineering Science Reference • © 2023 • 293pp • H/C (ISBN: 9781668499795) • US $300.00

Environmental and Technological Aspects of Redox Processes
Gheorghe Duca (Institute of Chemistry, Moldova State University, Moldova) and Ashok Vaseashta (International Clean Water Institute, USA)
Engineering Science Reference • © 2023 • 435pp • H/C (ISBN: 9798369305126) • US $295.00

For an entire list of titles in this series, please visit:
http://www.igi-global.com/book-series/advances-environmental-engineering-green-technologies/73679

IGI Global
PUBLISHER of TIMELY KNOWLEDGE

701 East Chocolate Avenue, Hershey, PA 17033, USA
Tel: 717-533-8845 x100 • Fax: 717-533-8661
E-Mail: cust@igi-global.com • www.igi-global.com

Table of Contents

Detailed Table of Contents

for just 6.4% of the earth's surface. On a global scale, the wetlands and freshwater ecosystems are under pressure and are seeing declining species diversity. Freshwater biodiversity has not been part of the ongoing mainstream conservation efforts and has been largely ignored. This is in part due to invasive human activities, pollution, climate change, deforestation, habitat destruction and degradation, the introduction of alien competing species, and excessive consumption. This chapter gives a broad overview of the current scenario of declining freshwater species biodiversity and threats to vulnerable aquatic ecosystems. Delineation of biodiversity hotspots, adoption of a proactive approach, and focused conservation efforts towards freshwater ecosystems are some of the suggested remedial measures to address this global problem.

The 15th meeting of the Conference of the Parties (CoP 15) to the Convention on Biological Diversity held in Montreal, Canada began a new era of global biodiversity leadership. This summit brought awareness that the member states have not been able to satisfactorily achieve all 20 Aichi targets. Though it varies from state to state, progress to protect biodiversity around the world is too slow, and the level of depletion is critical. Considering the concerns raised, a new agreement has been signed. This agreement aims to protect 30% of the planet's land, coastal areas, and inland waters by 2030; halve food waste; and achieve the vision of living in harmony with nature by 2050. A monitoring and evaluation framework has also been set out. COP 15 launched a new era for a bright future for biodiversity that needs renewed leadership. The new vision, however, is born with a lag in funding. The renewed leadership also requires retailored governance functions: narrative and high-level direction-setting, integration, mainstreaming, coordination, stakeholder engagement, accountability, etc.

The rapidly increasing world population creates more need for shelter, transportation, energy, and nutrition, leading to the construction of new settlements and agricultural areas, making habitats fragmented, endangering many species and a significant portion of local populations. Biodiversity is adversely affected by the deterioration of the functioning, and the integrity of habitat ecosystems. Insects, which constitute

Shorebirds, or Waders, are called sentinels of global environmental change. Many of these species breed in the arctic and subarctic and migrate back to tropical and temperate regions in their non-breeding seasons. Presently, more than half of shorebird populations with documented trends are in decline and migratory species are more inclined to be impacted. The extent and taxonomic range of the shorebird populations' decreases, along with facts showing that the Arctic is experiencing a higher rate of temperature rise than the rest of the world, indicate global warming as the probable cause of these declines. Primary factors responsible for their decline are degraded habitat quality, which has been influenced by anthropogenic activities and climate change. This highlights the necessity to pinpoint the drives of challenges in order to implement successful conservation measures. In this chapter, the authors aimed to understand the perspective on how and where the habitats of shorebirds are changing and factors responsible for their declines.

Preface

As editors of *Perspectives on Global Biodiversity Scenarios and Environmental Services in the 21st Century,* it is our pleasure to introduce this comprehensive reference book to readers from diverse backgrounds and interests. Our journey into this work began with an acknowledgment of the profound transformations our planet is undergoing, particularly in the context of biodiversity and the environmental services it provides. Our combined expertise, both in academia and our roles as environmental educators, inspired us to curate this volume, drawing from the knowledge of countless experts from around the world.

In an era defined by unprecedented ecological shifts, the term "biodiversity" resonates as a critical and multifaceted concept. It encompasses the intricate relationships among genotypes, species, communities, ecosystems, and biomes, as outlined by the Convention on Biological Diversity. The intricate web of life on Earth, which supports not only our ecological systems but also the well-being and resilience of human societies, is now facing challenges of unprecedented magnitude.

The past half-century has witnessed a significant decline in biodiversity, and the implications of this decline are far-reaching. It affects our ability to achieve sustainable development, threatens societal resilience, and has direct consequences for the health and well-being of individuals. In *Perspectives on Global Biodiversity Scenarios and Environmental Services in the 21st Century*, we explore these issues in depth, providing insights into the interplay between ecological services and biodiversity.

This book is a collaborative effort, pooling the knowledge and perspectives of a vast network of biodiversity experts, researchers, academicians, and students who have dedicated themselves to studying diverse taxa and habitats across terrestrial, freshwater, and marine ecosystems. Their collective wisdom and expertise have allowed us to compile a resource that not only informs but also inspires proactive policy and management plans. Our goal is to guide our readers through the complex landscape of global biodiversity and its multifaceted scenarios, offering a comprehensive overview of the challenges and opportunities that lie ahead.

Perspectives on Global Biodiversity Scenarios and Environmental Services in the 21st Century is intended for a broad audience. It serves as a valuable resource for students at all levels of environmental studies, from school to postgraduate courses. Additionally, it caters to the needs of professionals in environmental industries, environmentalists, biodiversity experts, researchers, and academicians. In fact, we believe that anyone seeking a deeper understanding of the changes in global biodiversity projected for the 21st century will find this book both informative and thought-provoking.

The chapters in this book cover a wide array of topics, each delving into specific facets of global biodiversity scenarios and environmental services. From historical perspectives and cxtinction rates to the effects of climate change and the pressures on biodiversity, our aim is to provide a comprehensive view of the challenges and opportunities ahead. We invite you to explore the rich tapestry of insights offered within these pages, with the hope that it will empower you to engage with these critical issues and contribute to the ongoing global dialogue on biodiversity conservation and sustainability.

ORGANIZATION OF BOOK

Chapter 1

Freshwater resources are vital for human well-being, supporting agriculture, energy production, transportation, and more. In this chapter, we explore the close connection between these essential ecosystem services and biodiversity. We delve into how increasing human population and economic development have transformed natural water systems and the profound implications for both biodiversity and sustainable human life.

Chapter 2

Freshwater, a precious but limited resource, sustains life on our planet. Yet, it constitutes only a small fraction of Earth's water. This chapter underscores the critical state of freshwater biodiversity, often overlooked in mainstream conservation efforts. We discuss the threats facing aquatic ecosystems, including pollution, habitat destruction, climate change, and invasive species, and suggest proactive measures to address this global concern.

Chapter 3

The 15th Conference of the Parties (CoP 15) to the Convention on Biological Diversity marked a turning point in global biodiversity conservation. In this chapter, we analyze the outcomes of this pivotal summit, which highlighted the inadequacy of progress in protecting biodiversity worldwide. We explore the new agreement aiming to safeguard 30% of the Earth's land and waters by 2030 and offer insights into the leadership required for its success.

Chapter 4

Rapid population growth and urban development are fragmenting habitats, endangering species, and disrupting ecosystems. Insects, a crucial part of global biodiversity, play essential roles in the food chain. In this chapter, we focus on the unique insect diversity found in Anatolia, examining the factors contributing to its richness and the threats it faces, which reflect broader global challenges.

Chapter 5

This chapter explores the dynamic relationship between environmental regulation and the promotion of socio-ecological and green technological innovation within organizations. By examining the impact of regulatory incentives, we provide insights into how environmental policies influence the development of sustainable practices and technologies.

Chapter 6

Forests, as the natural habitat of diverse plant and animal species, are dwindling worldwide. This chapter underscores the importance of forest conservation in maintaining life on Earth, providing clean air, shelter, and a sanctuary for a vast array of species. We explore the vital role forests play in safeguarding terrestrial biodiversity and improving socioeconomic aspects of regions.

Chapter 7

Shorebirds, often referred to as sentinels of environmental change, are facing declines in their populations. This chapter investigates the reasons behind these declines, emphasizing the role of climate change, habitat degradation, and anthropogenic activities. By understanding the challenges faced by these migratory birds, we pave the way for effective conservation measures.

Chapter 8

This chapter explores the intersection of urban anthropology and Anthropocene urban socio-ecology planning resilience. We examine how urban anthropology can inform and strengthen urban planning efforts in the Anthropocene era. Through ethnographic interpretations and a review of theoretical and empirical literature, we provide insights into the foundation for resilient, ecologically conscious urban development.

We, the editors, wish to express our sincere gratitude to all the contributors whose dedication and expertise have made this work possible. It is our hope that *Perspectives on Global Biodiversity Scenarios and Environmental Services in the 21st Century* will serve as a valuable resource and a source of inspiration for all who share our commitment to preserving our planet's biodiversity for future generations.

Naveen Chourasia
Government College, Bichhua, India

Kavita Chahal
Government College, Bichhua, India

Chapter 1
Changes in Freshwater Systems Management and Governance:
A Case Study From Mandla Madhya Pradesh

Bhavna Bawra
https://orcid.org/0009-0001-3020-2557
Government Degree College, Bhua-Bichhiya, India

Manohar Pawar
Foundation for Ecological Security (FES), India

ABSTRACT

Freshwater resources provide a broad set of environmental and ecosystem services which are essential for healthy living of humans and their food and energy production, social and economic stability. This system is also critical for biodiversity, for example providing habitat for many invertebrate and vertebrate species, aquatic animals, and fish species. Along with the increasing human population and socioeconomic development levels have resulted in rapid development of water resources and the replacement of naturally occurring and functioning systems with highly modified and human-engineered systems. Humans need the freshwater provisioning services such as irrigation, domestic water, power production, temperature maintenance and transportation which come from inland water ecosystems such as rivers, lakes, ponds and wetlands. Freshwater ecosystem is directly associated with terrestrial ecosystem through recreation, scenic values, fisheries and biodiversity preservation and ecosystem function, which is essential for the sustainable development of human life.

DOI: 10.4018/978-1-6684-9034-1.ch001

1. INTRODUCTION

Fresh water is a basic and essential need of all the living organisms. it is a renewable, variable resource and predominant for all the ecological services it offers. This system is also an important part of biodiversity, for example providing habitat for many invertebrates and vertebrate species, aquatic animals and fish species. Climate change affects the water availability in many regions and can harm the health of people and ecosystems. For example, rising temperatures, drought, and reduced snowfall are putting more pressure on water supplies in which can put water infrastructure (such as dams, sewers, and water treatment facilities) at risk (Lall, U., et al. 2018).

Human activities like emission of greenhouse gases that contributed the Global warming which has direct and widespread impact on the hydrological cycle and on the aquatic ecosystems in different parts of the world (Huntington 2006; Oki and Kanae 2006). Freshwater ecosystems are highly stressed due to these climatic changes like eutrophication, species invasion, change in rainfall pattern and rising temperatures. Concerning environmental change, attention has increasingly shifted from localized problems and phenomena to global and regional issues requiring global collaboration. Changes in the environment and sustainability are inextricably linked to social, economic, and political developments. Population growth, economic development and globalization, industrialization, and urbanization are the underlying driving forces of environmental change, all of which have profound effects on the physical environment and the organization of human space (MEA 2005). These factors appear in a variety of ways at the global, regional, national, and local levels. Environmental issues are increasingly trans-boundary, with seemingly local phenomena having global implications and global events having regional and local ramificationsAs a result, environmental management frequently necessarily requires international collaboration (USGCRP 2014). Due to the changing temperature and heat the quantity and quality of water is also affecting which is severely influencing the aquatic habitat and some species are not able of adopt in this changing water environment.(Shuter and Meisner 1992), the uncertain or excess rainfall, increases the velocity of water flow and rise in water level causes the displacement of the sediments and mix other unusefull in aquatic system and cause the disturbances in natural habitat of aquatica animals, making them vulnerable and losses of many fishes. (Pujolaretet al.2011).

India is gradually moving away from being a water-rich country due to increasing population pressure and urbanization. It currently supports 18% of the world's population while using only 4% of global water resources. As a result, water resource management has taken on new significance. Today, water resource availability is a major issue and a major challenge for our country (Water resources in India n.d). Water bodies are an integral part of fresh water resources. The landscape of India is dotted with large number of water bodies. Traditionally, these water bodies have played an important role in supply of drinking water, for domestic use and agriculture purposes etc. They have been the major source of minor irrigation system for agriculture in India. Through the ages, these water bodies, either natural or manmade such as lakes, tanks, ponds and similar structures have sustained the Indian agriculture. In urban areas also water bodies play an important role as a source of drinking water, absorption of flood water and a conduit for ground water recharge. Therefore, it is essential to assess where freshwater resources exist, how they are used, and how climate, technology, policy, and people can play important role in conservation/ restoration of these water bodies for healthy and sustainable development.

In Madhya Pradesh the main source of replenishable ground water resources is recharge from rainfall, which contributes to nearly 77% of the total annual ground water recharge. A major part of the state receives rainfall mainly during SW Monsoon season spread over the months of mid-June to September. Over 99% of the annual rainfall is received in the four rainy months for June to September only thereby leading to large variations on temporal scale. Agriculture is the main stay of the people of Madhya Pradesh(GWRA MP Final 2020).

2. MANDLA MADHYA PRADESH AND NARMADA RIVER

Mandla District is situated in the east part of Madhya Pradesh. The district lies almost entirely in the catchment of river Narmada & its tributaries. Mandla District extends over the highest plateaus of the Satpura ranging from 500 meters to 500 meters above mean sea level (Geography Mandla district n.d)

2.1 Physiography of Narmada River in MP

The Narmada basin, which includes the north-eastern portion of the sub-basin and the south-western (Wainganga sub-basin) and south-eastern extend (Seonath sub-basin) of the Godavari basin, occupies a significant portion of the district. The region is a generally easterly gradient. The central Indian highlands to the north, the Satpura-Maikal hill ranges to the south, and the eastern plateau to the southeast define the terrain of the entire statetogether with the Narmada-Son valley.

The Maikal ranges, which are the eastern, broader, and semicircular portion of the Satpuras, serve as a watershed between western and eastern India. The Amarkantak plateau, where the Narmada and Son rivers originate, is one of these ranges.

The Narmada basin has varying topography, with the uppermost part of the terrain being hilly, giving rise to the largest west flowing river Narmada at Amarkantak. As the river flows through the middle section, the terrain changes to a fertile plain.

The coastal region of the basin is where the river meets the Arabian Sea. (NVDA 1985;Narmada Basin n.d)

The basin is divided into three sub-basins. The Narmada Upper, Middle, and Lower Sub-basins have a total of 150 watersheds. Along with the main river Narmada, (CGWB report Mandla n.d) important tributaries drain an area of 92,672.42 Sq.km, or 3% of the country's total geographical area.

- **Climate**

Because the district is located in Satpura's highest plateaux, which range in altitude from 443 to 887 m, it is cooler than the low-lying plains of Raipur and Jabalpur to the north and south, respectively. The hot summers, high rainfall of 1,400–1,600 mm, and generally dry winters make up the district's climate.

- **Forest**

Two main types of forest can be seen in the Bichhiya block of the East Mandla Forest Division: teak forests and moist and dry mixed deciduous forests. Patches of Sal woodland can be seen in close to the Reserve's central area. There are mixed deciduous forests which are located in the Buffer Zone of the Kanha Tiger Reserve, but barren of Teak and Sal, with a few areas of moist and dry mixed deciduous trees along stream banks.

- **Narmada basin**

The Narmada basin is divided into four Agro-climatic regions: Eastern, Central, Western plateau hill region, and Gujarat plains. Agro-ecologically, the basin is divided into two zones: hot sub humid eco-regions with red and black soils, and hot semi-humid arid eco-regions with medium and deep black soils. The basin is divided in 3 Sub-basins viz. Narmada Upper, Narmada Middle and Narmada Lower Sub basin.

The drainage network of Narmada River consists of 41 major and minor tributaries. The Narmada Upper Sub-basin covers maximum area of the basin accounting to 46.61% of the basin area. The Narmada Middle and Narmada Lower Sub-basin covers 43.78% and 9.61% respectively of the total geographical area of the basin. Total 71 watersheds coming under upper catchment of the river and in Mandla we are studying 4 sub basin areas.

In Narmada basin, Surface water bodies spreads over 2919.32 Sq.km, which is around 2.95% of the total geographical area of the basin (Source: River Basin Atlas of India,2012). There are a total of 13994 surface water bodies. The number of Tanks is 12529 which are 91.83% of the total surface water bodies in the basin. Several large reservoirs of considerable size have also been constructed in the Narmada Basin. Some of the large reservoirs in this basin are Indira sagar, Bargi, Tawa, Sardar Sarovar, Jobat, Kolar Reservoir etc (District Statistical Handbook, Mandla).

2.2 Upper Catchment of Narmada River

The Vindhyanchal Baghelkhand region includes the upper hilly areas and plains. The region is a hill-valley complex that includes the Madhya Pradesh districts of Annupur, Shahdol, Dindori, Mandla, Balaghat, Jabalpur, Narsimhapur, and Chhindwara.

Figure 1. Location of the area

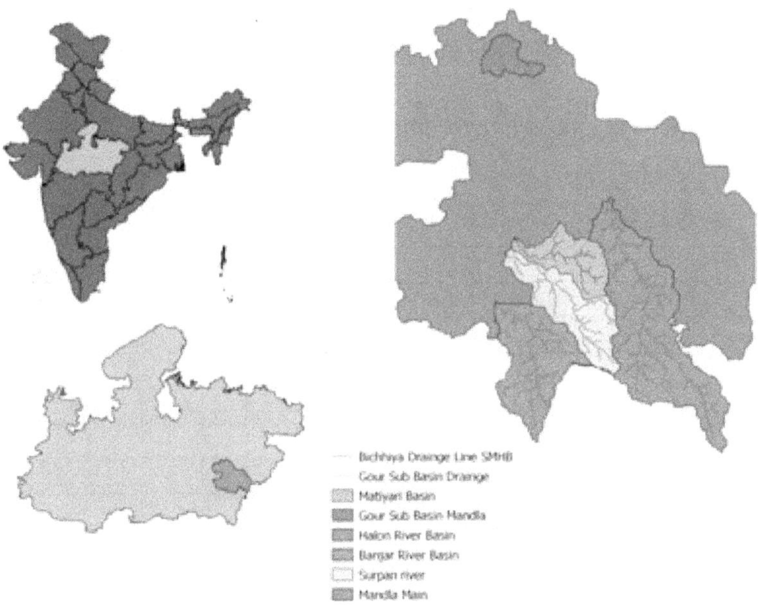

Bichhiya Dam on Kathar Nala, Halon Dam on Halon River, Matyari Dam on Matiyari Nala, and Mand tank are the names of the three main irrigation tanks (Figure: 1). These large bodies of water are crucial sources for agricultural irrigation. In addition to these water features, there are several small water bodies in villages that are mostly utilized for drinking water, cattle drinking, domestic use, wildlife, and very little for irrigation. Many rivers that are upper catchments of the Narmada River of Bichhiya block have its origins in the Kanha Tiger Reserve Forest. These preserved green spaces have a considerable impact on the tributaries of Matiyari, Halon, Banjar, and Surpan, as well as surface water bodies and minor streams.

Catchments that had more upstream forest cover were shown to be able to maintain flow longer into the dry season. According to the "infiltration 97 evapo-transpiration trade-off" hypothesis proposed by Bruinjzeel, which states that differences in evapo-transpiration and infiltration between land cover and groundwater recharge are interlinked, less forest cover resulted in low-flows in dry season, so hydrologic responses response correlated with catchment area forest cover and habitat (Bruinjzeel 2004). Although significant recharging occurred under all three habitats and natural forests provided the most temporally stable groundwater recharge, which supports base flow in rivers of higher order downstream.

As per 2007 land records data of total land area of Bichhiya block is67735 haand forest area is 20996 ha, which is 31% of the total land area as compares to other blocks of the district. The details of total area and forest area are given below in Table-1.

Table 1. Block wise and land wise total area of Mandla district

	Total Area	Land under Forest	Net Sown Area	Non-Agricultural Land	Barren land	Other Uncultivable Land	Untapped Cultivable Land
Nainpur	69115	10010	44662	7995	10451	3030	3698
Mandla	54218	10544	39730	6667	4439	2428	2189
Mohgaon	32780	5168	33238	3830	3597	1805	946
Ghughri	65477	15494	24046	6054	11173	2433	2732
Bichhiya	67735	20996	38639	5030	5106	4070	1409
Mawai	53104	9378	31370	5183	7357	2750	1932
Niwas	40052	4225	20652	3682	7746	1224	2711
Bijadandi	44821	11621	23192	7826	5505	1205	2366
Narainganj	39854	7248	26564	7066	6458	781	1722
Overall District	**965559**	**593082**	**282093**	**53333**	**61832**	**19726**	**19705**

Surface water bodies have traditionally played an important role in the lives of common people in India by way of irrigation, drinking water supply, ecology, tourism and domestic uses.

Figure 2. Drainage map showing the different river basin with Kanha Tiger Reserve

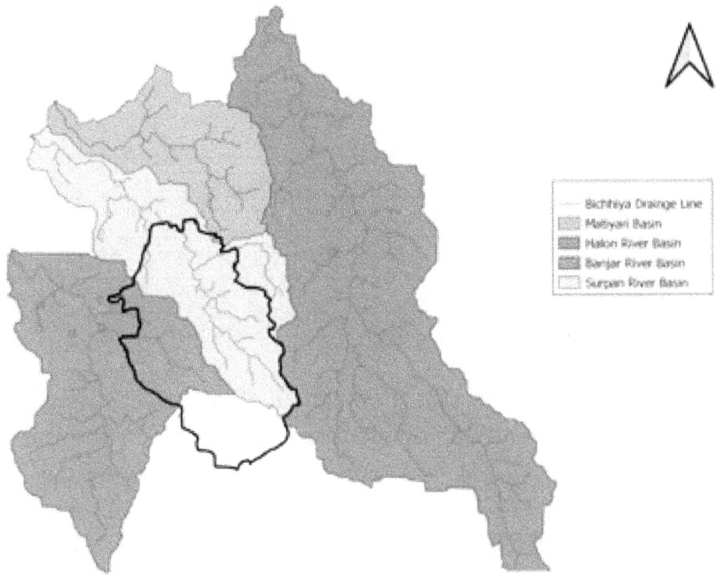

Due to undulating terrains and hills the distribution of water of irrigation tanks,it is bit difficult to reach in the all area of its catchments and communities living in areas where water resources are scarce, such as those dependent on small farming fields, are particularly vulnerable to water-related challenges. These communities often rely heavily on agriculture for their livelihoods and may not have access to alternative sources of income. Climate change and other environmental factors can exacerbate water scarcity, leading to reduced crop yields and increased competition for water resources.

Additionally, these communities may lack the infrastructure and resources necessary to manage their water resources effectively, such as irrigation systems or water storage facilities. This can further limit their ability to adapt to changes in water availability and can lead to increased vulnerability to water-related shocks and stresses. Mandla district receives 1200-1400 mm rainfall yearly butdue to hilly terrains it flows down quickly. Adequate harvesting structures need to catch this water. In the study area almost every village have a small harvesting structureand water bodies that fulfills the year around domestic and other needs. Efficient managements of these water bodies are ensuring water availability throughout yearTable- 2.

Table 2. Total Number of water bodies in Mandla district

Size Range (Area in ha.)	Number of water bodies	%
0.1 - <=0.5	977	51.94
0.5 - <=1	552	29.35
1 - <=2	215	11.43
2 - <=10	89	4.73
10 - <=50	42	2.23
50 - <=100	1	0.05
> 100	5	0.27

https://www.indiawris.gov.in/

As per National Water Informatics CentreOut of the 65941 surface water bodies recorded in Mandla, 1881 hold 2.85% of the total resources for surface water harvesting. Small water bodies are 977 (51.94%) and contribute significantly in range of 0.1 to 0.5 hector, whereas 0.5 to 1 hector contribute 552 (29.35%) water bodies in the district. Small water bodies are 977 (51.94%) and contribute significantly in range of 0.1 to 0.5 hector whereas 0.5 to 1 hector contribute 552 (29.35%) water bodies in the district.(Figure- 3)

The data shows that number of small water bodies is higher which contribute more for domestic use, cattle care and drinking purposes. The management of these small water bodies is very crucial to supply the fresh water throughout year, so the governance and institutionalsetup play very important role for fresh water supply at village level.

Figure 3. Range distributions (area in hectors) of fresh water bodies in the Mandla district

A wide and varied collection of case studies from Bichhiya Block, Mandla District, Madhya Pradesh are described here, despite the fact that many patches still require special attention. The community's involvement in managing and restoring freshwater resources was examined in these case studies.

3. INSTITUTIONAL GOVERNANCE, MANAGEMENT, AND RIVER BASIN ORGANIZATIONS (SMALL AND LARGE WATER BODIES)

- **Case study: 1. Management practices of Kurwahi Village**

Kurwahi is a small village, like many other tribal villages in the Kanha Tiger Reserve's buffer zone. During the summer months, and even in the early summer, the village faces severe water shortages. The village has a small pond that is traditionally used for irrigation and other domestic purposes. Due to the undulating terrain and limited resources, the water available in the pond is insufficient to meet the needs of the villagers. In 2015, the village institution (Panchayat and ad hoc committees) took over governance of this water body and established some rules for the use of water and its resources. After one year, it was decided by the villagers to begin fisheries, and management rules helped to not only secure water availability until monsoon, but also to provide food security and incentives from the small pond. In this way the villagers and institutions were encouraged to conserve and manage their water bodies in better way. (Figure –4)

Figure 4. Community fisheries not only securing food but also ensure fresh water availability year-round

- **Case Study: 2. Management practices Changariya Villages**

Changariya village is located in Bichhiya block of Mandla district in Madhya Pradesh, India. It is situated 17km away from sub-district headquarter Bichhiya (Block office) and 47km away from district headquarter Mandla. As per 2011 Census, Changariya is the gram panchayat of Changariya village.

The total geographical area of village is 483.66 hectares. Changariya has a total population of 681 peoples, out of which male population is 351 while female population is 330. Literacy rate of Changariya village is 46.70% out of which 54.42% males and 38.48% females are literate. There are about 149 houses in Changariya village.

Figure 5. Restoration of fresh water resources for the benefit of the entire village

Until 2016, the larger pond was primarily used for irrigation by two nearby farmers and by a few others for households and domestic purposes. The rest of the villagers had limited access to this fresh water resource. Due to the negligence and lack of maintenance, over a period of time, the ponds got heavily infested with *invasives and slit*. This ignorance was lead to reduction of the water holding capacity of the pond considerably.

The Natural Resource Management Committee (a village level ad hoc committee of Panchayat) sets up for the governance and management of these fresh water resources collectively. All members of the community bear responsibility for the pond's upkeep and use. The institution decided to conduct a community fishery in the pond, and they also established rules for fish harvesting. Every year, the village institution engages in community fishing in the pond. The smaller pond has also come to be governed and maintained in a similar manner. The efforts have resulted in the restoration of the ponds and the enrichment of the people's diet. It has also revived the community together as a collective and ensured that the fresh water resources can be maintain for a longer time to give environmental, social and financial benefit to the entire village.

- **Case Study: 3. Management practices of Matiyari Dames**

In hilly or uneven terrain, it was difficult to use water efficiently. After government initiative anearthen dam was proposed to fulfill the requirement of irrigation and other purposes in command area villages. The large earthen dam was built in near Kharpariya villagee but the communities in lives in catchment area are heavily dependent previously on it for livelihood had serious concern, so the community members of nearby village (Changariya, Chilpi, Kharpariya, Gadiya and Umariya) decided to caught fishes from the Matiyari dam with some rules and regulation.This step was not only to resolve the conflict between communities, but also to make it beneficial for wildlife as well. This freshwater resource provides not only economic and livelihood benefits but also helping to reduce the pressure on catchment area forest resources. The Matiyaridam is connected with many small canals and used to fill many small tanks which additional supports and provide economic incentives with irrigation and wildlife support in their Command area. The governance and management of these small tanks will further help and support for sustaining fresh water ecosystems.

- **Case Study: 4. Management Practices of Kumharrah Village**

Kumharra village is situated in the Mandla district's Bichhiya block. It is located 37km from Mandla, the district headquarters, and 31km from Bichhiya, the sub-district headquarters and block office. According to statistics from 2009, Kanskheda is the gram panchayat for Kumharra village. The village's entire geographic area is 358.88 hectares. Kumharra is home to 232 individuals in total, 111 of them are men and 121 are women. Kumharra village has a 56.90% literacy rate, with 72.97% males and 42.15% females being literate. In the village of Kumharra, there are roughly 52 houses.

The tribal dominated village primarily depended on the agriculture and forest resource for their livelihood. Like other villages Kumharrah village have small pond in downstream of forest. The rich forest is supporting and fulfilling the requirement of water of the villagers but due to climate change and erratic rainfall the pond was not filled with full capacity. There were no management rules and it was covered with lantana camera and other aquatic invasives. The villagers realize the situation and collectively approach to Panchayat authority for conservation and management of the pond. Later the pond was deepened by gram panchayat officials and the fish culture was allowed. So, by the Community efforts the dry pond became alive and also useful for various domestic needs, irrigation and wildlife.

4. GOVERNANCE: INSTITUTIONS FOR MANAGING SHARED WATERS

Integrated water resource management is a challenge in terms of governance, economics, and obtaining adequate scientific knowledge.There are many conflicting between uses and interests that frequently accompany changes in polices, objectives and management practices.

Incorporation also requires the formation of appropriate and effective institutional arrangements that incentivize stakeholders to cooperate and adhere to management plans at village level and landscape level. Because water is so vital to human society, problems in water management often reflects weak points of the governance and serve as a springboard for broader policy reforms while these organizations have helped to raise awareness of fresh water-related issue and solution.This is a general acknowledgement that there is a significant gap between framework of rules and actual practices. The continue effort to support more effective on-the-ground implementation is always useful, particularly if placed within a structured framework for learning and adaptive management.

Probably, the most significant methodological challenge is developing the capacity for a landscape or watershed-based approach to assessment, which is required to identify ecosystem functions that support the provision of valued ecosystem services in a specific context, as well as to select feasible and appropriate institutional framework.It can also be considered an institutional challenge because the effects on livelihoods and other aspects of well-being are easily overlooked in the absence of effective participation and recognition of the rights of the various stakeholders and local communities on which people depend heavily and identify as potential decision makers.

The desire to determine ecosystem water requirements can be driven by anecdotal observations of a decline in certain ecosystem characteristics and functions or by the results of applied analysis spawned by observation. A change in condition may be detected by local communities (for example, a decline in native fish production and a change in vegetation structure and land use changes) or by government-sponsored surveillance monitoring programs carried out as a legislative requirement.Many of the services provided by fresh water ecosystems are supported by natural variability in spatial and temporal patterns in water distribution, abundance, and quality, as well as interactions between the basin, climate, geology, topography, and vegetation.(Poff et al. 1997; King and Brown 2003, Whiting 2002). Water resource development, such as dam construction and irrigation systems, can provide benefits to provisioning services such as water supply for drinking and agriculture. However, it can also negatively impact the regulating services, supporting services, and cultural services provided by water ecosystems. For example, altered water flow can disrupt the natural regulation of water quality and quantity, leading to decreased biodiversity and ecosystem health. Additionally, the loss of wetlands and other natural habitats can impact cultural services such as recreation and spiritual connections to nature.

Institutions involved in determining ecosystem water requirements are typically those with the necessary governance and technical capacity. These are primarily government policymaking or non-profit organizations seeking to inform advocacy positions. The process will typically take place at the catchment level, but will necessarily involve the active participation of actors ranging from the local to the regional levels.Actual or potential beneficiaries' willingness to pay for freshwater services is determined not only by demand, but also by their confidence in the effectiveness of strategic planning and institutional arrangements required ensuring that the service is delivered and that they will have access to the stream of benefits. Recognition of community right will strengthen the institution that stands behind it. While in some cases, the government has a very strong influence, it does not operate alone.

In many cases, customary or religious law may be more influential. Because of this complexity, changes in state law alone do not automatically translate into changes in property rights on the ground (Meinzen- Dick and Pradhan 2002).In many cases, protection of watersheds has been used by governments or other local authorities as justification for centralized control of land and water resources and regulatory conservation policies that exclude upland communities. This approach tends to be resisted by local populations using various means such as fire, which further reduce the effectiveness of regulatory approaches (Swallow et al. 2001; Blaikie and Muldavin 2004).

5. RIVER BASIN ORGANIZATIONS (RBO)

The concept of river basin level community organization is not new but the Recognition among multiple interests and users of fresh water provided by river basins they lead a trend toward the formation of river basin organizations and strengthen as a vehicle for basin-wide assessment, planning, management, and conflict resolution.

River Basin Organizations may be formed to manage basins within individual villages or Panchayat as well as in block or district or watershed context, which presents an added layer of management challenges and gives rise to sovereignty considerations.Pivotal concerns for basin-wide organizations are typical issues of water allocation among multiples and often conflicting uses.

Decisions about infrastructure development such as dams, reservoirs, and navigation canals are critical because these decisions modify and diverts flow of water that makes possible changes in agricultural irrigation and urban development, and therefore they are a major drivers of land use change. They also block flows of sediment that maintain downstream river channels. Other key concerns at the basin level are with cumulative impacts of land use changes and inputs of pollutants that are detectable over long distances and after a long peroiod of time.

However, such organizations are not entirely new. Among the better-known early examples are the Tennessee Valley Authority, a regional river basin authority formed in 1933, and the French system of water management, organized around river basins, which was established in 1964. The Mekong Committee, predecessor of the Mekong River Commission, was created in 1957, one of the earliest formal River Basin Organizations (RBO) in the developing world (Ratner 2003).

While India has been making efforts towards integrated river basin management, there are still challenges in implementing it on a large scale(Amarsinghe et al 2005). The country has established river basin organizations, such as the Brahmaputra Board and the Ganga River Basin Authority, to coordinate management efforts across different states and sectors. However, there are still gaps in policy implementation and enforcement, as well as a lack of adequate resources for effective management. Additionally, transboundary river basin management remains a challenge due to political tensions and disputes between neighboring village, block, District and state Authorities(Saleth, R.M. 2004). Therefore, while India is making progress towards integrated river basin management, there is still room for improvement and further efforts are needed to address the existing hydrological, ecological, and socio-economic problems in a holistic manner(Shah, Tushar; et. al. 2002)(Raju, KV & Taron, Avinandan. 2019).

River Basin Organizations range is a type of farm who have the authority to plan, promote, and enforce their plans, to those that lack authority but play important roles in an advisory capacity in the assessment needed to inform planning, priority setting, stakeholder negotiations, and decision-making. An example of the Communities in Banjar River basin faces issues with invasive species.

Banjar basin organization collectively include the 36 villages of around Kanha tiger reserve approach to forest department and clean the road site invasive which was further help to reduce human wildlife conflict and also support forest regeneration and ultimately support the freshwater ecosystem. River basin committees, and, at sub-basin tributary levels, by local water committees, come together and preparation of landscape level planning and manage fresh water resources effectively.

In an alternate model, which can be considered an adaptive approach, many RBOs have evolved from a narrow and sectoral focus or working on other issues are also come together and create a multi actor platform which will further help to resolving a fresh water related issues in a comprehensive way. Capacity building thus tends to occur as needed, in response to recognition of new types of problems that are beyond existing response capacities, or in response to extreme events that expose structural weaknesses and bring problems to broad public attention more rapidly.

Regulatory authority tends to rest with existing agencies and political jurisdictions, which may need to be politically compelled to effectively participate, or offered some form of incentive.This kind of RBO tends to play an intermediary role by activities such as convening interested parties for purposes of strategic planning, management, and conflict resolution, preparing master plans, proposing policies, creating an information system, and coordinating monitoring efforts.

There are many emerging local community forums/organizations that evolved in a similar fashion in recent like fisher man groups, as its focus shifted from fisheries to other watershed/ basin level issues in absence of legal authority or failure of system created for their management.

6. MAJOR CHALLENGES

Temperature

In many regions due to the greenhouse effect the temperature rise in air and surface temperatures also increases which leads to variability in temperature and an enhancement in glacial melting (Arnell et al. 2001). High temperature accelerates the rate of evaporation from land, vegetation and oceans and rivers or local water bodies. So, the makes the land dry and reduce the water table. Unfortunately, some areas in India gets less precipitation and rain becomes for short period sometimes cause drought and reduce the ground water for the basic need like drinking and irrigation. This will trigger a vicious cycle of water scarcity.

Precipitation and Water Quantity and Flow Changes

The water cycle is also called the hydrologic cycle. In the water cycle, water from oceans, lakes, swamps, rivers, plants and even from people and animals, is converted into water vapors and released into the atmosphere. When the water droplets in the clouds get too heavy, they fall back on to the Earth - called precipitation. This includes rain, snow and hail, but most precipitation is in the form of rain. More evaporation is causing more precipitation, With more evaporation, there is more water in the air so storms can produce more intense rainfall events in some areas. This can cause flooding – a risk to the environment and human health.

Climate change also leads to significant changes in groundwater recharge. Smaller change in temperature has been observed on those freshwater ecosystems that receive input from groundwater than those dependent on precipitation. seasonal heavy rainfall events already surpass the natural infiltration rates of soil, leading to high sediment input and dangerous levels of pesticide runoff from agricultural lands (Pringle 2000).13

Threat From Invasive Species in India's Forests

Invasive species have changed the forests across India (Singh 2005, Kohli et al. 2006, Hiremath and Sundaram 2005, Kannan et al. 2013). Many tiger reserve are also infested with invasive species. These are not only impacting wildlife and habitat but also effect the fresh water ecosystem which crucial for long term water availability and supply. Sub humid forests are where this is very noticeable. By evapo-transpiration, species like Lantana which account for a sizeable portion of the biomass on a per-unit-area basis may have altered soil moisture regimes. It is vital to look at how lantana affects the relationships between linked plants' carbon and water needs, as well as how it affects the hydrological responses of forests.

Climatic Variability, Climate Change, and Indian Forests

With changes in the significance of ocean-atmosphere processes like ENSO and the Indian Ocean Dipole, the Indian Monsoon systems are undergoing change. Also, during the 1950s, a decline in the monsoon has been noted, despite an increase in rainfall occurrences (Krishnaswamy et al. 2014) According to the some report climate change may impact Kanha Pench Corridor area more than other part of the state. This will affect the ecosystem services provided by the forest adversely.

CONCLUSION

The social and economic aspects of fresh water issues are very significant. In order to manage freshwater resources, an integrated ecosystem approach is required. It is obvious that comprehensive management of watersheds is required given the connections between water resources, land degradation, soil erosion, and deforestation. Frameworks for incentives and policy that are appropriate must be implemented for this goal.Along with new and improved technologies, the necessary tasks also involve institutional development, policy changes, research, education, and capacity building. Improved access to trustworthy data and the development of monitoring technologies that enable us to evaluate the state of affairs and trends in fresh water are other vital tasks. Demand management needs to be prioritized because there are very few opportunities to create significant additional freshwater resources (Schindler et al. 2014; Hornung et al. 2019). However, behavioral change also needs to be a big part of it.The markets can play a significant role in this process. The world's poorest individuals and communities lives in upper catchments of basin are most at risk from water-related issues. While pollution and excessive use may also have an impact on fresh water resources, they are better equipped to handle the issues through technological solutions, regulatory changes, and educational initiatives. Cooperation between researchers and citizen scientists who create and manage the programs to encourage people to test the health of their local waterways and share their findings is important. Improve communication, education and knowledge transfer are key to engage the communities of river basin and help to decision making and management of their fresh water resources efficiently.

ACKNOWLEDGEMENTS

We also thankful to Foundation for Ecological Security Anand for providing necessary facilities during the field work. We also acknowledge the support of the Madhya Pradesh Forest Department Mandla and Govt. Degree College Bichhiya, Mandla.

All maps in this study were created using QGIS version 3.16.8.

REFERENCES

Amarsinghe. (2005). Spatial Variation in Water Supply and Demand across river basins in India, IWMI. *Research Reports (Montgomery), 83*, 2005. http://www.iwmi.cgiar.org/pubs/pub083/RR83_distribution.pdf

Blaikie, P. M., & Muldavin, J. S. S. (2004). Upstream, downstream, China, India: The politics of environment in the Himalayan region. *Annals of the Association of American Geographers, 94*(3), 520–548. doi:10.1111/j.1467-8306.2004.00412.x

Bruijnzeel, L. A. (2004). Hydrological functions of tropical forests: Not seeing the soil for the trees? *Agriculture, Ecosystems & Environment, 104*(1), 185–228. doi:10.1016/j.agee.2004.01.015

Central Water Commission. (2014). *Narmada Basin*. CWC.

Central Water Commission. (n.d.). *Home*. CWC. http://cwc.nic.in/main/webpages/rm/activities.html

CGWB. (n.d.). *Mandla Report*. CGWB. http://cgwb.gov.in/District_Profile/MP/Mandla

CWC. (n.d.). *Climate change and water scarcity in India*. CWC. https://globalwarming.com/articles/climate-change-and-water-scarcity-in-india/

Government of India. (2011). *Census*. Government of India. https://censusindia.gov.in/2011census/dchb/DCHB.html

GWRA MP. (2020). *Assessment*. CGWB. http://cgwb.gov.in/GW-Assessment/GWR-2020-Reports%20State/2022-03-10

Hiremath, A. J., & Sundaram, B. (2013). Invasive plant species in India protected Areas: conserving biodiversity in cultural landscape. In L. C. Foxcroft, P. Pyšek, D. M. Richardson, & P. Genovesi (Eds.), *Plant Invasions in Protected Areas: Patterns, Problems and Challenges* (pp. 241–266). Springer. doi:10.1007/978-94-007-7750-7_12

Hornung, L. K., Podschun, S. A., & Pusch, M. (2019). Linking ecosystem services and measures in river and floodplain management. *Ecosystems and People (Abingdon, England)*, *15*(1), 214–231. doi:10.1080/26395916.2019.1656287

Huntington, T. G. (2006). Evidence for intensification of the global water cycle: Review and synthesis. *Journal of Hydrology (Amsterdam)*, *319*(1-4), 83–95. doi:10.1016/j.jhydrol.2005.07.003

Kannan, R., Shackleton, C. M., & Uma Shaanker, R. (2013). Reconstructing the history of introduction and spread of the invasive species, Lantana, at three spatial scales in India. *Biological Invasions*, *15*(6), 1287–1302. doi:10.100710530-012-0365-z

King, J. M., & Brown, C. (2003). Environmental flows: Case studies. In R. Davis & R. Hirji (Eds.), *Water Resources and Environment: Technical Note C.2*. World Bank.

Lall, U. (2018). Water. In *Impacts, risks, and adaptation in the United States: Fourth national climate assessment* (Vol. II, p. 147). U.S. Global Change Research Program.

MEA. (2005). *Millennium Ecosystem Assessment. Ecosystems and human well-being: biodiversity synthesis*. World Resources Institute.

MEA. (2005). *Millennium Ecosystem Assessment*. Freshwater Ecosystem Services, World Resources Institute.

MEA. (2005). *Millennium Ecosystem Assessment. Fresh water.* World Resources Institute.

Meinzen-Dick, R. S., & Pradhan, R. (2002). *Legal Pluralism and Dynamic Property Rights, Consultative Group on International Agricultural Research system- wide program on collective action and property rights no. 22*. International Food Policy Research Institute.

Narmada Valley Development Authority. (2014) *Government of Madhya Pradesh, Narmada Basin, Narmada Water Dispute.* Nvda.nic.in..

Oki, T., & Kanae, S. (2006). Global hydrological cycles and world water resources. *Science, 313*(5790), 1068–1072. doi:10.1126cience.1128845 PMID:16931749

Poff, N. L., Allan, J. D., Bain, M. B., Karr, J. R., Prestegaard, K. L., Richter, B. D., Sparks, R. E., & Stromberg, J. C. (1997). The natural flow regime. *Bioscience, 47*(11), 769–784. doi:10.2307/1313099

Pujolar, J. M., Vincenz, I. S., Zane, L., Jesensek, D., De Leo, G. A., & Crivelli, A. J. (2011). The effect of recurrent floods on genetic composition of marble trout populations. *PLoS One, 6*(9), e23822. doi:10.1371/journal.pone.0023822 PMID:21931617

QGIS Development Team. (2023). *QGIS Geographic Information System.* Open Source Geospatial Foundation Project.

Raju, K., & Taron, A. (2007). River Basins Organizations in India: An Overview. Conference. *IWMI-TATA, 2007*, 613–646.

Ratner, B. D. (2003). The politics of regional overnance in the Mekong River Basin. *Global Change, 15*, 59–76.

Regional Remote Sensing Center West. (2012). *NRSC, ISRO.* River Basin Atlas of India.

River Basins Organizations in India – Institutional Frameworks and Management Options. (1999). A Case for Fundamental Review, published by South Asian Network of Rivers, Dams and people. *Thematic Reviews: Institutional and Governance Issues.* http://www.zef.de/module/register/media/ 0f22_Saravanan-RiverBasin_wcd.pdf

Saleth, R. M. (2004). *Strategic analysis of water institutions in India: Application of a new research paradigm.* IWMI.

Schindler, S., Sebesvari, Z., Damm, C., Euller, K., Mauerhofer, V., Schneidergruber, A., Biró, M., & Essl, F. (2014). Multifunctionality of floodplain landscapes: Relating management options to ecosystem services. *Landscape Ecology*, *29*, 229–244. doi:10.100710980-014-9989-y

Sehgal, J., Mandal, D. K., Mandal, C., & Vadivelu, S. (1992). Agro-Ecological Regions of India. *Tech. Bull. No. 24.*

Shrivastava, R. N., Nema, R. K., & Awasthi, M. K. (2018). Strategy of Water Resource Utilization in Upper Narmada Basin. *International Journal of Current Microbiology and Applied Sciences*, *7*(05), 591–602. doi:10.20546/ijcmas.2018.705.073

Shuter, B. J., & Meisner, J. D. (1992). Tools for assessing the impact of climate change on freshwater fish populations. *GeoJournal*, *28*(1), 7–20. doi:10.1007/BF00216402

Swallow, B. M., Garrity, D. P., & van Noordwijk, M. (2001). *The Effects of Scales,Flows and Filters on Property Rights and Collective Action in Watershed Management*. International Food Policy Research Institute.

Uitto, J. (2001). World Forests, Markets and Policies. VL - 3 Publisher: Kluwer Academic Publishers pp.65-76. DO - doi:10.1007/978-94-010-0664-4_3

USGCRP. (2014). Water Resources. Climate Change Impacts in the United States: The Third National Climate Assessment. U.S. Global Change Research Program, 69-112.

Vorosmarty, C. J. (2002). Global change, the water cycle, and our search for Mauna Loa. *Hydrological Processes*, *16*(1), 1335–1339. doi:10.1002/hyp.527

Vorosmarty, C. J., Green, P., Salisbury, J., & Lammers, R. (2000). Global water resources: Vulnerability from climate change and population growth. *Science*, *289*(5477), 284–288. doi:10.1126cience.289.5477.284 PMID:10894773

Vorosmarty, C. J., & Sahagian, D. (2000). Anthropogenic disturbance of the terrestrial water cycle. *Bioscience*, *50*(9), 753–765. doi:10.1641/0006-3568(2000)050[0753:ADOTTW]2.0.CO;2

Whiting, P. J. (2002). Streamflow necessary for environmental maintenance. *Annual Review of Earth and Planetary Sciences*, *30*(1), 181–206. doi:10.1146/annurev.earth.30.083001.161748

WRI (World Resources Institute), UNEP, UNDP, & World Bank. (1998). World Resources 1998–99: Environmental Change and Human Health. Oxford University Press, New York, NY.

Chapter 2
Declining Freshwater Species Biodiversity

Santhosh Kumar Rajamani
https://orcid.org/0000-0001-6552-5578
MAEER MIT Pune's MIMER Medical College, India & Dr. BSTR Hospital, India

Radha Srinivasan Iyer
https://orcid.org/0000-0001-7387-4401
SEC School for Independent Living, India

ABSTRACT

Freshwater is a precious resource that is essential for human survival and liquid surface freshwater accounts for just 0.2% of total volume of water on Earth. Wetlands, which are home to one of the most biodiverse species of plants and animals, account for just 6.4% of the earth's surface. On a global scale, the wetlands and freshwater ecosystems are under pressure and are seeing declining species diversity. Freshwater biodiversity has not been part of the ongoing mainstream conservation efforts and has been largely ignored. This is in part due to invasive human activities, pollution, climate change, deforestation, habitat destruction and degradation, the introduction of alien competing species, and excessive consumption. This chapter gives a broad overview of the current scenario of declining freshwater species biodiversity and threats to vulnerable aquatic ecosystems. Delineation of biodiversity hotspots, adoption of a proactive approach, and focused conservation efforts towards freshwater ecosystems are some of the suggested remedial measures to address this global problem.

DOI: 10.4018/978-1-6684-9034-1.ch002

NEED FOR CONSERVATION OF FRESHWATER ECOSYSTEMS

Fresh water ecosystems are one of the most productive ecosystems in the world and home to an estimated 1,26,000 species of plants and animals. Freshwater is a precious resource that is essential for human survival that accounts for just 2.3% percentage of total planetary water, remaining quantity is saline or locked up as ice and glaciers unusable from a human perspective. Wetlands which are the home to one of the most biodiverse species of plants and animals account for just 6.4% of area of the earth (Rawat et al., 2015).

Freshwater is a rare and precious natural resource, forming constituting a small fraction, about 1 / 1,00,000th of total water of this planet. As the population continues to grow the global demand for freshwater is expected to increase exponentially. This demand for fresh water is expected to overtake supply by 2030 leading to severe water Scarcity (Harrington & Fisher, 2014).

This chapter gives a broad overview of current scenario of declining freshwater species biodiversity and threats to the vulnerable aquatic ecosystems. Delineation and conservation of Biodiversity hotspots, adoption of a proactive approach, focussed and concerted conservation efforts towards freshwater ecosystems, are some of the suggested remedial measures to address this global problem. Many freshwater bodies like rivers span many countries which means that disease or extinction can rapidly spread across many countries via waterbodies. Many species of freshwater amphibians, reptiles and crustaceans are also under threat of extinction. Riverine species of plants and animal seem to be most susceptible to human activities, followed by marsh and lake species (Harvey, 2013).

DECLINING FRESHWATER BIODIVERSITY

According to the available study findings, freshwater biodiversity is in serious trouble and is declining far more rapidly than the most severely impacted terrestrial ecosystems. On a global scale the wetlands and freshwater ecosystems are under pressure and are seeing declining species diversity with one-third of species threatened with extinction globally, according to some estimates. Fresh water biodiversity has not been part of the ongoing mainstream conservation efforts and has been largely ignored by conservationists (Belgrano et al., 2015).

Freshwater ecosystems are among the most endangered on the planet and are experiencing biodiversity losses on a par with or even larger than those of terrestrial systems. With around one-third of the 28,000 freshwater species examined for the International Union for Conservation of Nature (IUCN) Red List threatened with extinction, the global loss of freshwater species continues unabated. There are now

25% more extinct fish species in North America than there were in 1989, and up to 86 more species could become extinct by 2050. The loss of habitat, overfishing, invasive species, pollution, and hydrological changes are all blamed for the decline in freshwater biodiversity (Eldredge, 2002).

The given research findings show a worrying decline in freshwater biodiversity. Compared to terrestrial and marine environments, freshwater ecosystems have larger proportions of extinct and threatened species, with populations of vertebrates declining by 83% between 1970 and 2018. With just 2.3% of the Earth's surface is covered by freshwater habitats, where almost one-third of all vertebrate species reside. In North America, the rate of extinction in freshwater habitats is thought to be 4% per decade, and 32% of all amphibian species are thought to be endangered (Berlatsky, 2014).

Freshwater biodiversity is declining because of several reasons, including hydrological changes, habitat loss and degradation, overfishing, invasive species, pollution, and the several effects of climate change. This is in part due to invasive human activities, pollution, climate change deforestation, habitat ruin and degradation, introduction of alien competing species, and undue human consumption (José Galizia Tundisi & Takako Matsumura Tundisi, 2016).

Figure 1. Percentage of various compartments of water in our biosphere along with their turnover periods in the box. Surface freshwater in form of lakes and rivers that can be directly accessed by humans and animals totals about 0.0091% of the total water. Freshwater sources and ecosystems are a precious and rare resources that cannot be squandered away to pollution and human greed.
(Graph redrawn in plotly.com by authors based on data from the book Freshwater Ecology: Concepts and Environmental Applications of Limnology by Dodds, W. K., & Whiles, M. R. 2000.) produced with plotly.com.

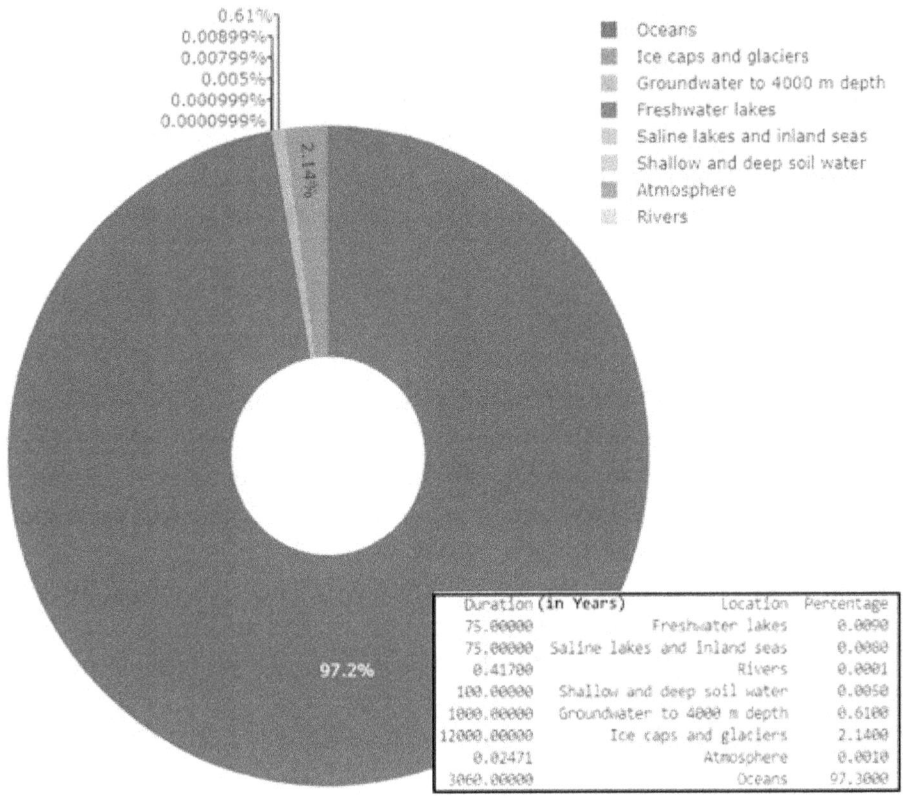

Duration (in Years)	Location	Percentage
75.00000	Freshwater lakes	0.0090
75.00000	Saline lakes and Inland seas	0.0080
0.41700	Rivers	0.0001
100.00000	Shallow and deep soil water	0.0050
1000.00000	Groundwater to 4000 m depth	0.6100
12000.00000	Ice caps and glaciers	2.1400
0.02471	Atmosphere	0.0010
3060.00000	Oceans	97.3000

The loss of variety and richness in the many kinds of living things and their interactions within an ecosystem is referred to as decreased ecosystem diversity. Many factors, including human activity, climate change, natural catastrophes, and pollution, can contribute to this loss. Reduced diversity in an ecosystem can cause imbalances in the natural processes necessary to maintain the ecosystem's health. For instance, the extinction of one species of plant can have an impact on other organisms' ability to feed, which might cause a drop in population. This may have repercussions across the ecosystem and eventually cause the whole system to become unstable (Berlatsky, 2014).

FRESHWATER ECOSYSTEMS ARE AN UNTAPPED GENETIC LIBRARY

The gene pool of aquatic fauna and flora is a precious natural resource that is untapped to its full potential and is base material for genetic manipulation. Freshwater home to a global 12% of plant species and 40% of fish species (Harvey, 2013). For example, thermally stable *Taq* DNA polymerase (DNA polymerase I) used in DNA amplification or Polymerase Chain Reaction technology was discovered in an aquatic bacterium living in hot springs called *Thermus aquaticus*. It was found in hot freshwater springs of Yellowstone National Park (Levin, 2001).

WETLAND ECOSYSTEMS

Wetland ecosystems offer both humans and the environment a variety of beneficial services and advantages, such as bettering water quality, storing carbon, providing habitat for wildlife, and supporting fisheries and leisure activities. The amount of biomass produced by the plants and other animals that dwell in wetland environments is a measure of their productivity (Dodds & Whiles, 2020).

Since they contain large concentrations of nutrients that encourage the growth of plant life, such as nitrogen and phosphorus, wetlands are extremely productive environments. The plants in wetlands produce vast amounts of biomass through photosynthesis, which provides food and habitat for a range of other creatures, including birds, fish, amphibians, and invertebrates (Keddy, 2016).

Figure 2. The six types of wetlands are namely, 1. swamp, 2. marsh, 3. fen, 4. bog, 4. wet meadows, and 6. shallow water or aquatic
(Drawn by authors from page 444, Life on Earth An Encyclopaedia of Biodiversity, Ecology, and Evolution, Vol-1, Eldredge N. 2002)

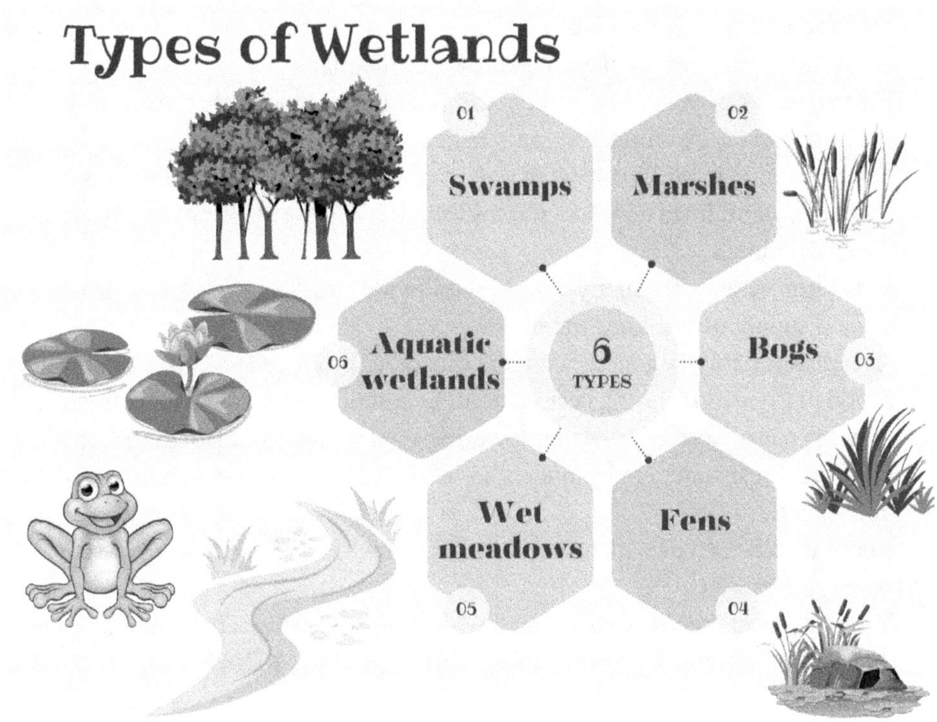

ECOSYSTEM SERVICES

Freshwater ecosystems provide numerous important services to humans and animals in form of food for example crabs, shrimps and fishes, recreation like swimming and boating, temperature regulation, carbon sequestration, flood protection and portable drinking water. The following services to rendered humans by freshwater ecosystems (Levin, 2001).

Biodiversity: A wide variety of plant and animal species, many of which are unique to freshwater habitats, can be found there. Most fauna and flora of wetland have developed many unique adaptations like for survival in a low oxygen and changing aquatic habitat.

Cycling of nutrients: Freshwater environments are crucial to this process, which is necessary for ecosystems to function.

Nutrition: Freshwater ecosystems are source of nutrition for both humans and animals. Fish harvested in freshwater forms a bulk of inland human diet. Fish is a rich source of essential fatty acids like Omega 3 Fatty acid, and vitamin D essential for human survival in temperate and arctic climates.

Storage of Carbon dioxide: Freshwater ecosystems absorb and store carbon dioxide, which helps to regulate the climate of the planet.

Filtration: Freshwater ecosystems can function as inherently effective systems for purifying water by eliminating impurities and pollutants.

Scientific study: Freshwater ecosystems offer significant opportunities for study that help us comprehend how ecosystems function and the effects of human activity on the environment.

Drinking water: For many people worldwide, freshwater habitats provide as their main supply of drinking water.

Irrigation: Crop irrigation, which is crucial to produce food, uses freshwater.

Flood prevention: During periods of high rainfall, wetlands and other freshwater ecosystems can absorb extra water, lowering the likelihood of flooding.

Habitat: Freshwater environments serve as vital homes for a variety of fish and wildlife species.

Recreation: Activities like swimming, boating, fishing, and hiking are possible in freshwater habitats.

Thus, freshwater ecosystems offer a wide range of services that are vital to both the survival of humans, and other animals. (G. Timothy, 2020).

BIODIVERSITY OR BIOLOGICAL DIVERSITY

Biodiversity or Biological diversity refers to species richness. Biodiversity refers to the variety of species that are found a particular locale. Species biodiversity is the abundance of various species found a particular locale. Genetic biodiversity is the variations in the single gene of a species. The accelerated rate of loss of Biodiversity loss observed in all ecosystems of our biosphere is deemed as **Biodiversity crisis**. There three main types of biodiversity, namely genetic diversity, species richness, and ecosystem diversity (P Legendre & L Legendre, 1988).

Species-area Curves are logarithmic plot of number of species to area of an ecosystem. A trivial reduction in area will cause drastically(logarithmic) decline in the species diversity for example Reduction by a factor 2 will lead to loss of 10^2 $=100$ species loss. This explains the dramatic loss of diversity due to habitat loss (P Legendre & L Legendre, 1988).

Whittaker's biodiversity markers are designed as alpha, beta, and Gamma. Alpha is number of species in an ecosystem. The comparison of species diversity as we transition from one ecosystem to another is done using Beta. Gamma diversity is a measure of diversity in a large area (P Legendre & L Legendre, 1988).

THREATS TO AQUATIC BIODIVERSITY

Biotic Pollution

Biotic pollution is introduction of foreign species that competes with the native species and alters the ecological balance of a place. Invasive Species is when the alien species is environmentally or economically damaging. Introduction of Water hyacinth or Water lettuce into water bodies of India is example of Biotic pollution (Dodson, 2005). These invasive species are a great threat to endemic, slow-growing and sensitive species like Cryptocoryne sp. or River Crinum Lilly.

Habitat Fragmentation

Habitat fragmentation is the process of breaking up a habitat into numerous smaller islands by human intervention. Typical examples include building dams, canals and bridges and embankments on rivers. These fragments of habitat support far fewer species than original habitat. Native species become stressed due to loss of their homes and this favors establishment of alien species (Keddy, 2016).

Aquaculture as Threat to Aquatic Biodiversity

Aquaculture is the practice of farming aquatic flora and fauna for commercial purposes, this excludes natural harvest like fishing and trapping. Aquaculture has impact on the local biodiversity of a locality. The uneaten fodder, urinary and other waste of aquaculture may contribute to Eutrophication. Hormones and chemicals like Antifungals used in aquaculture may reach natural waterbodies through runoffs and affect the amphibian, crustacean, and fishes. Species may escape cultivation and become potentially invasive in new habitats. Expanding aquaculture may lead to habitat fragmentation, resource depletion, altered food web and fall in indigenous biodiversity (Richman et al., 2015).

Loss of Biodiversity Due to Deforestation

Deforestation is loss of Biodiversity due excessive logging of trees of Forrest for human use of the land for example, agricultural purposes or building houses. Loss of such Forrest can lead to increased soil erosion, global warming, lowland floods, landslides, mud slides, and silt loading of downstream waterbodies (G. Timothy, 2020).

Damage to Aquatic Ecosystems and Loss of Biodiversity by Eutrophication

Plant macronutrients namely nitrogen, phosphorus, and potassium (NPK) are essential for growth of macrophytes and increases productivity of aquatic ecosystem. Paradoxically, excessive nutrients from agricultural runoff and sullage from homes containing soap and detergent use often, often reduce the aquatic diversity (G. Timothy, 2020). Higher levels of Phosphate (lesser extent ammonium), primarily trigger excessive growth and proliferation of algae and few pollution tolerant species like cattail or lesser bulrush (*Typha angustifolia*), water lettuce (*Pistia stratiotes*), water hyacinth (*Eichhornia crassipes*). This process is called eutrophication. Contamination of a waterbody may be from point sources (sewage from town) or from diffuse sources (soil erosion from deforestation or excessive cattle grazing) (Levin, 2001).

Change in Seasonal Patterns of Waterflow

The construction of dams and canals on rivers leads to alteration of water flow pattern. Excessive diversion of water by canals may reduce quantity of water causing disturbances in ecosystems. Dams may lead to redistribution of water by flooding certain areas and drying out some tributaries. All these activities disturb the fragile aquatic ecosystems built and balanced in a time frame of millions of years (Rawat et al., 2015).

Excessive Commercial Harvesting

Excessive commercial harvesting is also affecting the biodiversity of freshwater ecosystems. Travancore dwarf pufferfish (*Carinotetraodon travancoricus*) of South-west India, is a voracious fish which keeps populations of freshwater snails and molluscs in check. They are popular in aquarium trade due to bright coloration. This fish is now threatened due to excess harvesting for aquarium trade and habitat loss. Loss of this voracious species is likely to lead to increased populations of freshwater snails, which in turn might have implications on human health. Freshwater snails are known to be intermediate hosts of numerous animal parasites like *Schistosomiasis*, or bilharzia. (Sigee, 2005).

EXCESSIVE FISHING AND DISRUPTIVE METHODS OF FISHING

Exploitation of marine resource like excessive fishing, disruptive methods of fishing like using dynamite and cyanide, bottom trawling for fishing have threatened the costal ecosystems and endangered a precious resource that is source of nutrition for humans. Coral reefs that are home to many varieties of marine organisms are being destroyed at an alarming rate by disruptive fishing practices, increasing ocean temperatures and acidification of ocean waters (O'Sullivan & Reynolds, 2008).

CONCEPT OF PROTECTED AREA- ECOREGION

To accomplish certain conservation goals, a protected area is a geographically delimited area that is recognized and administered. It could be a conservation area, national park, wildlife sanctuary, or some other kind of reserve. Conserving biodiversity and preserving natural and cultural treasures for future generations are the main goals of a protected area. Governments, non-governmental organizations, and community-based organizations all create and maintain protected areas.

Ecoregion is defined by WWF as "a relatively large unit of land or water containing a geographically distinct assemblage of species, natural communities, and environmental conditions" (EPA, *Ecoregions | US EPA* 2020).

BIODIVERSITY HOTSPOTS

Biodiversity hotspots are regions of high biodiversity that must be protected. 36 such regions are declared throughout the world and includes Western Ghats and Sri Lanka. A total of 2.3% of world area is taken up by these 36 Biodiversity hotspots but is home to around 43% of world population of fauna. Most of the world's biodiversity can be found in these hotspots, which are frequently found in tropical and underdeveloped nations. The Western Ghats in India, the Cape Floristic Area in South Africa, and the Atlantic Forest in Brazil are a few examples of hotspots for biodiversity (National Geography Society, *Biodiversity hotspots* 2021).

BIODIVERSITY HOTSPOTS IN INDIA

There are numerous biodiversity hotspots, or areas with a high concentration of rare and endangered species, in India. The main biodiversity hotspots in India are as follows:

Western Ghats: One of the world's 34 biodiversity hotspots is the Western Ghats Mountain range, which runs parallel to India's west coast. Many endemic plant and animal species, including as the lion-tailed macaque (*Macaca silenus*), Nigiri tahr (*Nilgiritragus hylocrius*), the clouded leopard (*Neofelis nebulosa*) and the Asiatic black bear (*Ursus thibetanus*) and Malabar large-spotted civet (*Viverra civettina*), can be found there (Chitale et al., 2014).

Eastern Himalayas: The states of Arunachal Pradesh, Assam, Manipur, Meghalaya, Mizoram, Nagaland, Sikkim, and Tripura are all part of this hotspot, which is situated in northeastern India. The red panda is one of the many plant and animal species that call it home (Joseph et al., 2010).

Indo-Burma: This region includes sections of Myanmar, China, Thailand, Laos, and Vietnam in addition to northeastern India. Several endemic plant and animal species, such as the Siamese crocodile, the Burmese python, and the Irrawaddy dolphin, can be found there (Chitale et al., 2014).

Sundaland: This region includes parts of the Philippines, Malaysia, Indonesia, Brunei, and India. Many rare plant and animal species, like as the Bornean orangutan, Sumatran rhinoceros, and Javan rhinoceros, can be found there (Joseph et al., 2010).

Nicobar Islands: The Sundaland hotspot includes the Nicobar Islands, which are situated in the Bay of Bengal. Several endemic plant and animal species can be found there. a few of which include the Nicobar megapode, the Nicobar tree shrew, and the Nicobar pigeon (Chitale et al., 2014).

For the long-term survival of many species and the preservation of India's natural heritage, it is imperative that these hotspots and their biodiversity be protected.

FRESHWATER BIODIVERSITY HOTSPOTS OF THE WORLD

Regions with a high concentration of endemic freshwater species that are threatened by human activities including habitat destruction, pollution, and climate change are known as freshwater biodiversity hotspots. Following are some instances of hotspots for freshwater biodiversity:

Mekong river basin: More than 1,000 species of freshwater fish, including the Mekong giant catfish, one of the biggest freshwater fish in the world, may be found in the Mekong River watershed. Other rare and threatened species found in the area are the Siamese crocodile and the Irrawaddy dolphin.

Lake Victoria: East African Lake Victoria, the second-largest freshwater lake in the world, is home to more than 500 species of cichlid fish, many of which are endemic. Several additional aquatic and terrestrial species, many of which are threatened by pollution and overfishing, are supported by the lake and the marshes that surround it.

Amazon river system: The largest river system in the world, the Amazon River basin is home to more than 3,000 species of freshwater fish, many of which are unique to this planet. The Amazon River dolphin and the gigantic otter are just two of the numerous other aquatic and terrestrial animals that call this area home.

Great lakes of North America: Around 200 fish species, as well as a broad variety of other aquatic and terrestrial species, call the Great Lakes of North America, which include Lake Superior, Lake Michigan, Lake Huron, Lake Erie, and Lake Ontario, home. A sizable commercial fishing business is supported by the lakes, which are also a substantial supply of drinking water (Dudgeon et al., 2006).

WESTERN GHATS: FRESHWATER BIODIVERSITY

The Western Ghats are home to globally most biodiverse fauna and flora many of which are threatened with extinction. About 325 species of plants and animals found in IUCN Red Data List species, are found in Western Ghats (Reddy et al., 2016).

Seasonal mass flowering of wetland endemic, composite, *Senecio bombayensis*, or sonki (hindi sona = golden) covers entire section of hills in an orange-golden sheen. There are streams and medows filled with *Cryptocoryne spiralis* and rocky-marshes of *Hygrophila pinnatifida* or Dwarf hygrophila. Wet meadows full of short lived Habenaria grandiflora Ground white orchids flowers poking above the grass canopy. It is estimated that 50 to 60% of species found in Western Ghats are endemic.

Figure 3. IUCN red data list of threatened species occur in the Western Ghats in percentage
Redrawn by authors based on data available on from IUCN website, produced with plotly.com

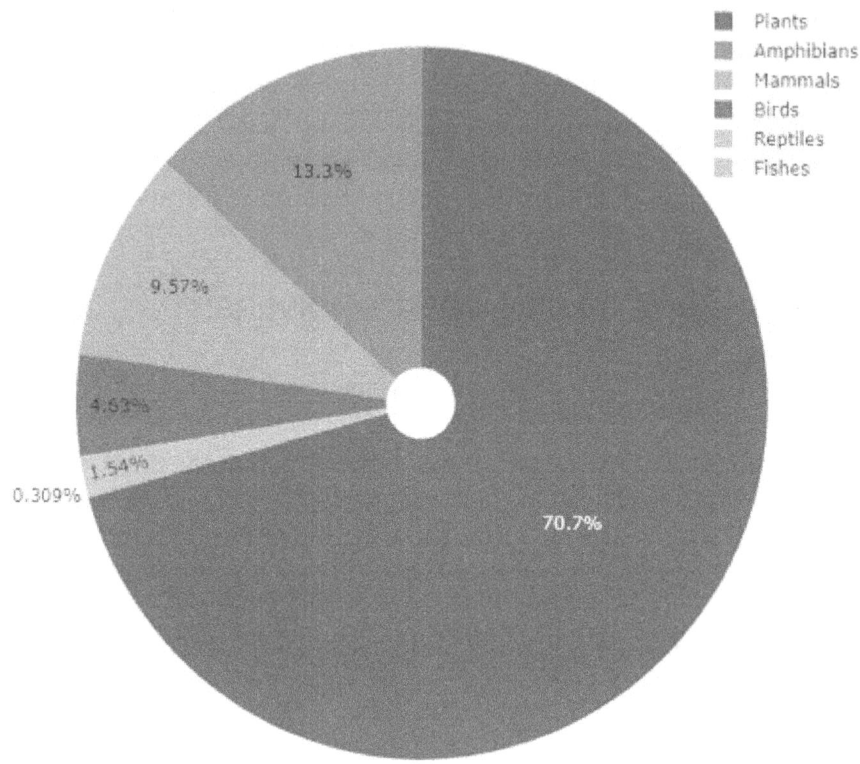

Almost 300 species of freshwater fish, many of which are endemic, can be found in the Western Ghats, a mountain range in southwestern India. Many other aquatic and terrestrial species, including some endangered turtle and amphibian species, call this area home. An extensive variety of amphibian species can be found in the Western Ghats. Many of these species are in danger of extinction because of things like habitat loss, pollution, climate change, and other issues (*IUCN SSC Western Ghats*).

ENDANGERED AMPHIBIANS OF WESTERN GHATS

Western Ghats endangered amphibians are being protected, but more must be done to save their habitats and stop further population reduction ("Home: ENVIS center on biological diversity," 2023).This is the list few of the Western Ghats' critically endangered amphibians:

Purple Nose Frog: The IUCN has designated the rare and elusive purple nose frog (*Nasikabatrachus sahyadrensis*), which is exclusively found in the Western Ghats, as an endangered species. It spends much of its existence underground and only comes out to reproduce during the monsoon season.

Malabar tree frog: The Western Ghats' forests are home to the Malabar tree frog (*Rhacophorus malabaricus*), an arboreal frog that is considered endangered owing to habitat loss and degradation.

Gundia Tree frog: Only located in the Western Ghats of Karnataka, the Gundia Indian tree frog (*Polypedates gundia*) is classified as highly endangered due to habitat loss.

Beddome's leaping frog: The Western Ghats' streams are home to the Beddome's leaping frog (*Indirana beddomii*), which is considered endangered owing to habitat loss and pollution.

Dancing frogs of the Western Ghats: *Micrixalidae* are family of dancing frogs that are endemic to western Ghats. They are named so because of their habit of dancing or waving their feet at female frogs as mating ritual.

Dancing cave frog *Micrixalus spelunca*: Dancing cave frog *Micrixalus spelunca* discovered in 2014 inhabits dark caves of Tamil Nadu. DNA Fingerprinting technology has discovered several new species dancing frogs, all of which are vulnerable due to habitat destruction and pollution.

Kerala's Silent Valley Frog: Due to habitat degradation and fragmentation, the tropical frog known as the silent valley frog (*Micrixalus thampi*) that can only be found in Kerala's Silent Valley National Park is classified as severely endangered.

Kottigehar dancing frog: Due to habitat loss and pollution, the Kottigehar dancing frog (*Micrixalus kottigeharensis*), a rare frog that inhabits streams in the Western Ghats, is classified as severely endangered (Daniels, 1992;Vijayakumar et al., 2016).

ENDANGERED FRESHWATER FISH SPECIES OF WESTERN GHATS

A variety of freshwater fish species can be found in the Western Ghats, a mountain range in India, many of which are threatened by habitat loss, pollution, overfishing, and other reasons. Here are some instances of Western Ghats freshwater fish that are in danger of extinction:

Humpback mahseer: Only the Chalakudy River in Kerala is home to the large freshwater fish known as the humpback mahseer (*Tor remadevii*). Due to habitat destruction, overfishing, and the building of dams and weirs, it is critically endangered.

Kerala Hill stream Loach: Kerala Hill stream Loach (*Pseudogastromyzon myersi*): The streams of the Western Ghats are home to this little, vibrant fish. Due to habitat loss and degradation brought on by deforestation and pollution, it is critically endangered.

Travancore Anchovy: This little, silvery fish, known as the Travancore Anchovy (*Stolephorus travancoricus*), is found in the rivers and estuaries of the Western Ghats. Overfishing, habitat loss, and pollution have made it endangered.

Malabar Pufferfish: Malabar Pufferfish (*Carinotetraodon travancoricus*) Western Ghats' streams and rivers are home to this diminutive freshwater pufferfish. Because to habitat degradation and overfishing for the aquarium trade, it is critically endangered. This is a popular aquarium fish in the west known for its puffing habit and its voracious appetite of predating on the pest snails.

Vellayani Peacock Eel: The Vellayani Peacock Eel (*Macrognathus aral*) is a fish that resembles an eel that is found in Kerala's Vellayani Lake. Because to habitat loss, pollution, and overfishing, it is critically endangered.

These are only a few of the numerous rare freshwater fish species that can be found in the Western Ghats. To prevent the continued reduction of these distinctive and significant species, conservation actions are required (Jumani et al., 2018; Patil et al., 2018; Raghavan et al., 2007).

MY CASE STUDIES ON ENDANGERED MACROPHYTES OF WESTERN GHATS

The following section is my own survey and analysis of the situation of threat to aquatic plants that are found in the western ghats ("checklist of plants of Maharashtra," 2022).

BLUE WATER LILY OR *NYMPHAEA NOUCHALI*

The blue water lily, or *Nymphaea nouchali* (N.L. Burman) is an aquatic plant that grows in ponds, lakes, and still streams. It has stunning blue blossoms and is employed in conventional medicine. Nonetheless, this species faces serious risks from pollution and habitat destruction ("Wetlands," 2020).

Figure 4. Nymphaea nouchali (Burm. f.) Red-blue water lily, known as Neelkamal in Hindi. This species is native to India. Habitat fragmentation and filling up marshes for agricultural activities has pressurized this species. This is Nymphaea nouchali var. nouchali identified by its blue coloured flowers and under surface of floating leaves are coloured an texture of blue-purple

ROTULA AQUATICA (LOUR.)

Rotula aquatica (Lour.) famous for medicinal properties in Ayurvedic stream of medicine, this wetland plant is now endangered due to excessive harvesting and loss of habitat. This species in sensitive to pollution and has a relatively slow growth rate. Specimens thrive in rainy seasons, flower in winters, and become quiescent in hot summers. Specimens cannot survive on lateritic soils and need very high humidity for cultivation ("Wetlands," 2020).

CRINUM VIVIPARUM (LAMK., R. ANSARI & V.J. NAIR)

This live bearing aquatic crinum lily is a close cousin of garden variety of white lily, having adapted to a life in water (Rajamani & Iyer, 2023a). This is a very slow growing species with special adaptation to survive among the rocky, shallow, waterlogged banks of western ghat streams. Due to habitat destruction and slow growth, this species is now threatened. Specimens can survive on garden soils hitherto flower once in two or three years (Dudgeon et al., 2006; Bastmeijer, 2016a).

CRYPTOCORYNE RETROSPIRALIS (ROXBURGH KUNTH)

Cryptocoryne retrospiralis is another endemic species found in Maharashtra, Karnataka, and Kerala. This species closely resemble another common ubiquitous species found everywhere in India *Cryptocoryne spiralis var spiralis.* The leaves are stiff, and Lance shaped. *C. crispatula var. crispatula* is another species closely resembling this species but not distributed in South-west India but found in Indo-Burma-China belt (Bastmeijer, 2016a).

CRYPTOCORYNE SIVADASANII (BOGNER)

Cryptocoryne sivadasanii is a rare endemic species localized to Karnataka and Kerala. The leaves of this species grow to long lengths and are devoid of a leaf blade. The species is available on some aquarist websites and collection from natural habitat has threatened this species. The spathe of this species does resemble Cryptocoryne retrospiralis as does its annual cyclical growth pattern. Because of Pollution and garbage disposal in Kerala and Karnataka stream this species is under threat (Bastmeijer, 2016a).

CRYPTOCORYNE COGNATA (SCHOTT)

Cryptocoryne cognata is a very rare and endemic species of Cryptocoryne, found only in Ratnagiri district of Maharashtra. It was considered a lost species for 150 years but rediscovered in 1990s. Due to endemic nature of this species, I believe this species is highly threatened (Rajamani & Iyer, 2023b). Most rivers and streams of Ratnagiri district are witnessing phenomenal pollution, garbage littering and habitat destruction due to population growth and growth of tourism in this region. This problem is compounded with low tolerance of Cryptocoryne to pollution and climate change.

LAGENANDRA SPECIES

Lagenandra species are rare aquatic species confined to the streams and rivers of Western Ghats. The species is known to occur naturally in the province of Karnataka, Kerala and Sri Lanka. *Lagenandra* species that are found in India/Sri Lanka include *Lagenandra meeboldii* (Engl.C.E.C. Fisch.), *L. nairii* (Ramamurthy & R.Rajan), *L. ovata* (L. Thwaites), *L. toxicaria* (Dalzell), and *L. undulata* (A.R.K. Sastry).

This genus is threatened and on decline in the wild. Their tolerance to pollution is low and plants seem to be unable to grow outside high humidity zones found near waterbodies. There may be yet undiscovered species in these native habitats as the genus has not been well studied. Some species like *Lagenandra meeboldii* bleeding heart are prized by aquarists for their unique leaf coloration. This species is also found frequently aquarium trade of the West. Excessive collection from wild is putting pressure on this exotic genus (Bastmeije, 2016b).

ERIOCAULACEAE OR PIPE WORT FAMILY

Most members of the *Eriocaulaceae* family are small aquatic or semiaquatic plants that thrive in moist soil or shallow water. They are typically little plants; several species only reach a height of a few millimetres (Rajamani & Iyer, 2022). They stand out for their characteristic rosettes of leaves and their tiny, insignificant flowers, which are frequently encircled by bracts. *Eriocaulaceae* is a family of flowering plants that includes around 1200 species

India is home to many species of Eriocaulon, which are commonly known as pipeworts. With over 130 species, India possesses one of the world's richest biodiversity of *Eriocaulaceae* floras. Some of the Eriocaulon species found in India include:

1. *Eriocaulon cinereum* - Assam, Meghalaya, West Bengal, and Sikkim.
2. *Eriocaulon quinquangulare* - Kerala, Karnataka, Tamil Nadu, and Maharashtra.
3. *Eriocaulon madayiparense* - Kerala.
4. *Eriocaulon karaikkalense* - Tamil Nadu.
5. *Eriocaulon koipilliense* - Kerala.
6. *Eriocaulon nilgiriense* - Tamil Nadu.
7. *Eriocaulon quinquenerve* - Kerala, Karnataka, Tamil Nadu, and Maharashtra.
8. *Eriocaulon robustum* - Assam, Meghalaya, and Arunachal Pradesh.
9. *Eriocaulon sedgewickii* - Assam, Meghalaya, and Arunachal Pradesh.
10. *Eriocaulon subulatum* - Assam, Meghalaya, and Arunachal Pradesh.

Eriocaulon quinquangulare, Eriocaulon sexangulare, Eriocaulon sedgewickii, Eriocaulon robustum, and Eriocaulon euryspermum are among the most prevalent *Eriocaulaceae* species in India. Some of these plants are often used in aquariums as beautiful plants. These plants are frequently utilized in Indian traditional medicine. Although the conservation status of many *Eriocaulaceae* species in India is unknown, many species are thought to be vulnerable or endangered because of habitat loss (*Eriocaulaceae - Efloraofindia*, n.d.).

AMPHIBIAN BIODIVERSITY OF NORTH-EAST: NEW SPECIES OF SIJU CAVE CASCADE FROG OR AMOLOPS SIJU DISCOVERED IN 2023

In 2023, team of scientists working for Zoological Survey of India (ZSI) made discovery of a new species of cave dwelling frog called *Amolops siju.* This frog was found deep inside Siju cave in Meghalaya state's, Southern Garo Hills district. Siju cave is a system of limestone caverns 4 kilometers long. Previously they had discovered 3 new species of cascade frogs of genus *Amolops* namely: *A. chanakya, A. tawang, and A. terraorchis.* This is another example of undiscovered freshwater amphibian biodiversity of Northeastern part of India.

India is home to a diverse array of aquatic plants, many of which are endangered due to habitat loss, pollution, and other human activities. These are just a few examples of the many endangered aquatic plants of India. Conservation efforts such as habitat restoration, pollution control, and the regulation of invasive species are necessary to protect these important species and the ecosystems they inhabit (Abellán et al., 2007).

PORTULACA LALJII (P. SIVARAMAKRISHNA & P. YUGANDHAR SP. NOV.)

In 2020 a new and rare species of wild sun rose plant called *Portulaca laljii (Portulacaceae* family) has been discovered in the Eastern Ghats, Prakasam district Andhra Pradesh province. This species is unique in possessing thick tuberous roots, that are succulent. This is an adaptation for its arid, rocky, and hilly habitat. This new species is named in honour of Indian Botanist, Dr. Lal Ji Singh. The species seems to be endemic, confined to a narrow range and hence vulnerable.

MAJOR LOSS OF AMPHIBIAN BIODIVERSITY

Amphibian's populations are in decline in not just areas around human habitats like villages and towns. Amphibians counts in even remote areas without any direct interference from human are showing a declining trend. The IUCN Global Amphibian Assessment estimates that as many as 168 amphibian species have vanished and are now thought to be extinct. A third of all amphibian species are in decline, and if they are not brought into captivity, more than 40% could go extinct by the year 2060 (Briggs et al., 2010).

Agricultural runoff containing Pesticides and herbicides, leading to contamination of ponds and lakes has been linked with fall in frog populations. Scientists at university of Southern Florida discovered that Atrazine herbicide interacted with Phosphate in NPK fertilizers to produce weakness of frog immune system. Such immunosuppressed frogs could not fend off trematode and other infestation.

Global increase in temperature has been linked to growth and Expansion of Chytrid fungus that causes major mortality among frog populations. Lastly, incidence of deformities like extra limbs or malformed limbs or missing eyes, is on rise. This has been linked with pollution and pesticide contamination of water (Lips et al., 2006).

CHYTRID FUNGUS PANDEMIC OF BATRACHOCHYTRIUM DENDROBATIDIS- GLOBAL LOSS OF AMPHIBIAN BIODIVERSITY

An important problem that has been linked to a chytrid fungus pandemic is the global loss in the biodiversity of amphibian populations. The infection causes swelling, sloughing, and ulceration in the skin of frogs, which can be fatal. It has been determined that chytridiomycosis, a fungus that is spread by the chytrid fungus ***Batrachochytrium dendrobatidis*** (Bd), is the main reason amphibian populations are declining in many parts of the world (Fisher et al., 2009).

Environmental persistence: Even in the absence of amphibians, the chytrid fungus can persist for long periods of time in the environment. As a result, it is challenging to stop the fungus's growth and it continues to pose a threat to amphibian populations.

Global distribution: Every continent where amphibians are found has been confirmed to have chytrid fungus, which has led to population decreases and extinctions in many regions of the world. It poses a danger to amphibian biodiversity because of its widespread global distribution and capacity to influence several amphibian species.

Treatment ineffectiveness: Wild amphibians infected with the chytrid fungus are now untreatable. In captive breeding operations, antifungal treatments have been employed, but these techniques are not workable for wild populations.

The illness affects amphibians' epidermis (outer skin layer), which is crucial to their life because it is involved in amphibian respiration, maintaining hydration, and defense against external pathogens (Fisher et al., 2009).

LIFE CYCLE OF CHYTRID FUNGUS IN AMPHIBIANS

Chytridiomycosis is a disease that affects frogs and is brought on by the chytrid fungus (*Batrachochytrium dendrobatidis*). The free-swimming zoospore stage and the encysted stage are the two primary phases of the chytrid fungus' life cycle in amphibians.

Zoospore stage: The chytrid fungus begins its life cycle as a flagellated, motile zoospore. From fully developed sporangia, which are reproductive structures, zoospores are released into the environment. They employ enzymes to pierce the top layer of skin because they are drawn to the skin secretions of amphibians.

Figure 5. Lifecycle of Chytrid fungus Batrachochytrium dendrobatidis (BD) that is causing global decline in Amphibian biodiversity. The fungus exists in two stages namely Zoospores that are motile and find fresh hosts or reinfect the original host (labelled A). They are actively motile by flagella. Zoosporangia, (shown under labels B. and H.) that form once they attach and encyst in the outer soft tissue of an amphibian host, most commonly a frog (Labelled E). The zoosporangia produce numerous roots like rhizoids (labelled 2 in H) and acquire a cap (labelled 1 in H) and possess a unique cellular body under the sporangium called apophysis that seems to act as a bridge between sporangia and feeding rhizoids. C. Intracellular development of zoosporangia, with eventual death of the frog in about 2 weeks from cardiac failure (labelled G). The gestating zoosporangia lose their cap (labelled E) and release numerous motile Zoospores on contact with water. This disease has emerged as a panzootic disease in the 21st century maybe due to global warming or effect of pesticides in weakening the immune systems of frogs. This has caused major decline in the worldwide freshwater biodiversity.

An original illustration of the authors

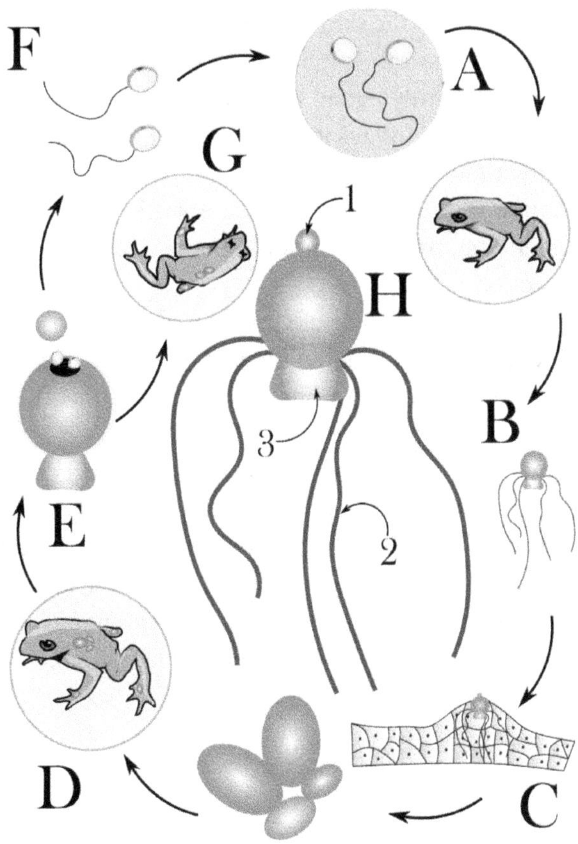

Encysted stage: The zoospore changes into an encysted form, which is capable of asexual reproduction once it has entered the amphibian. More zoospores are released into the environment as the encysted form develops and creates new sporangia.

The chytrid fungus may complete its life cycle in a matter of days, which enables it to spread quickly throughout amphibian populations. Even in the absence of amphibians, the encysted stage can survive in the environment for a very long time (Longcore et al., 1999).

PATHOPHYSIOLOGY OF THE CHYTRID FUNGUS IN AMPHIBIANS

Amphibian chytrid fungal pathology is complicated and multifaceted. The following characteristics define the pathophysiology of the chytrid fungus in amphibians:

Hyperkeratosis: When the fungus infects the keratinized skin of amphibians, it results in an expansion of the skin's thickness. As a result, the exchange of gases, liquids, and electrolytes across the skin may be reduced, which may cause mortality.

Osmoregulation disruption: Amphibians' skin plays a key role in controlling the equilibrium of their electrolytes and water. By harming the skin and reducing its ability to function, the chytrid fungus upsets this equilibrium and causes electrolyte imbalances and dehydration.

Ulceration: As the illness worsens, the skin suffers increasing skin damage, which results in ulceration. Amphibians suffer both systemic illnesses and eventual death from this infection.

Immune system disruption: The chytrid fungus can also impair amphibians' immune systems, rendering them more vulnerable to other diseases.

Heart failure: If the infection is serious enough, it may travel to the heart, causing heart failure and eventual death.

Chytrid fungus poses a serious threat to frog populations due to its pathogenic characteristics, thus efforts to stop its spread and create efficient therapies are essential for conservation efforts. Chytrid fungus has been linked to the decline and extinction of numerous species, and it poses a serious threat to amphibian populations all over the world. It is essential to comprehend the chytrid fungus' life cycle to create efficient management and control plans (Scheele et al., 2019).

RESTORATION OF WATER CHEMICAL QUALITY

The act of returning water's chemical composition to its original or desired state is referred to as "water chemical restoration." Usually, this is done to raise the standard of water used for drinking, irrigation, and industrial uses.

Pesticides, herbicides, fertilizers, and industrial pollutants are just a few of the substances that can contaminate water and endanger both human health and the environment. Chemical restoration may entail several procedures that can eliminate or lower the concentration of dangerous compounds in water, including filtration, disinfection, oxidation, and ion exchange.

Water chemical restoration may entail adding chemicals to water to change its pH, alkalinity, or hardness, improving its appropriateness for uses, in addition to eliminating impurities. For instance, adding lime or chlorine to water can raise the pH and lessen acidity while also disinfecting the liquid. Ultimately, water chemical restoration is a significant component of water management and is essential to ensure that everyone has access to clean water in a sustainable manner (Dodds & Whiles, 2020).

By removing any impurities or toxins using various chemical techniques like filtration, distillation, or reverse osmosis, water can be chemically restored. With these procedures, the water is returned to its cleanest state by being separated into its constituent molecules and other impurities. Water is filtered by running it through a porous substance, like activated carbon, which collects and eliminates contaminants. Water is boiled in the distillation process, and any contaminants that do not evaporate are left behind after collecting the condensed vapor in a different container. When water is forced through a semipermeable membrane, it selectively allows water molecules to pass through, leaving behind any impurities that are too large to pass through the membrane. This process is known as reverse osmosis (Shahidi & Fereidoon Naczk, 1992).

Water impurities can also be eliminated via chemical processes including ozonation, UV radiation, and chlorination. By adding chlorine to the water, which reacts with pollutants and destroys bacteria and other germs, chlorination is accomplished. Strong oxidants are used through ozonation and UV light to degrade and eliminate water pollutants (Welch & T Lindell, 1992).

In general, the procedure used to restore water chemically relies on the kind and quantity of contaminants present as well as the required level of purity (Welch & T Lindell, 1992).

MATRIX ASSEMBLY CLUSTER SOURCE (MACS), AN ECO-FRIENDLY PROCESS TO REMOVE CHEMICALS FROM WATER

The Matrix Assembly Cluster Source (MACS), a recently developed machine, was used to develop a ground-breaking solvent-free water treatment process. Toxic chemical removal from water now has a new, eco-friendly method according to Swansea University researchers. The study was conducted by Professor Richard Palmer at Swansea University's Centre for Innovative Materials, Processing, and Numerical Technologies (IMPACT), with funding provided by the EPSRC (Cai et al., 2020).

This represents a revolution in the way that water treatment and other catalytic processes are approached. (Cai et al. 2020) Professor Palmer stated that the MACS method of designing functional materials at the nanoscale "opens up whole new horizons across a wide range of fields, from physics and chemistry to biology and engineering."

WATER POLICY REFORM-LEGISLATION

Water managers face an urgent need for water policy reform that can support a sustainable future for people and nature, as climate change estimates indicate even bigger imbalances between water supply and demand in the coming decades. The water flows of western rivers have been increasingly harnessed over the past 150 years to supply water for farming, public water supplies, electricity generation, and other purposes. This has allowed for rapid growth in cities and the regional economy as well as the production of agricultural goods that are exported around the world. Nevertheless, this growth has come at a high cost to the environment (Samiha Ouda et al., 2020).

INTEGRATED CATCHMENT MANAGEMENT (ICM)

A basin of water is a defined geographic area where natural resources are managed using a comprehensive strategy called integrated catchment management (ICM). This is a technique to managing land, water, and other natural resources within a catchment area in a coordinated and sustainable way. The area of land from which water flows to a certain body of water, such as a lake or river, is referred to as the catchment area. A catchment is a piece of land where water from precipitation such as rain and other types collects before flowing into a river, lake, or other body of water (Riddiford, 2021).

Figure 6. The steps involved in implementation integrated catchment management (ICM) for effective and sustainable utilisation of water resources like streams, lakes, and rivers

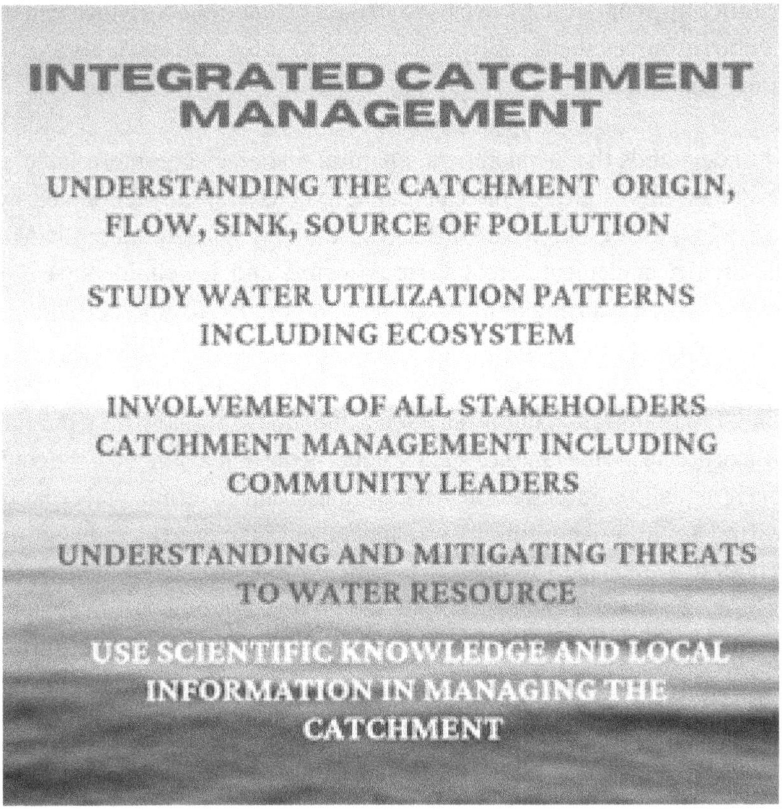

ICM aims to strike a balance between the need to use natural resources like water, land, and forests responsibly and the desire to conserve them for future generations. It entails teamwork among numerous stakeholders, including government agencies, regional communities, non-profit groups, and businesses (Samiha Ouda et al., 2020).

ICM's guiding principles include:

1. **Involve all stakeholders:** All parties involved in the planning and decision-making process work together to make sure that all points of view are considered.
2. **Responsive:** ICM acknowledges that natural systems are complex and undergoing ongoing change. As a result, management plans should be responsive to changing conditions.

3. **Ecosystem-based approach:** ICM adopts an ecosystem-based approach, acknowledging the interconnectedness of natural systems and the need for all management decisions to take the ecosystem into account.
4. **Balance approach:** ICM strives to strike a balance between the need to utilise natural resources sustainably and the need to safeguard and conserve them for future generations.

ICM understands that a catchment's natural resources are interrelated and that decisions made in one area of the catchment may have an impact on nearby areas. To ensure that these resources are utilised sustainably and in concert, ICM aims to integrate their management across various sectors and stakeholders (*A Beginners Guide to Integrated Catchment Management and the EPA's Role*, 2016).

The following crucial steps are often included in ICM:

Mapping watershed areas: Knowing the watershed entails mapping the area, identifying the important natural resources, and analyzing how they interact. ICM aims to incorporate a wide range of stakeholders in the management of the catchment, including regional communities, farmers, business, and governmental organizations.

Creating a management plan: A management plan is created that outlines the steps required to manage the natural resources in a sustainable and coordinated manner based on the stakeholders' input and an understanding of the catchment.

Modelling water flow: We can identify and target the best solutions to solve an issue by simulating the flow, terrain, and species of a watershed. We can select and target the best solutions to solve a problem by simulating the flow, terrain, and species of a watershed. For native fish, large timber significantly increases habitat complexity (G. Timothy, 2020).

Implementation and monitoring: The management plan is put into practice, and development is tracked to make sure the actions are having the desired effects. Stakeholders may improve results for the environment and the populations that depend on it by cooperating (Abellán et al., 2007).

ICM is becoming more widely acknowledged as a successful strategy for managing natural resources in a sustainable and integrated way. ICM has been used successfully to enhance the management of natural resources in various regions of the world, including Europe, Asia, and Africa (*A Beginners Guide to Integrated Catchment Management and the EPA's Role*, 2016).

STEPS TO PEVENT DECLINE OF FRESHWATER BIODIVERSITY

Our planet's health and sustainability may be significantly impacted by the loss of freshwater biodiversity. The following suggestions can stop this decline:

Habitat protection: Protecting the habitats where freshwater species dwell is one of the best methods to stop the loss of biodiversity in the freshwater ecosystem. This can be done by taking steps to conserve some places, control how land is used, and avoid polluting freshwater habitats.

Pollution reduction: Pollution reduction is essential for safeguarding freshwater ecosystems. Increased usage of ecologically friendly products, stronger laws on industrial and agricultural operations that contribute to water pollution, and a decrease in the amount of plastic garbage that enters rivers are some approaches to accomplish this.

Improve Research: Funding scientific investigation can aid in our understanding of freshwater ecosystems and the dangers they confront. This may result in the creation of better freshwater resource management techniques and more successful conservation initiatives.

Public awareness and interest: Raising public understanding of the value of freshwater biodiversity can aid in increasing public support for conservation initiatives. This can be accomplished through media efforts, educational initiatives, and public relations campaigns that highlight the importance of freshwater ecosystems and the dangers they confront.

Encourage environmentally friendly water usage: Promoting environmentally friendly water use can assist to protect freshwater resources and lessen the strain on ecosystems. This can be accomplished by taking steps like raising public awareness of water conservation, promoting water-saving technologies, and encouraging the adoption of irrigation techniques that require little water.

Encourage sustainable methods: Encourage sustainable fishing methods, as they can significantly reduce freshwater biodiversity. Sustainable fishing approaches, such as the use of selective fishing techniques and catch quotas that prevent overfishing, should be promoted to stop this (José Galizia Tundisi & Takako Matsumura Tundisi, 2016;Dmitriĭ Mikhaĭlovich Bezmaternykh, 2013).

IN SITU CONSERVATION ACTIVITIES

In situ conservation, as opposed to captive environments like zoos or botanical gardens, refers to the preservation of species within their native habitats or ecosystems. There are numerous initiatives being made to conserve species in situ since this strategy is frequently more successful at conserving a species' genetic diversity and ecological interactions. In situ conservation activities include, for example:

Protected areas: To preserve natural habitats and ecosystems, national parks, animal reserves, and other protected areas have been created. These areas are managed to reduce human disturbance and safeguard species from dangers like habitat loss and poaching.

Restoration: Restoration of degraded ecosystems can help native species thrive by creating better conditions for their survival. This can entail getting rid of unwanted species, growing native vegetation, or repairing natural streams. Community-based conservation involves collaborating with regional groups to advance ethical land-use decisions, curtail unlawful hunting and poaching, and create conservation zones that are under local control.

Wildlife corridors or Habitat corridors: By connecting fragmented ecosystems with wildlife corridors, inbreeding is reduced, and genetic diversity is increased. This allows for the mobility of animals and encourages genetic exchange. Habitat corridors are bridges that connect various island created by habitat fragmentation.

Relocation of species: Species may occasionally need to be temporarily removed from their natural habitats and reproduced in captivity to secure their survival. They can be returned into their natural habitats once their population has grown.

These are only a few instances of the worldwide in situ conservation initiatives that are being made. To ensure that endangered animals and their habitats are protected for future generations, numerous organizations, governments, and individuals are cooperating (Eldredge, 2002).

EX SITU CONSERVATION ACTIVITIES

Ex situ conservation is the practice of protecting animals away from their natural habitats, frequently in well-regulated environments like zoos, botanical gardens, or seed banks. Examples of ex situ conservation initiatives include as follows:

Captivity breeding: Programs that breed endangered species in captivity with the intention of returning them to the wild are known as captive breeding programs. Programs for captive breeding are carried out in venues like zoos, aquariums, and wildlife refuges.

Botanical gardens and seed banks: For the goal of conservation, these organizations gather and store seeds and other plant materials. By doing this, you can protect genetic diversity and guarantee the survival of plant species that could be under danger of extinction in their native habitats. These institutions gather and preserve genetic material from threatened species in their DNA and tissue banks. This can be used to support conservation efforts and is beneficial for study.

Cryopreservation: Cryopreservation is the process of storing genetic material, including as sperm, eggs, and embryos, in liquid nitrogen for use in conservation initiatives in the future. This method is frequently applied to endangered species conservation.

Reintroduction programs: These entail returning animals raised in captivity or who have undergone rehabilitation to their original habitats. The goal is to rebuild declining wild populations of endangered species.

Ex situ conservation activities are crucial for safeguarding and preserving threatened species and their genetic diversity overall (Eldredge, 2002).

CONCLUSION

These studies suggest that stopping the decline of freshwater species biodiversity can be achieved through restoring natural hydrological regimes and water chemical status, creating clean-water ponds for biodiversity, and implementing whole-catchment management and conservation actions. These studies suggest that stopping the decline of freshwater species biodiversity can be achieved through restoring natural hydrological regimes and water chemical status, creating clean-water ponds for biodiversity, and implementing whole-catchment management and conservation actions.

REFERENCES

Abellán, P., Sánchez-Fernández, D., Velasco, J., & Millán, A. (2007). Effectiveness of protected area networks in representing freshwater biodiversity: The case of a Mediterranean river basin (south-eastern Spain). *Aquatic Conservation*, *17*(4), 361–374. doi:10.1002/aqc.778

Bastmeijer, J. D., & Wongso, S. (2016a). Indian Cryptocoryne. *The Crypts pages*. Www.cryptocoryneworld.org. https://www.cryptocoryneworld.org/india.php

Belgrano, A., Woodward, G., & Jacob, U. (2015). *Aquatic functional biodiversity: an ecological and evolutionary perspective*. Elsevier: Academic Press.

Berlatsky, N. (2014). *Biodiversity*. Greenhaven Press.

Biodiversity hotspots. (n.d.). *Home - National Geographic Society*. Biodiversity. https://www.nationalgeographic.org/encyclopedia/biodiversity-hotspots/

Briggs, C. J., Knapp, R. A., & Vredenburg, V. T. (2010). Enzootic and epizootic dynamics of the chytrid fungal pathogen of amphibians. *Proceedings of the National Academy of Sciences of the United States of America*, *107*(21), 9695–9700. doi:10.1073/pnas.0912886107 PMID:20457916

Cai, R., Cao, L., Griffin, R., Chansai, S., Hardacre, C., & Palmer, R. E. (2020). Scale-up of cluster beam deposition to the gram scale with the matrix assembly cluster source for heterogeneous catalysis (propylene combustion). *AIP Advances*, *10*(2), 025314–025314. doi:10.1063/1.5142836

Chitale, V. S., Behera, M. D., & Roy, P. S. (2014). Future of Endemic Flora of Biodiversity Hotspots in India. *PLoS One*, *9*(12), e115264. doi:10.1371/journal.pone.0115264 PMID:25501852

Daniels, R. J. R. (1992). Geographical Distribution Patterns of Amphibians in the Western Ghats, India. *Journal of Biogeography*, *19*(5), 521. doi:10.2307/2845771

Dmitriĭ Mikhaĭlovich Bezmaternykh. (2013). *Sustainable Development of Asian Countries, Water Resources and Biodiversity Under Climate Change*. Siberian Branch of the Russian Academy of Sciences.

Dodds, W. K., & Whiles, M. R. (2020). *Freshwater ecology: concepts and environmental applications of limnology*. Academic Press.

Dodson, S. I. (2005). *Introduction to Limnology*. McGraw-Hill Science, Engineering & Mathematics.

Dudgeon, D., Arthington, A. H., Gessner, M. O., Kawabata, Z.-I., Knowler, D. J., Lévêque, C., Naiman, R. J., Prieur-Richard, A.-H., Soto, D., Stiassny, M. L. J., & Sullivan, C. A. (2006). Freshwater biodiversity: Importance, threats, status and conservation challenges. *Biological Reviews of the Cambridge Philosophical Society*, *81*(02), 163. doi:10.1017/S1464793105006950 PMID:16336747

US EPA. (n.d.). *Ecoregions*. EPA. https://www.epa.gov/eco-research/ecoregions

Eid, E. M., & Shaltout, K. H. (2017). Growth dynamics of water hyacinth (Eichhornia crassipes): A modelling approach. *Rendiconti Lincei. Scienze Fisiche e Naturali*, *28*(1), 169–181. doi:10.100712210-016-0589-4

Eldredge, N. (2002). Life On Earth. Vol. 1: A-G: An Encyclopedia of Biodiversity, Ecology, and Evolution. Abc-Clio.

Eriocaulaceae - efloraofindia. (n.d.). Google. https://sites.google.com/site/efloraofindia/species/a---l/e/eriocaulaceae

Fisher, M. C., Garner, T. W. J., & Walker, S. F. (2009). Global Emergence ofBatrachochytrium dendrobatidisand Amphibian Chytridiomycosis in Space, Time, and Host. *Annual Review of Microbiology*, *63*(1), 291–310. doi:10.1146/annurev.micro.091208.073435 PMID:19575560

Gomez Isaza, D. F., Cramp, R. L., & Franklin, C. E. (2020). Living in polluted waters: A meta-analysis of the effects of nitrate and interactions with other environmental stressors on freshwater taxa. *Environmental Pollution*, *261*, 114091. doi:10.1016/j.envpol.2020.114091 PMID:32062099

Harrington, L. W., & Fisher, M. J. (2014). *Water Scarcity, Livelihoods and Food Security: Research and Innovation for Development*. Taylor and Francis. doi:10.4324/9781315851662

Harvey, B. (2013). *Blue Genes*. Taylor & Francis. doi:10.4324/9781849770606

Home: ENVIS center on biological diversity. (2023). BSI Envis. https://bsienvis.nic.in/Home.aspx

IUCN SSC Western Ghats Plant Specialist Group. (n.d.). *IUCN*. IUCN. https://www.iucn.org/our-union/commissions/group/iucn-ssc-western-ghats-plant-specialist-group

Joseph, S., Sudhakar Reddy, Ch., Thomas, A. P., Srivastava, S. K., & Srivastava, V. K. (2010). Spatial interpolation of carbon stock: A case study from the Western Ghats biodiversity hotspot, India. *International Journal of Sustainable Development and World Ecology*, *17*(6), 481–486. doi:10.1080/13504509.2010.516107

Jumani, S., Rao, S., Kelkar, N., Machado, S., Krishnaswamy, J., & Vaidyanathan, S. (2018). Fish community responses to stream flow alterations and habitat modifications by small hydropower projects in the Western Ghats biodiversity hotspot, India. *Aquatic Conservation*, *28*(4), 979–993. doi:10.1002/aqc.2904

Keddy, P. (2016). *Wetland ecology: Principles and conservation*. Cambridge University Press.

Legendre, P., & Legendre, L. (1988). *Numerical Ecology*. Elsevier.

Levin, S. A. (2001). Encyclopedia of biodiversity. Academic Press and Elsevier Inc.

Lips, K. R., Brem, F., Brenes, R., Reeve, J. D., Alford, R. A., Voyles, J., Carey, C., Livo, L., Pessier, A. P., & Collins, J. P. (2006). From The Cover: Emerging infectious disease and the loss of biodiversity in a Neotropical amphibian community. *Proceedings of the National Academy of Sciences of the United States of America*, *103*(9), 3165–3170. doi:10.1073/pnas.0506889103 PMID:16481617

Longcore, J. E., Pessier, A. P., & Nichols, D. K. (1999). Batrachochytrium Dendrobatidis gen. et sp. nov., a Chytrid Pathogenic to Amphibians. *Mycologia*, *91*(2), 219–227. doi:10.1080/00275514.1999.12061011

Lovejoy, T. (1994). The quantification of biodiversity: An esoteric quest or a vital component of sustainable development? *Philosophical Transactions of the Royal Society of London. Series B, Biological Sciences, 345*(1311), 81–87. doi:10.1098/rstb.1994.0089 PMID:7972359

Maheswaran, U., Rajamani, S. K. S. K., Deivasigamani, S., Joshi, K., Verma, D., & Kumar Rajamani, S., & ROSS, D. S. (n.d.). (2022 *Computational Intelligence Approach to Improve The Classification Accuracy of Brain Tumor Detection 1192–1196*https://doi.org/ doi:10.1109/SMART55829.2022.10047792

O'Sullivan, P., & Reynolds, C. S. (2008). *The Lakes Handbook*. John Wiley & Sons.

Ouda, S., Zohry, A. E.-H., & Noreldin, T. (2020). *Deficit irrigation: A remedy for water scarcity*. Springer. doi:10.1007/978-3-030-35586-9

Patil, T., Jamdade, R. A., Patil, S., Govindwar, S. P., & Muley, D. V. (2018). DNA barcode based delineation of freshwater fishes from northern Western Ghats of India, one of the world's biodiversity hotspots. *Biodiversity and Conservation, 27*(13), 3349–3371. doi:10.100710531-018-1604-0

Raghavan, R., Prasad, G., Ali, P. H. A., & Pereira, B. (2007). Fish fauna of Chalakudy River, part of Western Ghats biodiversity hotspot, Kerala, India: Patterns of distribution, threats and conservation needs. *Biodiversity and Conservation, 17*(13), 3119–3131. doi:10.100710531-007-9293-0

Rajamani, S. K., & Iyer, R. S. (2022). Development Of an Android Mobile Phone Application for Finding Closed-Loop, Analytical Solutions to Dense Linear, Algebraic Equations for The Purpose of Mathematical Modelling in Healthcare and Neuroscience Research. *NeuroQuantology : An Interdisciplinary Journal of Neuroscience and Quantum Physics, 20*, 4959–4973. doi:10.6084/m9.figshare.c.6156024.v1

Rajamani, S. K., & Iyer, R. S. (2023a). A Scoping Review of Current Developments in the Field of Machine Learning and Artificial Intelligence. In D. Samantha (Ed.), *Designing and Developing Innovative Mobile Applications* (pp. 138–164). IGI Global. doi:10.4018/978-1-6684-8582-8.ch009

Rajamani, S. K., & Iyer, R. S. (2023b). Machine Learning Based Mobile Applications Using Python and ScikitLearn. In D. Samanta (Ed.), *Designing and Developing Innovative Mobile Applications* (pp. 282–306). IGI Global. doi:10.4018/978-1-6684-8582-8.ch016

Rawat, M. Sumit Dookia, & Chandrakasan Sivaperuman. (2015). Aquatic Ecosystem: Biodiversity, Ecology and Conservation. Springer.

Reddy, C. S., Jha, C. S., & Dadhwal, V. K. (2016). Assessment and monitoring of long-term forest cover changes (1920–2013) in Western Ghats biodiversity hotspot. *Journal of Earth System Science, 125*(1), 103–114. doi:10.100712040-015-0645-y

Richman, N. I., Böhm, M., Adams, S. B., Alvarez, F., Bergey, E. A., Bunn, J. J. S., Burnham, Q., Cordeiro, J., Coughran, J., Crandall, K. A., Dawkins, K. L., DiStefano, R. J., Doran, N. E., Edsman, L., Eversole, A. G., Füreder, L., Furse, J. M., Gherardi, F., Hamr, P., & Collen, B. (2015). Multiple drivers of decline in the global status of freshwater crayfish (Decapoda: Astacidea). *Philosophical Transactions of the Royal Society B: Biological Sciences, 370*(1662), 20140060. doi:10.1098/rstb.2014.0060

Riddiford, J. (2021). *Current integrated catchment management policy and management settings in the Murray–Darling Basin*. Murray-Darling Basin. doi:10.1016/B978-0-12-818152-2.00009-7

Scheele, B. C., Pasmans, F., Skerratt, L. F., Berger, L., Martel, A., Beukema, W., Acevedo, A. A., Burrowes, P. A., Carvalho, T., Catenazzi, A., De la Riva, I., Fisher, M. C., Flechas, S. V., Foster, C. N., Frías-Álvarez, P., Garner, T. W. J., Gratwicke, B., Guayasamin, J. M., Hirschfeld, M., & Canessa, S. (2019). Amphibian fungal panzootic causes catastrophic and ongoing loss of biodiversity. *Science, 363*(6434), 1459–1463. doi:10.1126cience.aav0379 PMID:30923224

Shahidi, F., & Fereidoon Naczk. (1992). *Ecological Effects of Wastewater*. CRC Press.

Sigee, D. C. (2005). *Freshwater Microbiology*. John Wiley & Sons.

Sivaramakrishna, P., & Yugandhar, P. (2020). A new species of the genus Portulaca L. (Portulacaceae) from the Eastern Ghats, India. *Journal of Asia-Pacific Biodiversity, 13*(4), 755–761. doi:10.1016/j.japb.2020.08.010

Soria, M., Leigh, C., Datry, T., Bini, L. M., & Bonada, N. (2017). Biodiversity in perennial and intermittent rivers: A meta-analysis. *Oikos, 126*(8), 1078–1089. doi:10.1111/oik.04118

The IUCN Red List of Threatened Species. (n.d.). *IUCN Red List of Threatened Species*. IUCN. https://www.iucnredlist.org/resources/molur2011

G., T. (2020). *Ecosystem-Based Management, Ecosystem Services and Aquatic Biodiversity: Theory, Tools, and Applications*. Springer International Publishing, Imprint Springer.

Tundisi, J. G., & Tundisi, T. M. (2016). *Limnologia*. Oficina de Textos.

Vijayakumar, S. P., Menezes, R. C., Jayarajan, A., & Shanker, K. (2016). Glaciations, gradients, and geography: multiple drivers of diversification of bush frogs in the Western Ghats Escarpment. *Proceedings of the Royal Society B: Biological Sciences, 283*(1836), 20161011. 10.1098/rspb.2016.1011

Welch, E. B., & Lindell, T. (1992). *Ecological effects of wastewater: applied limnology and pollution effects*. E & Fn Spon. doi:10.4324/9780203038499

Chapter 3
Global Leadership on Biodiversity With Regard to the Montreal Conference of the Parties 15 Summit

Isidore E. Agbokou
https://orcid.org/0000-0003-1225-9311
Centre d'Etudes Diplomatiques et Stratégiques de Paris (CEDS), France

ABSTRACT

The 15th meeting of the Conference of the Parties (CoP 15) to the Convention on Biological Diversity held in Montreal, Canada began a new era of global biodiversity leadership. This summit brought awareness that the member states have not been able to satisfactorily achieve all 20 Aichi targets. Though it varies from state to state, progress to protect biodiversity around the world is too slow, and the level of depletion is critical. Considering the concerns raised, a new agreement has been signed. This agreement aims to protect 30% of the planet's land, coastal areas, and inland waters by 2030; halve food waste; and achieve the vision of living in harmony with nature by 2050. A monitoring and evaluation framework has also been set out. COP 15 launched a new era for a bright future for biodiversity that needs renewed leadership. The new vision, however, is born with a lag in funding. The renewed leadership also requires retailored governance functions: narrative and high-level direction-setting, integration, mainstreaming, coordination, stakeholder engagement, accountability, etc.

DOI: 10.4018/978-1-6684-9034-1.ch003

GLOBAL LEADERSHIP ON BIODIVERSITY WITH REGARD TO THE MONTREAL CONFERENCE OF THE PARTIES 15 SUMMIT

Diversity within species, between species, and of ecosystems contributes in various ways to sustaining life on earth. Diversity ensures the food of the constantly growing world population is linked to the evolution of medicine through, among other things, the improvement of agricultural production with pollinating agents. Nearly four billion people mainly use natural health remedies, and about 70% of the drugs used to treat various diseases, including cancer, are derived from nature (IPBES, 2019). Energy needs are met from biomass. Both terrestrial and marine ecosystems are carbon sinks. They sequester more than half of global anthropogenic emissions, which helps to limit global warming and make life on earth pleasant. The fight to keep temperature trends below 1.5°C above pre-industrial levels would be more difficult to achieve without these ecosystems. Nature maintained by biodiversity plays a crucial role in the life and future of humanity.

Despite the importance of biodiversity, it is constantly threatened with extinction by various factors, especially those of an anthropogenic nature. The current rate of its deterioration is the fastest that it has ever been, and it takes place under the combined effect of several factors, both direct and indirect. Direct drivers include land use and land-use change, forest degradation and deforestation, overexploitation of seas and oceans, climate change, direct exploitation of organisms, invasive alien species, and pollution. Indirect factors induce the degradation of biodiversity through unsustainable modes of production and consumption, the upward trend in demography, technological innovations, and modes of governance.

People and organizations are becoming aware of the importance of biodiversity and the rate of its destruction and have taken measures to remedy the situation. Thus, during the United Nations Conference on Environment and Development, also known as the Earth Summit, held in Rio de Janeiro in 1992, a Convention on Biological Diversity (CBD) was adopted by the states parties. Nations Unies (1992) states its objectives:

The conservation of biological diversity, the sustainable use of its components and the fair and equitable sharing of the benefits arising from the use of genetic resources, in particular through satisfactory access to genetic resources and appropriate transfer of relevant techniques, taking into account all rights to these resources and to the techniques, and with adequate funding.

This convention aims to encourage actions that lead to a sustainable future for humanity, to leave a habitable planet for future generations. It concerns different types of biological diversity ranging from ecosystems to genetic resources, and it embraces all areas of life including agriculture, science, culture, politics, education, research, trade, and business.

A governing body has been created to monitor the implementation of the biodiversity convention consisting of the 196 states that have ratified the convention. This body is called the Conference of the Parties (COP), with its secretariat based in Montreal, Canada and headed by an executive secretary. Among other functions, it supports governments in the implementation of the convention and its work programs, organizes meetings, drafts documents, and ensures coordination with other international organizations. It ensures the dissemination of information related to the convention and its meetings. The COP is held every two years to review the progress made in the implementation of the convention and the decisions of the COPs. It establishes priorities and makes decisions on substantial work plans. Several countries have already hosted this conference, the first of which was held in the Bahamas from November 28 to December 9, 1994. The last, COP 15, was held in Montreal, Canada from 7 to 19 December, 2022.

After four postponements related to the COVID-19 pandemic, COP 15 on biodiversity was finally opened on Wednesday, December 7, 2022. It was preceded by a phase held online, following COVID-19 precautions, with limited face-to-face participation, in October 2021, in the city of Kumming in China. The Secretary General of the United Nations, Antonio Guterres, and the Canadian Prime Minister, Justin Trudeau, took part in the official opening alongside the president of the conference, Runqiu Huang, Chinese Minister of Ecology and the Environment. COP 15 saw the participation of the 196 Parties to the Conference. Four speeches were delivered including one from Inger Andersen, Executive Director of the United Nations Environment Program; Steven Guilbeault, Minister of Environment and Climate Change of Canada; Elizabeth Mrema, Executive Secretary of the Convention on Biological Diversity; and the President of the COP 15, Runqiu Huang.

The COP 15 is important to the future of biodiversity but its forecasts are limited as it comes after the COVID-19 pandemic and amid the Russo-Ukrainian conflict, which both have variously impacted humanity. This insight aims to take stock of the progress noted on the path of biodiversity conservation as emerged from the COP 15 and the high-level commitments in order to secure a better tomorrow for humanity.

ISSUE, SCOPE, AND METHODOLOGY

Biological diversity is constantly evolving. It is enriched by genetic modifications and evolutionary processes and also impoverished by threats under anthropogenic action. Climate change affects all sectors of life and poses a threat to biodiversity. The Paris Climate Agreement is therefore of significant importance to the survival of all animal and plant species. Its objective is to keep the temperature rise of the planet below 1.5°C above pre-industrial level. Doing so would reduce harmful global effects and bolster biodiversity. The depletion of biodiversity can lead to an ecosystem imbalance, and consequently, to the disruption of the benefits gained from it. Food chains will be broken and a gradual extinction of species will occur. To alleviate this disaster, and in view of the usefulness of biodiversity, it is important to ensure its conservation through concrete measures and actions. As humanity has become aware of the importance of biodiversity, to promote its conservation, an international framework has been drawn up. This convention on biodiversity was opened for signature on June 5, 1992, at the United Nations Conference on Environment and Development, or Earth Summit, in Rio de Janeiro. A total of 196 parties have ratified it. The convention's secretariat is based in Montreal, Canada. To assess the implementation of the convention and assess the progress on the biodiversity path, the parties meet regularly. Since ratification, several sessions on biodiversity have been organized. The COP 15 took place in two main stages. As stated, the first part was held in China, with remote attendees, in October 2021. It was a high-level meeting that worked on the development of the post-2020 global biodiversity framework. It resulted in the Kunming Declaration, which includes 17 commitments for biodiversity. The second stage that took place in Montreal enabled the adoption of the Kunming Declaration.

To advance the convention, each decade, the conference decides on a framework for global action in favor of biodiversity. Thus, during the 2010 COP 10 held from October 18 to 29 in Nagoya, Japan, the strategic plan for biological diversity for the period 2011-2020, including the 20 Aichi targets, was adopted. The COP 15 should make it possible to review the results obtained over the past decade, assess the strengths and weaknesses, internalize and share the lessons learned, and make progressive adjustments in the decade to come. Like the other COPs, for an efficient conduct of work, a chair was appointed in that of Runqiu Huang, Chinese Minister for Ecology and the Environment. To allow the study of the assorted items on the agenda, two collaborative working groups have been set up. The first working group is chaired by Ms. Rosemary Paterson from New Zealand. It considered agenda items of the COP 15 and the COP 4, the meeting of the Parties to the Nagoya Protocol, which focused on access to genetic resources and the fair and equitable sharing of benefits arising from their use.

Some of these items on the agenda are 1) review of progress made in implementing the convention and the strategic plan for biological diversity 2011-2020 and in achieving the Aichi biodiversity targets, 2) post-2020 global biodiversity framework, 3) digital sequencing information on genetic resources, 4) resource mobilization and funding mechanism, 5) capacity building and technical and scientific cooperation, and 6) planning, reporting and review mechanisms. The second group chaired by

Helena Jefferey Brown of Antigua and Barbuda also considered some of the items on the agenda of the COP 15, the COP 10, serving as the meeting of the Parties to the Cartagena Protocol on Biosafety, and the COP 4. Among other issues addressed by this group are 1) review of the effectiveness of the processes of the convention and its protocols, 2) marine and coastal biodiversity, 3) invasive alien species, 4) sustainable management of wildlife, 5) biodiversity in relation to climate change, and 6) agriculture and health.

At the difference of the previous COPs that were essentially conservation sector gatherings with delegates from ministries of environment, COP 15 recorded more than 1,400 participants registered from the finance and private sectors, with many country delegations including delegates from non-environment ministries such as health, commerce, and finance. The finance and business sector, health, human rights, and other non-conservation players were significantly more prominent at this COP. This is a sign of high-level global awareness of biodiversity depletion concern and reflects the growing understanding that nature underpins our lives, societies, and economies, and that biodiversity loss is everyone's business. This witnesses the global leadership expression that without a 'whole-of government and whole-of-society approach', the world will not be able to effectively tackle the nature crisis or interlinked climate, poverty, inequality, and fragility crises.

GLOBAL BIODIVERSITY ISSUES

Biological diversity or biodiversity refers to all forms of life on earth and their natural characteristics. It therefore includes plants, animals, and microorganisms, and it provides several services to humanity. Three types to be distinguished are 1) ecological diversity (ecosystems), 2) specific diversity (species), and 3) genetic diversity (genes). It is involved in the functioning of ecosystems and provides ecosystem services essential to human well-being. The food security of human beings is almost ensured by biodiversity, particularly through animals and plants. From a health perspective, plants are essential in the manufacture of pharmaceuticals. In view of its importance and the threats it faces, an international day has been dedicated to biodiversity to keep front of mind how vital it is to the development of humanity and the need for its conservation. The International Day for Biodiversity is celebrated on 22 May every year, in reference to the adoption of the text of the Convention on Biological Diversity by the General Assembly of the United Nations.

In the context of the conservation of biodiversity, the United Nations equipped itself in 1992 with an instrument capable of ensuring its sustainable management. This convention on biodiversity was ratified by the 196 parties. The objectives of this convention are 1) conservation of biological diversity; 2) sustainable use of components and the fair and equitable sharing of the benefits arising from the use of genetic resources through satisfactory access to genetic resources and the transfer appropriate technology, considering all rights to these resources and to the technology; and 3) appropriating adequate funding. Alongside this convention, it appeared necessary to have a consultation framework for the conservation of biological diversity. The COP is the governing body of this convention, and it is responsible for monitoring its implementation and ensuring its progress through the decisions taken during its periodic meetings. The first meeting of the COP to the Convention on Biological Diversity was held in the Bahamas from November 28 to December 9, 1994. It is a rotating COP. The COP 3 took place in Buenos Aires, Argentina, from November 4 to 15, 1996, and the fifth was held in Nairobi, Kenya, from May 15 to 26, 2000. The COP 9 was held in Bonn, Germany from 19 to 30 May 2008. During this conference, a special plenary session was organized to celebrate the International Day for Biodiversity.

In the world's march toward sustainable development, all states signed the 2030 agenda for sustainable development in 2015. That followed on from the Millennium Development Goals (MDGs) of 2000-2015. MDG assessments have shown varying progress across countries. The design of the Sustainable Development Goals (SDGs) considered the track record of the MDGs, its success and failure factors, with the purpose of capitalizing on the march undertaken by humanity toward a development that leads to a better future. The conservation and sustainable use of biological diversity is an essential factor in this development. Given the importance of biodiversity for life itself, several of these objectives are focused on its conservation. These include the SDGs 12, relating to sustainable production and consumption; 13, which deals with the fight against climate change and its repercussions; 14, relating to the conservation and sustainable use of aquatic resources; and the SDG 15 to preserve and restore terrestrial ecosystems.

DETERMINANTS OF GLOBAL LEADERSHIP ON BIODIVERSITY AT THE COP 15 IN MONTREAL

The COP 15 on Biological Diversity Following the COP 27 on Climate

COP 14 on biological diversity was held in Sharm El-Sheikh, Egypt in November 2018 with the central theme of investing in biodiversity for people and the planet. At the end of the work, several commitments were made by the 196 signatory states of the Convention on Biological Diversity including convening of a summit on biodiversity for heads of state by 2020, the acceleration of actions for the achievement of the 20

Aichi biodiversity targets by 2020, the establishment of a global and participatory process aimed at developing a post-2020 global biodiversity framework, the launch of the agenda for nature and people, pollinator issues, sustainable wildlife management, health, invasive alien species, and the mobilization of financial resources.

Climate change and biodiversity are interdependent. The negative impacts of climate change act on all sectors of socioeconomic life, ecosystems, and all living beings. Climate change leads to the degradation and loss of terrestrial and marine biodiversity, thus promoting its extinction. Biodiversity has a beneficial effect on climate change as it has the power to adapt and mitigate its adverse effects. Thanks to the biodiversity of the oceans, 30% of greenhouse gas emissions due to anthropogenic actions are absorbed (Gruber et al., 2019). Actions to maintain and restore ecosystems and biodiversity are therefore beneficial for humanity in reducing the impacts of climate change. The richer and more varied the biodiversity, the more plant species will be available for carbon sequestration. Climate change has no beneficial effect on biodiversity, and its harmful effects include the disruption of the life cycle of fauna and flora through the uncontrolled variation of precipitation and the rise in temperature leading to the extinction of species, and the acidification of the oceans which harms aquatic life.

At the COP 21 held in Paris in 2015, which is sanctioned by the historic climate agreement, concerns related to ecosystems and more broadly to biodiversity have been addressed. The COP 21 acknowledged the importance of biodiversity for the slowing of climate change and decided that it be considered in the Paris Agreement which is adopted by the state's parties in its preamble and articles seven and eight in the Global Biodiversity Framework from Kunming-Montreal COP 15 on biodiversity—climate change is included as a threat to the maintenance and development of biodiversity. It is proposed to ensure that climate change is limited so that biodiversity can flourish. The crises of climate change and the progressive loss of biological diversity are linked and can only be solved in an integrated way. The various authorities dealing with these two areas are required to work together. Thus, at the COP 26, held in Glasgow, the foundations for taking ecosystems into account were laid. During this COP, the issue of biodiversity was raised through the need to ensure the integrity of all ecosystems. During the COP 27, held in Sharm el-Sheikh in Egypt, biodiversity was also the subject of many side events and informal discussions. At the end of the session, the nature-based solution was approved as one of the approaches to combat the adverse effects of climate change. Some participants wanted the COP 15 on biodiversity to be the equivalent of the COP 21 on climate, sanctioned by the Paris Agreement. Target eight of the Global Biodiversity Framework from Kunming-Montreal clearly addresses climate change as an element of biodiversity extinction and calls for minimizing its impact. It calls for the adoption of nature-based solutions and ecosystem approaches for biodiversity enhancement.

Lessons Learned from Global Leadership on Biodiversity After a Decade of Actions

The COP 15, after a review of the results obtained at the end of the decade of biodiversity 2010-2020, had the main objective of reaching a historic agreement to halt and reverse the loss of biodiversity. The assessment shows significant gaps in the achievement of the Aichi objectives with an advanced loss of biodiversity. Finding a consensual and lasting solution is essential. The COP should therefore provide an effective response that limits the loss of biodiversity. To mitigate loss, many actions are necessary, some of which have economic impacts for certain countries. These range from reducing the use of pesticides in all their forms, combating invasive species, reforming or removing subsidies that harm the environment, and increasing funding for the protection of nature drawn from public and private sources.

Expectations of the International Community

The expectations of the international community with regard to the COP 15 were announced at the G20 summit held on November 15 and 16, 2022 in Bali, Indonesia. This group, genuinely concerned about the future of the planet, led the heads of state who took part in this summit to adopt an ambitious, balanced, pragmatic, effective, robust, and transformative post-2020 strategy. In 2010, in Nagoya, Japan, the Strategic Plan for Biodiversity 2011-2020 was adopted. This strategic plan was developed around the twenty Aichi targets for 2020, divided into two strategic goals: 1) manage the underlying causes of the loss of biological diversity by integrating biological diversity into the whole government and society and 2) enhance the benefits to all of biodiversity and ecosystem services. It should make it possible to halve the rate of loss of natural habitats in the world and safeguard the biodiversity of 17% of land areas and continental waters (Aichi Biodiversity Targets 2011-2020).

The COP 15 was the perfect opportunity for actors and stakeholders in biodiversity conservation to act. Its activities were conducted in accordance with the note by the executive secretary on the proposed organization of work (CBD/COP/15/1/Add.4/Rev.1–CBD/CP/MOP/10/1/Add. 4/Rev.1–CBD/NP/MOP/4/1/Add.4/Rev.1). Thus, the two working groups chaired by Rosemary Paterson and Helena Jefferey-Brown were established. They alternated between group and plenary work which enabled several delegations to speak on behalf of the members of their group to explain and defend their position, and express their expectations of the COP.

Need to Make Progress in the Implementation of the United Nations Convention on Biological Diversity

Like previous meetings of the parties to the convention, the COP 15 aimed to advance the biodiversity convention. As such, several issues are associated with its holding. Solutions must be found to the gradual loss of biodiversity. The main factors of biodiversity degeneration must be addressed through 1) the change in the exploitation of the seas, 2) the change in land use, 3) the cessation of the overexploitation of organisms for sustainable exploitation, 4) the fight against the adverse effects of climate change, 5) the cessation of pollution, and 5) the control of invasive non-native species.

Need to Change the Approach to Implementing the Convention

Challenges arise with finding ways to put an end to overconsumption, unsustainable production, and food waste. These factors are linked to agriculture, which is itself interdependent with biodiversity. Biodiversity is the basis of agriculture and agriculture contributes to the conservation and sustainable use of biodiversity (Secretariat of the Convention on Biological Diversity, 2008). But agriculture is a major driver of biodiversity loss. Agriculture and urbanization are the cause of 80% of the loss of biodiversity in many regions of the world (La Galaxie de Futura, n.d.). This factor must therefore be made known to the population of cities and agricultural actors in order to remedy the problem. Agricultural actors are also the guardians of agricultural biodiversity. The involvement in the conservation of biological diversity is required of the whole community including the financial and business sectors, the private and public sector, NGOs, and civil society. The challenge is to invoke action from the different strata of society in such a way that everyone is engaged because the actions of everyone matter. Indigenous peoples' and local communities' involvement as actors, not simply as consumers of solutions in decision-making processes relating to nature, as well as the recognition of their rights to land are essential for a better future for the preservation of biodiversity. Biodiversity conservation has a cost in that agreements will have to be concluded under the arbitration of the international community in relation to the financial resources that will go from developed countries to developing countries in the form of aid to finance the conservation of biodiversity.

DIVERGENCE OR DIVISIVE ISSUES AT THE COP 15

The implementation of the previous agreement on the decade of biodiversity has not worked to achieve the agreement's objectives. The 196 signatories of the COP

15 agreement adopted a gradual approach with well-measured ambitions. The new agreement provides for the protection of 30% of terrestrial and freshwater areas and 30% of coastal and marine areas by 2030 (Global Biodiversity Framework from Kunming-Montreal 2022, target 2 and 3). But it needs to be recognized that protection must concern all biodiversity. It is an approach that considers the known failures in the implementation of the steps to accomplish objectives of the decade of biodiversity and that leads the actors to be cautious in setting the targets to be achieved. The first step to achieving the 30% biodiversity protection goal relates to the financing of the agreement. Indeed, it will be necessary to reduce nature-destroying subsidies by at least $500 billion dollars by 2030 and mobilize at least $20 billion dollars per year until 2025, then $30 billion dollars in annual international aid by 2030 in favor of biodiversity (Global Biodiversity Framework from Kunming-Montreal, 2022. target 18). The issue of what share should go to developing countries has been debated between developed and developing countries. The latter preserve their biodiversity for humanity; although, they could have used it to reduce poverty. So, it is important that compensatory measures are found. A global fund dedicated to biodiversity was not established by the conclusion of the COP15. It has, however, been decided to establish one in 2023 within the Global Environment Facility (GEF), a branch dedicated to biodiversity.

The agreement did not provide a binding mechanism to encourage countries to follow their commitments. The application of the convention is left free to each actor. Such a construct is not advantageous to achieving the objectives of biodiversity conservation. There are no numbers or deadlines for certain goals. The agreement provides for limited education and awareness for decision-makers and civil society without formulating any obligatory requirements. This agreement could be compared to the tragedy of the commons. It is proactive and each actor, depending on its assessment of the agreement, may do nothing to achieve the objectives of the agreement. It should have a mechanism for periodic evaluation of achievements.

INNOVATIONS AND SCIENTIFIC ISSUES AT THE COP 15

The COP 15 aims to take action to achieve the 2050 vision of living in harmony with nature (Global Biodiversity Framework from Kunming-Montreal, 2022. P4). But it is also aware of the abuses of organisms that could result from the various manipulations of the agreement. In this sense, the COP 15 provides for the full application of the Cartagena Protocol on Biosafety through the development, adoption, and application of a plan for its effective implementation and one for strengthening capacities of actors. The protocol aims according to its article 1:

To contribute to ensuring an adequate degree of protection for the safe transfer, handling and use of living modified organisms resulting from modern biotechnology which may have adverse effects on the conservation and sustainable use of biological diversity, also taking into account risks to human health, with particular emphasis on transboundary movements (Montreal, 2000).

The COP places at the center of the debates, the adoption of approaches to ecosystem vitality and nature-based solutions that will, among other things, make it possible to combat the loss of biological diversity, restore degraded ecosystems, support sustainable food production, and promote health.

Prior to the CoP 15, many worlds class knowledge providers organizations issued authoritative contributions that informed the decisions at this historic summit. Special tribute should be paid to the United Nations agencies (UNDP, UNEP, UNESCO, FAO, etc.), World Wildlife Fund (WWF), International Union for Conservation of Nature (IUCN), world committed Non-governmental Organizations etc.

Their works should be extended to the universities across the world and robust partnership should be envisaged between those organizations and universities. Such partnership will help to leverage with accuracy and timely those universities potential for better, exhaustive, and extensive understanding of the biodiversity depletion and the better measures to be deployed to address their irreversible issues. This should also lead to reshaping the education curricula and introducing systematically biodiversity sensitivity in the spirit of the current and future generations. Foreseen partnership could take the form of Global Universities Alliance for Biodiversity Safeguard. All education doctrine should embody biodiversity concern, protection and worth.

RECURRING AND EMERGING BIODIVERSITY CHALLENGES AT THE COP 15

The loss of biological diversity has a direct impact on human health, the planet's food and nutritional security, and economic activities, among other things. It is therefore important to ensure its preservation. The various successive the COPs have worked to raise the international community's awareness of the dangers facing humanity in the face of the erosion of biodiversity. The alert, launched more than a dozen years ago, remains topical and continues to intensify. There is indeed a close link between biodiversity, climate change, the quality of life on earth, and the quality of water and air. During the so-called decade of biodiversity (2011-2020), progress in achieving the Aichi goals has been slow. Thus, a review made in 2019 by the experts, whose report is published by the convention, pointed out that none of the 20 Aichi targets has been fully achieved, and only six objectives were partially achieved—the halving of the loss of natural habitats, the end of subsidies for products harmful to biodiversity and the protection of 17% of land areas and 10% of marine areas have remained as forecasts despite being considered achievements at the time of the review (Secretariat of the Convention on Biological Diversity, 2020). The report rejects massive support of $500 billion dollars per year for activities harmful to biodiversity in favor of $78-91 billion dollars allocated each year to the protection of biodiversity (Secretariat of the Convention on Biological Diversity, 2020).

OUTSTANDING QUESTIONS

The adopted framework for the conservation of biodiversity post-2020 adopted at dawn on December 19, 2022, by the COP 15 involved lively debates at the venue in Montreal. Its objective is to accelerate actions in favor of the conservation of biodiversity. Although adopted by the COP, there are certain items that remain to be addressed further. Consensus could not be reached on certain points of the agreement which could not be considered in the final text. For example, in the field of fishing, the destruction of marine ecosystems under the influence of industrial fishing could not be resolved. Industrial fishing can continue on its current trajectory, which constitutes a danger for the survival of marine biodiversity. The IUCN Red List of Endangered Species includes several species of fish. Many stocks are overfished or have become extinct, which reduces the gene pool of these species. Overfishing as well as pollution degrade marine ecosystems. Oceans and marine life provide half of the atmospheric oxygen and absorb approximately 30% of anthropogenic carbon dioxide emissions into the atmosphere per year, linking the oceans to the proper functioning of the planet (Gruber et al., 2019). In view of this, one of the objectives of the 2030 agenda for sustainable development, in particular SDG 14 relating to aquatic life, is dedicated to oceanic ecosystem conservation. So, it is important that actions that negatively impact aquatic life and marine biodiversity are addressed. But industrial fishing is an economic activity for many countries party to the convention. It contributes to the economic well-being of their population. However, industrial fishing should be regulated in line with sustainability which will benefit oceanic biodiversity and the industry economically.

Livestock is also one of the main factors in the erosion of biodiversity. It participates in the modification of habitats, the overexploitation of species, global warming, pollution, and biological invasions. Livestock development involves a change in land use and allocation, deforestation, and forest degradation. To produce animal feed, crop areas are ever-increasing to the detriment of the protection of natural resources. Unanimity was not achieved on potential actions that should be taken by the livestock sector to encourage its participation in the maintenance of biodiversity.

As part of the fight against certain diseases or crop pests, scientists are increasingly using gene drives. It is a technique that consists of fixing a gene of interest, called a transgene, in a population using genetic engineering. It increases the probability of passing the desired genetic information from one generation to the next. Thus, with gene drives, it is possible to move toward the standardization of species or to make certain species disappear. Unlike natural mutations where the mutated sequences are not passed on to offspring, in gene drives transmission, all the information is transmitted from one generation to the next. This constitutes a threat to biodiversity. According to the Food and Agriculture Organization of the United Nations (FAO), from the beginning of the twentieth century to the present, 75% of the genetic diversity of agricultural crops has been lost (Wildfong, 1999). This erosion is linked to genetically modified organisms, which are being developed more in the agricultural sector. This is a product of the agro-food multinational companies that are investing in the field and financing research. The desire to obtain plants resistant to diseases and other hazards leads to the reduction of biological diversity at the level of the plant kingdom.

These items were passed over in silence at COP 15 partly due to industrial and economic lobbies. It is important that these different technologies are used intelligently so as not to lead to the degradation of biodiversity to the detriment of humanity.

RUSSO-UKRAINIAN CONFLICT AND ITS IMPLICATIONS ON BIOLOGICAL DIVERSITY

The Russo-Ukrainian Conflict escalated in February 2022. It has profound consequences on the environment. Referring to the estimates made by the various actors in the fight against environmental degradation, this conflict has harmful effects on the environment and leaves a bleak future for humanity on the path to sustainable development. According to the Ukrainian authorities, almost 900 protected natural areas in the country are affected by military activities. They represent around 30% of all protected areas in the country (OECD, 2022). Marine species are threatened by sonars of warships, and they and other species are affected by other engines of conflict as well. The United Nations Environment Program (UNEP), with its mandate within the United Nations system, and with the participation of partner organizations, carried out an initial monitoring of the effects of the conflict, the conclusions of which are unequivocal. Significant damage has been observed on both the urban and rural environment. For example, UNEP's rapid review reveals incidents at nuclear power plants and facilities, energy infrastructure including tankers, oil refineries, drilling rigs, gas installations and distribution pipelines, mines, and industrial and agro-industrial sites. All of this has a direct impact on the quality of water, air, soil, ecosystems, livelihoods, food and nutrition, security, and more generally, on life. Containers of hazardous substances such as solvents, ammonia, and plastics have been destroyed and their contents dumped into nature, contributing to the contamination of surface water and groundwater. Several nature reserves and protected and forest areas are under threat of fires as well. A ton of biomass contains about half a ton of carbon dioxide (Antoine et al., 2005), so forest fires and other biomasses will release significant greenhouse gases into the atmosphere. UNEP Executive Director, Inger Andersen describes the conflict as toxic.

According to the Convention on Biological Diversity, Ukraine contains 35% of the biodiversity of Europe, with nearly 70,000 animal and plant species identified (Fifth National Report of Ukraine to the Convention on Biological Diversity, 2015). Indeed, Ukraine is located at the heart of the migratory routes of many birds; it contains several rivers and a vast expanse of wetlands. The various bombardments cause fires in protected areas, nature reserves, and forest areas. This has a negative impact on these fragile ecosystems. But this violent conflict also has positive aspects on the environment. Because of the measures taken by Russia, several European countries are moving toward renewable energy. In the long term, this conflict will have environmental benefits. Overall, however, this conflict will prove to have many more harmful effects than beneficial to the environment. On this subject, leaders at the COP, without taking sides for one or the other of the states in conflict, asked for the end of the conflict.

The Ukrainian Recovery Conference held from 4 to 5 July 2022 in Lugano, Switzerland come to the conclusions and recommendations that 900 protected natural areas of Ukraine have been affected and an estimated 1.2 million hectares, or about 30% of all protected areas of Ukraine, suffer from the effects of conflict.11 Forests have been destroyed by fires from shelling and through misuse by the Russian forces, many being littered with destroyed or abandoned military vehicles (OECD, 2022).

Furthermore, according to WWF and BCG (2022, 22 p), the Ukrainian forests have been directly impacted by the conflict, with around 280,000 hectares subject to violent conflict-caused deforestation or mass felling. The same source also reported that it is estimated that 600,000 hectares of forest are currently located in conflict areas, putting them at a high risk of destruction and that, close to 3 million hectares of forests are or have been affected by the conflict, representing around 30% of the country's total forests. Last, forestry assets have also been damaged, with an estimated $440m of destructions inflicted on the State's forestry enterprises.

As of May 2022, The Ministry of Environmental Protection of Ukraine recorded more than 2000 crimes against the environment with more than 260 facts of ecocide. Restoration and development of natural resources will mitigate the negative impacts of climate change and improve quality of life. Among overall options issued by the conference to deal with the Ukraine case that become a global concern, the recovery process envisaged to focus on adaptation to climate changes as a response to the long-term threats posed by climate change on Ukraine's natural and human resources, including biodiversity, impact on ecosystems and migration. The future biodiversity leadership is expected to address the above-mentioned issues by undertaking the following measures:

- Develop forestry by managing forest clearance and planting new trees;
- Ensure the safety of Ukrainian territories by conducting mine clearance;
- Restore degraded lands and effectively use them;
- Build facilities for effective waste management and circular use of materials;
- Develop natural resources that strengthen the resilience of territories to natural disasters such as floods and fires;
- Preserve and develop natural resources of Ukraine.

The global biodiversity leadership at CoP 15 should have endorsed those conclusions and recommendations and enclosed them in its future agenda as a support to global biodiversity commitment. Ukrainian Biodiversity recovery should have been at the heart of the Summit for the great interest of the world and the future of the humanity.

AFRICAN AND EU POSITION AT THE COP 15

African nations participated in COP 15 with the firm will to bring about changes regarding protection and preservation of biodiversity. Senegal spoke on behalf of other nations in its capacity as current Chairman of the African Union. For Africa, the COP 15 offers a framework for the biodiversity community to redress its poor performance in achieving the targets set for the decade 2011-2020. For a better future for biodiversity, the COP should provide for the following elements:

- The establishment of adequate financial resources and the creation of a global fund for biodiversity of at least $100 billion dollars per year with an overall objective of 1% of global GDP in order to carry out ambitious conservation and management actions on biodiversity (Senegal's speech on behalf of African countries at the opening of COP15).
- The equitable sharing of benefits arising from the use of digital sequencing information and the establishment of a multilateral mechanism and the means for its implementation.
- Making substantial arrangements for capacity building, development, and transfer of technology to support the implementation of the convention.

- Consideration of the axes relating to integration, monitoring, and reporting which are of paramount importance for the effective implementation of the intersectoral political and legal instrument.
- The balanced consideration of the convention and its protocols during the COP. Thus, for the Protocol on Accountability and Preparedness, Africa is committed to launching a strong call for it to become operational. Considering the difficulty of the countries eligible for the framework of the realization of the replenishment of the 7th Global Environment Facility (GEF 7), to implement the program on biosafety, Africa calls for the creation of a window dedicated to implementation of the Cartagena Protocol.
- Regarding the low representation of African delegations to the COP linked to the difficult process of obtaining visas, it is imperative that relief be made for the next conferences. Africa remains committed to leading very constructive discussions and contributing to the success of the conclusions of the conference.

The European Union's 27 member states spoke with one voice through their spokesperson. They asked all the countries around the world to set up clear and ambitious biodiversity objectives and targets and to allocate substantial means to achieve them. This will make it possible to straighten the curve of the direct and indirect loss of biodiversity throughout the world. It is necessary to maintain objectives and targets, planning, and measurable indicators with a strong implementation mechanism and a resource mobilization framework. For the COP to be successful, it is essential that the integration of biodiversity issues be done in a global manner with the consideration of all actors, including the private sector, which is an operational arm in the financing of initiatives. The members of the EU are committed to taking their share of responsibility to meet the objectives, to put in place an ambitious framework through the mobilization of all possible resources. Thus, financial resources will be mobilized from different sources for the benefit of developing countries and for the implementation of the Nagoya Protocol. They ask for the continuation of discussions on the sequencing of information on genetic resources and the implementation of the various additional protocols such as Kuala Lumpur. Furthermore, they disapprove of the war in Ukraine and invite the Russian Federation to put an end to it immediately given the consequences borne by humans and the natural environment. The EU is committed to working inclusively; all the parties to the agreement constitute a force to be considered in the various actions. Representatives of women, young people, and other stakeholders must be consulted to put in place an ambitious post-2020 biodiversity framework for the benefit of the planet and its inhabitants.

FUTURE OF THE WORLD'S BIODIVERSITY AFTER THE COP 15: RESPONSIBLE ADJUSTMENT OF THE GLOBAL VISION

The 196 participating parties to the COP adopted the Kunming-Montreal agreement, which provides answers to the extension of biodiversity. Indeed, it includes 23 measures and actions, which, if well implemented, will reduce the extensive destruction of biodiversity. This includes protecting 30% of the planet and restoring 30% of degraded ecosystems (Global Biodiversity Framework from Kunming-Montreal, 2022. Target 2 & 3). Through this agreement, the extinction of wild species will be slowed and the risks caused by pollution will be avoided. An alternative will be found to the abusive exploitation of natural resources. Responsible production and consumption will be promoted in the various sectors. Funding that goes against the protection of nature will be nonexistent. The new financial framework for biodiversity will be put in place with the use of national, international, public, and private funds in the amount of $200 billion dollars (Global Biodiversity Framework from Kunming-Montreal, 2022. Target 19). Within the framework of biodiversity financing, a global aid fund for developing countries housed at the GEF will be set up and mobilize $30 billion dollars by 2030. But the COP did not see the participation of heads of state and government. Their presence would have enhanced the image of the session. However, it is first a technical debate that leads to the final agreement.

Progress is being made, but the old initiatives never flourished. The 2010-2020 decade in favor of biodiversity has only brought mixed results. None of the 20 Aichi targets has been achieved. A reason for that is because human beings may not be aware of the importance of biodiversity and nature in general. But life on earth is only possible thanks to the elements of nature such as microorganisms and pollinating insects whose loss would be unrecoverable. All measures remained at the promise stage. There was no binding action, and no monitoring method put in place. This raised fears of a bitter failure for the agreement and a more accelerated loss of biodiversity. It is for this reason that the results of COP 15 should become an awakening for all world leaders who must find ways to move from declarations of intent and conditional commitments to political will and courage, matched by decisive actions to reverse the trends of degradation and loss of biological diversity. Indeed, at the end of the COP 15, a historic multi-pronged agreement was signed. It is now possible to note progress in many areas relating to biodiversity. For the UNDP administrator, this agreement will have the merit of limiting the loss of biodiversity and protecting and restoring land and sea in a way that preserves the planet and respects the rights of indigenous peoples and local communities. Different measures will be used for this purpose. The aim is to promote the restoration of ecosystems degraded by human actions, to reduce the damage and loss of biodiversity linked to the excessive and uncontrolled use of pesticides in various sectors, and to strengthen the funding for the conservation of wild species. Agriculture is one of the sectors where pesticides, herbicides, and mineral fertilizers are used in copious quantities. Also, the establishment of new species without taking the necessary precautions is observed in several sectors with harmful consequences on biodiversity.

This conference also made it possible, on the second day of its opening, to proceed with the launch of the platform intended to help countries to accelerate the implementation of their Framework for Action on Biodiversity. Thus, 23 countries, led by Colombia and Germany, signed a declaration which sets up the Accelerator Partnership intended to help the governments of the countries to speed up the implementation of their Strategies and Plans of National Actions for Biodiversity (SPANB). Biodiversity is put back on the path to recovery for a better future. Progress is now possible in the governance of biodiversity. Capitalizing on the failures of the decade of biodiversity, the actors went beyond their individual divergence to conclude that humanity has more to gain in the combined efforts of parties. A monitoring framework has been adopted by the parties to the agreement to measure the progress made. An effective monitoring and evaluation system is essential to measure the impact of the agreement. A global assessment is also planned to be conducted in 2027.

While the COP 15 summit opened a new era for global biodiversity leadership, the renewed leadership will need retailoring of the governance functions: narrative- and high-level direction-setting, integration, mainstreaming, coordination, stakeholder engagement and alignment, oversight, accountability and enforcement, finance mobilization and channeling, robust knowledge, and expert advice and strategy articulation.

The implementation of the UNCBD deserves to benefit from the same dedicated attention as the UNFCCC at global level, and the global leadership should avoid all procrastination in the financial commitment as all cash disbursement delays can definitively have negative, dangerous, and irreversible impacts, accelerating depletion of biodiversity and therefore, equating to failure of global leadership.

Starting from now, all public and especially educational policies throughout the world and at all ages of education must always give a sacred and sublime place to biodiversity in times of war or armed conflict as well as in times of peace. Resource allocations for biodiversity mainstreaming and biodiversity research must be systematic and systemic and all governments around the world regardless of their strategic priorities must be committed to and accountable for. This is the new face of global biodiversity leadership required for the future. Visons, claims and declarations will not be sufficient to save our planet and guarantee the survival of humanity.

CONCLUSION AND RECOMMENDATIONS

The COP 15 on biodiversity, after several postponements, was held in Montreal, Canada from 7 to 19 December, 2022. It took place the day after COP 27 on climate change, which was held in Sharm El Sheikh in Egypt. It was attended by several delegations which are States Parties to the Convention on Biological Diversity. It is a COP that comes at a crucial time for humanity. Indeed, it stands on the ashes of the decade of biodiversity, which had mixed results. None of the Decade of Biodiversity targets have been met for a variety of reasons. The various delegations gave voice and showed their position at the COP 15. The work took place under the chairmanship of Runqiu Huang, Chinese Minister of Ecology and Environment. Several actors took part including the Secretary General of the United Nations, Mr. Antonio Guterres. At the end of the discussions, a new commitment was made. Thus, the Kunming-Montreal agreement was adopted by all states. Its purpose is to protect 30% of the planet's land, coastal areas, and inland waters by 2030. It also aims to halve food waste. Several measures have been proposed to enable the achievement of these objectives and, also to prepare the way for the 2050 vision of living in harmony with nature. The Accelerator Partnership to help country governments speed up the implementation of their National Biodiversity Strategies and Action Plans (NBSAPs) was set up by a coalition of countries committed to the cause of biodiversity. But the question of the financing of the agreement remains forefront. Specifically, the goal is to reduce subsidies harmful to nature by at least $500 billion dollars by 2030 by mobilizing at least $20 billion dollars per year until 2025 and then 30 billion dollars of annual international aid by 2030 (Global Biodiversity Framework from Kunming to Montreal, 2022. Target 19). This mechanism will manifest later with the Global Environment Facility. The access mechanisms are not yet known. It is urgent to accelerate this process in order, to provide states with the means necessary to conduct the global policy for the preservation of biodiversity.

The violent conflict in Ukraine has devastating consequences on the environment and therefore on biodiversity. According to the inventory made by the United Nations Environment Program and its partners, this conflict will have long-term environmental consequences; stopping it will benefit the environment. The agreement sealed between the various states party to the convention has a chance of succeeding because its implementation will consider the poor performance recorded during the decade of biological diversity.

Public policies mainstreaming biodiversity, with subsequent resource allocations regardless of the strategic priorities of the world governments without exception deserve attention, commitment, decisions, and accountability for an inspirational global biodiversity leadership.

REFERENCES

Anderson, C. B., Athayde, S., Raymond, C. M., Vatn, A., Arias, P., Gould, R. K., Kenter, J., Muraca, B., Sachdeva, S., Samakov, A., Zent, E., Lenzi, D., Murali, R., Amin, A., & Cantú-Fernández, M. (2022). Conceptualizing the diverse values of nature and their contributions to people. In P. Balvanera, U. Pascual, M. Christie, B. Baptiste, & D. González-Jiménez (Eds.), *Methodological assessment report on the diverse values and valuation of nature of the intergovernmental science-policy platform on biodiversity and ecosystem services.* IPBES Secretariat. doi:10.5281/zenodo.6493134

Andreas, B., Irene, L., Theodota, N., Tetiana, R., & Bohdan, V. (WWF), Mamdouh, A., Adrien, A., Tomislav, C., Fabien, H., Nicolas, K., (BCG). (2022). Ukraine: A sustainable Economic Recovery for People and Nature. *WWF&BCG, 67.* https://wwfeu.awsassets.panda.org/downloads/ukraine___a_sustainable_economic_recovery_for_people_and_nature___wwf_bcg___sept_2022_light.pdf

Antoine, A. A., José, C., Jérôme, D., Frédéric, D., Philippe, G., Ludovic, L., Didier, M., Joseph, M., Alfredo, N., Arthur, R., Daniel, R., Yves, R., Yves, S., Michaël, T., Marc, W., (2005). *Guide biomasse énergie*, Institut de l'énergie et de l'environnement de la Francophonie (IEPF) 2e Édition. Stavelot, Belgique 418P.

Balvanera, P., Pascual, U., Christie, M., Baptiste, B., Lliso, B., Monroy, A. S., Guibrunet, L., Anderson, C. B., Athayde, S., Barton, D. N., Chaplin-Kramer, R., Jacobs, S., Kelemen, E., Kumar, R., Lazos, E., Martin, A., Mwampamba, T. H., Nakangu, B., O'Farrell, P., & González-Jiménez, D. (2022). The role of the values of nature and valuation for addressing the biodiversity crisis and navigating towards more just and sustainable futures. In P. Balvanera, U. Pascual, C. Michael, B. Baptiste, & D. González-Jiménez (Eds.), *Methodological assessment report on the diverse values and valuation of nature of the intergovernmental science-policy platform on biodiversity and ecosystem services.* IPBES Secretariat., doi:10.5281/zenodo.6418971

Convention on Biological Diversity. (2022). *Ecological civilization-building a shared future for all life on Earth* [Conference session]. High-Level Segment - Part II of the 2020 UN Biodiversity Conference. 2P.

COP 15. (2022a). *Projet de rapport du groupe de travail I.* COP15.

COP 15. (2022b). *Projet de rapport du groupe de travail II.* COP15.

IPBES. (2019). *Rapport d'évaluation mondiale sur la biodiversité et les services écosystémiques.* IPBES.

Item 8 of the Provisional Agenda. (2018). *Pending assessments: thematic assessment of the sustainable use of wild species; methodological assessment regarding the diverse conceptualization of multiple values of nature and its benefits; and thematic assessment of invasive alien species* [Conference session]. Plenary of the Intergovernmental Science-Policy Platform on Biodiversity and Ecosystem Services, Sixth Session, Medellin, Colombia. 5P.

La Galaxie de Futura. (n.d.). *La COP15 s'attaque aux 5 causes de l'érosion de la biodiversité*. Futura. https://www.futura-sciences.com/planete/actualites/biodiversite-cop15-attack -5-causes-erosion-biodiversity-102186/

Ministry of Ecology and Natural Resources of Ukraine. (2015). Fifth National Report of Ukraine to the Convention on Biological Diversity. MENRU.

Nations Unies. (1992). *Convention sur la diversité biologique* [Conférence session]. 32P. UN.

Nicolas, G., Dominic, C., Brendan, R. C., Richard, A. F., Steven, V. H., Mario, H., Masao, I., Robert, M. K., Alex, K., Siv, K. L., Claire, L. M., Jeremy, T. M., Akihiko, M., Are, O., Fiz, F. P., Christopher, L. S., Toste, T., & Rik, W. (2019). The oceanic sink for anthropogenic CO2 from 1994 to 2007. *Science*, *363*(6432), 1193–1199. doi:10.1126cience.aau5153 PMID:30872519

OCDE. (2022). *Conséquences environnementales de la guerre en Ukraine et perspectives pour une reconstruction verte*. OCDE..

OECD. (2022). *Environmental impacts of the war in Ukraine and prospects for a green reconstruction, Policy Response, Tackling the Policy Challenge -Hub*

Point 9 de L'ordre du Jour Provisoire. (2018). *Dispositifs financiers et budgétaires pour la Plateforme* [Conférence session]. Plénière de la Plateforme Intergouvernementale Scientifique et Politique sur la Biodiversité et les Services Ecosystémiques, Sixième Session, Medellin, Colombie. 32P.

Quinzième Réunion (Partie I). (2021). *Déclaration de Kunming vers une civilisation écologique: bâtir un avenir collectif pour toute vie sur terre* [Conférence session]. Conférence des Parties à la Convention sur la Diversité Biologique, Kunming, Chine. 298P.

Quinzième Réunion (Partie II). (2022). *Projets de décisions pour la quinzième réunion de la conférence des parties à la convention sur la diversité biologique* [Conférence session]. Conférence des Parties à la Convention sur la Diversité biologique, Montréal, Canada.

Secrétariat de la Convention sur la Diversité Biologique. (2008). *Biodiversité et agriculture: Protéger la biodiversité et assurer la sécurité alimentaire.*

Secretariat of the Convention on Biological Diversity. (2020). Global biodiversity outlook 5, Montreal. Secretariat of the Convention on Biological Diversity.

Tropical Forest Alliance and Word Economic Forum (2021). *Forests, food systems and livelihoods: Trends, forecasts and solutions to reframe approaches to protecting forests insight report: September 2021*. Tropical Forest Alliance, WEF.

UNEP. (2022). *CBD/COP/15/2, 2022: Projets de décisions pour la quinzième réunion de la conférence des parties à la convention sur la diversité biologique.* UNEP. https://www.cbd.int/doc/c/c9de/d8ef/7225d7bb822f39ad3426c52b/cop-15-02-en.pdf

Wildfong, B. (1999). saving seeds. *Alternatives Journal*, *25*(1), 12–12.

Chapter 4
Insect Biodiversity of Turkey: Perspectives and Future Directions

Hasan Sevgili
Department of Molecular Biology and Genetic, Ordu University, Ordu, Turkey

ABSTRACT

The rapidly increasing world population creates more need for shelter, transportation, energy, and nutrition, leading to the construction of new settlements and agricultural areas, making habitats fragmented, endangering many species and a significant portion of local populations. Biodiversity is adversely affected by the deterioration of the functioning, and the integrity of habitat ecosystems. Insects, which constitute an important part of the animal kingdom, and exist all over the globe, affect human life both directly and indirectly, and are an indispensable part of the food chain; yet insect diversity is also under serious threat. Anatolia is a land where rich habitat and climate diversity are observed, being as rich as the whole of continental Europe in terms of plant and insect diversity on its own. In this review, the origins of, and the contributing factors to, insect diversity in Turkey is examined through data on a few specific insect groups that dwell steppe, aquatic, cave, and mountainous regions.

1. INTRODUCTION

Insects are one of, arguably, the most important group of animals in the ecosystem (Klein et al., 2007). They play key roles in many processes, such as the spread of seeds, and pollination. They have a major impact on agriculture, and even in non-agricultural areas, both positively and negatively (FAO, 2021). They affect the water quality, act as a vector in the transport and transmission of certain diseases, and their role in the food chain is being the most important food source for many vertebrates

DOI: 10.4018/978-1-6684-9034-1.ch004

(Reptiles, amphibians, birds, humans, etc.) (Prather et al., 2013). While the number of insect species discovered and published in the literature to date is only over one million, insects are the most common animal group in nature and are estimated to be represented by over 5 million species worldwide. While the calculations show that there exists a spreading of more than half a million insect species in the Palearctic region, which Turkey is a part of, it is predicted that there might exist 1.5 million species in the world belonging to the Coleoptera order, which would mean that this order has the most species diversity (Stork, 2018). Why are there so many species of insects? The most important reasons are: *a) As it is generally known, insects have a small body size. b) Some insects are able to spread widely as a result of their ability to fly. c) As their life cycles are short, their reproductive capacity is very high. Some species exhibit the facultative embryonic diapauses, giving birth to new generations several times a year under favorable temperatures, and conditions. Others necessarily spend the winter in eggs and give birth to a new generation per year, while some species spend the winter in the final stage of nymph or juvenile form, then quickly become reproductively active in early spring. Some species exhibit summer egg diapause (e. g. Anacridium aegyptium, Orthoptera). d) One of the reasons for the high diversity in Coleoptera, Lepidoptera, Hymenoptera and Diptera is the elimination of the competition for food between the nymphal and the adult period as there is a complete metamorphosis (Holometabolism). e) Since insects are coated by chitin, they are protected against factors such as dehydration and microbes cannot enter their bodies. f) With wide topographical and climatic variety, natural selection gives way to increasingly fitter populations. The insect species that can adapt to many different habitats exhibit a wider variety. g) There exists a correlation between plant and insect diversity, as these lifeforms co-evolve, meaning that richer the diversity of plants is, richer the diversity of insects will be.*

Before going into detail as to the main causes of the richness of species in insects, it will be useful to take a look at the general points on, and the main examples of the subject. The distribution of insects in an area is governed by factors such as food availability, suitability for reproductive activities, population density, and vulnerability to predation. These factors limit the distribution and population densities of the insects (le Roux et al., 2020). The body sizes of insects are related to these factors. Since insects have small body sizes, more than one insect species may survive and reproduce in small habitats, but due to competition and food requirements, habitats are often used by very few individuals of larger animal species (Bush & Butlin, 2004).

A significant number of insects are herbivores. Yet some insect species have a parasitic (e. g. mosquitoes) or parasitoid (e. g. wasps, flies, beetles) lifestyles, with their nutrition requirements and developmental processes adapted to their preferred hosts. These distinct lifestyles in insects have both encouraged speciation, and lead to niche differentiation and diversification in the same habitat. Closely related species of herbivore insects can feed on different parts of the same plant or appear on the same plant at different time intervals at different developmental stages. For example, in pollinator insects, a niche overlap occurs when two species feed on the same flowering plant. In such a case, if the resources are sufficient, the functionality of those species in the ecosystem will be more stable, but if the resources are limited, competition will arise (Blüthgen & Klein, 2011). Moreover, if two species pollinate different flowered plants or the variants of the same plant (e. g. color or fruit variations), the dynamics of the system may be disrupted with the elimination of one of the plants. Rich biological diversity enables a productive ecosystem and promotes speciation. For example, pollination of apical and basal stigma in strawberry flowers is split between small and large bees (Blüthgen & Klein, 2011). It has been reported that the visit and pollination of the same plant by different pollinators also affects fruit quality (Abrol et al., 2019; Wietzke et al., 2018). One of the species sharing the same habitat might hatch in early summer and disappear after a few months of breeding, while another species might emerge in a similar niche in the same habitat over the next few months. For example, *Isophya sikorai*, one of the green plump bushcricket, hatches in the lower parts of the Southeastern Anatolia (Şanlıurfa, Mardin) at the end of February and the beginning of March and lives during April on fresh and green vegetation (Uma & Sevgili, 2015).

Later, with the warming weather, after the breeding period, *I. sikorai* (Orthoptera) disappears completely around mid-May. The habitat used by *I. sikorai* is then occupied by *P. syriacus* belonging to the close genus *Poecilimon* (Orthoptera). *P. syriacus* is a species that is more drought resistant and continues to live on plants that are about to dry up. In late June, its habitat is occupied by the following species (Sevgili, 2016).

Turkey has a very rich habitat diversity (Figure 1). Approximately 26-28% (22 million ha) of Turkey is covered with forested areas (Güngöroğlu, 2017; Kaya & Raynal, 2001) and more than 50% of the land is ecologically suitable for forest development (Atalay et al., 2014). Forests and forest glades are favorable areas with rich and important variety of insects, animals, and plants. These areas, accompanied by deep valleys and high-altitude mountains intersecting with them, bolster diversification of insect populations over short distances, thus, numerous species have been discovered at them. In terms of particular groups of terrestrial and aquatic insects, levels of endemism in such areas are quite high (Demirsoy, 2008). For example, in the Black Sea region, the northern parts of the mountains are covered with broad-leaved trees at low altitudes, and spruce, beech, chestnut and hornbeam forests at slightly higher altitudes. Moreover, relict Mediterranean elements such as oak and maquis trees are seen on the southern slopes of the valleys further inland (e.g. Çoruh) (Atalay, 2006). The sharp climatic variability seen in this region has also led to adaptive radiation in insects over short distances (Figure 2). For example, bush cricket species of the genus *Phonochorion* (including western Georgia) inhabit the northern and higher parts of the Eastern Black Sea in Turkey, but there are no records of them at the southern parts of Kaçkar Mountains (Sevgili et al., 2010, 2015), while *Bucephaloptera bucephala* (Orthoptera), which is commonplace at the more southern part of Turkey, prefers dry and steppe-like grasslands, extends up to the Çoruh valley located in inner and southern part of the eastern Black Sea Region (unpublished data, HS). The distribution of some taxa is determined especially by climatic parameters and the mobility of the species. For example, the species of the genus *Saga* (Orthoptera: Saginae), have large body sizes, have short wings (which acts as a sound producing organ in males), and cannot fly. Thus, the mobility of these predatory bush crickets are very limited. *Saga* species can be found in shrubbery and tall grass vegetations in the Western, Southern, and Eastern regions of Anatolia, while in high mountains they spread especially in habitats with tragacanth (*Astragalus* spp.). However, the distribution of this group is limited in the northern parts of the Black Sea region (Şirin et al. 2019). The southeastern portion of Black Sea region of Turkey, sheltered by high mountains, has a humid subtropical climate, with abundant precipitation, warm winters, and humid summers. Therefore, this area is not suitable for *Saga* species that have adapted to habitats with relatively dry climates (Şirin et al., 2019). There are more than 12 thousand plant taxa in Turkey and about one-third of these plants (3649 taxa) are endemic species (Avcı, 2005; Şenkul & Kaya, 2017).

In Turkey, which has such a rich variety of flowering plants that it is quite natural that herbivorous and pollinator insect diversity is also as rich. Factors such as the heterogeneity of the topography, climatic variabilities within a short distance, and habitat diversity together provided suitable environments for plant and insect diversity and evolution in Anatolia. Plant-insect associations, herbivore host shifts, and host plant selection lead to diversification of the pollinator insects in particular (Barkan & Aytekin, 2013; Cameron et al., 2007).

Even though Anatolia is generally examined in terms of its seven different geographical climatic regions, each of these seven regions has sub-regions with distinct climatic characteristics. The northern parts of the Black Sea region and the inner regions facing south differ vastly from one another in terms of climatic characteristics. Thus, the biodiversity, gradually gets richer towards the inner regions, in accordance with a heterogeneous topographic pattern break up by numerous deep valleys. A similar situation is also valid for the Mediterranean region. While the southern parts of the Taurus mountains are covered with the typical Mediterranean vegetation, with scrub and pine trees, the northern regions are much cooler and exhibit steppe characteristics. These topographical and climatic differences resulted in an adaptive speciation in faunal and floral structure at both parts of the regions. Within the context of biological diversity of Anatolia, Eastern Anatolia should also be mentioned. Eastern Anatolia, where the heaviest snowfalls occur, is a region with the tallest mountains in Anatolia (such as Ağrı, Süphan, Munzur mountains) and it is a home to the deep valleys that were formed by large rivers such as the Euphrates and Tigris. As will be mentioned later, this region has a unique faunal and floral diversity as a consequence of the mountain ranges running diagonally in the north-south direction. Southeastern Anatolia, which includes the Upper Mesopotamia (Fertile Crescent) region, on the other hand, is the driest and the warmest part of Anatolia, with a very low average annual rainfall. This region, with lower species richness compared to other regions, is mostly represented by arid and semi-arid species.

2. REASONS WHY BIODIVERSITY MATTERS

Compared to many countries in the Palearctic zoogeographic region, Anatolia has a rich habitat diversity in terms of both genetic and species diversity. In order to understand the evolutionary process of biological diversity, disciplines such as paleobiology, paleoclimatology and phylogeography which enable us to analyze the factors in shaping biological diversity, are of great importance. Especially since the Pleistocene period (2.6 mya-11.700), the recurrent ice ages affected the distribution areas and routes of the species and consequently led to the evolution of populations across regions with a heterogeneous topography such as Anatolia, the Balkans and the Iberian Peninsula (Hewitt, 2011).

2.1 Tectonic Events

The tectonic evolutionary process of Anatolia, which is primarily responsible for the formation of the existing heterogeneous flora and fauna structure, has a very complex geological story (Okay, 2008; Okay et al., 2020). A section of the Alpine-Himalayan orogenic belt is located across Turkey. About 80 million years ago, Anatolia was not yet a land. The Anatolian plateau uplifted above the sea level since the Late Eocene (approximately 41 mya). Okay et al. (2020) published a study on the land forming processes of Pontides in the north, and taurids in the southern and the Eastern Anatolian mountains that formed the Central Anatolian Plateau (Figure 3), which illustrates Anatolia based on biostratigraphic, Cenozoic mammalian age, Cenozoic magmatic rock and thermochronological data. According to this publication, the formation of the sediments in Northern Anatolia and Central Anatolia that point to the Middle Eocene (ca. 41 Ma), are the youngest marine sediments. This area hasn't been a sea since the Middle Eocene. The formation of marine sediments in Eastern and southeastern Anatolia coincides with the Early-Central Miocene (Burdigaliyen, ca. 15 Mya). The Eastern Anatolian region had started becoming a land after this period. The Taurus Mountains, extending in the form of mountain ranges stretching from the Eastern Mediterranean to the Aegean region, uplifted above sea level and was transformed into a terrestrial environment much later than other regions of Anatolia. The age of sea deposits in these mountains is about 8 million years.

In the late Miocene epoch (about 10.4-5 million years ago) there was a vast Paratethys sea extending from Hungary (Western Paratethys) to Lake Aral and inland Iran (Eastern Paratethys) including the Black Sea. Modeling studies show that if present sea levels rise about 150 meters, a sea that could cover an area similar to old paratethys could form in the region (Esin et al., 2018). During the Messinian (Miocene) period, the uplift of the Caucasus disrupted the marine connection between the Paratethys domains (Blacksea + Caspian Sea). In the same epoch, the disconnection between the Paratethys (Serravalian period) and the Atlantic (Gibraltar strait blocked) ocean and the Mediterranean Sea caused the Messinian Salt Crisis (5.96 -5.33 million years ago) in the Mediterranean. Although it is stated that the influence of the evaporation process in the Mediterranean which emerged in this period on paleoclimatic changes in Anatolia is limited (Meijers et al., 2018), it is clear that the change in the earth's axis (precession of the equinoxes seen every 23 thousand years) corresponding to this period caused climate changes (Krijgsman et al., 1999). However, the drastic desiccation, and withdrawal of Mediterranean waters (Lago Mare) resulted in the emergence of terrestrial connections between Anatolia and the present-day Aegean islands, North Africa and the Balkans. Thus, faunal exchange became possible between Africa and Western Europe. Following the collision of the Arabian Plate with the Eurasian Plate during the Mid-Miocene, the Eastern Anatolian land part started to uplift above the sea and a heterogeneous topography formation came forth with subsidences and uplifting mountains within the region (Şengör et al., 2008). As a result of this collision, the 2500 km long Bitlis-Zagros suture, which occasionally causes major earthquakes in the region, was formed.

In the meantime, as a result of the persisting volcanic activities throughout the Neogene-Quaternary period, a large number of volcanic activities concentrated in the north of the Eastern Mediterranean (Erciyes, Melendiz, Karacadağ, Meke and Karadağ Mountains) and Eastern Anatolia (Karacadağ, Nemrut, Süphan, Tendürek and Ağrı mountains) led to the formation of the volcanic mountains of the region. This period ended 5.33 million years ago and with the rupture of the Gibraltar Strait, the waters of the Atlantic Ocean rapidly filled the Mediterranean (as the last remnant of the Tethys Sea) and thus, many terrestrial connections ceased to exist. As the Arabian plate raised Anatolia from the east and caused depressions in the Aegean region, the Aegean plate area was submerged under the Mediterranean waters and the sea extended to the north, leading to the emergence of many islands. On the other hand, large groundwater movements have unveiled the process of karstification and led to the formation of many submarine caves, especially on the rising Taurus

Mountains (Ilgar, 2015). Since the Tertiary, the forming of mountain ranges with Anatolia beginning to geomorphologically take shape, faunal exchange between Africa, Asia, Europe, and Caucasia as a result of natural passages emerging, getting submerged, and re-emerging had been the most significant processes in the shaping of Anatolian fauna. In addition, during the Late Pleistocene and Early Holocene (11,700-8,200 years ago), as the Bosphorus strait was a terrestrial passage between Europe and Anatolia that allowed faunal exchange between Balkan and Anatolian populations (Kerey et al., 2004). During the Late Pleistocene and Early Holocene, a topographical barrier is reported to exist on both sides of the south-central part of the Bosphorus (Kerey et al., 2004). This terrestrial connection lasted until recently (about 7000 years ago) when the sea levels rose, and the Marmara Sea and the Black Sea were united at the Istanbul Strait. The spread of bush locusts such as *Isophya rectipennis*, *I. speciosa*, *I. amplipennis* and *I. paveli* which all have short wings and extremely restricted mobility (Orthoptera, Phaneropterinae), can be put forward as an example of the aforementioned faunal transitions/transactions across both Thrace and the Balkans, as well as Marmara and Western Black Sea regions (Sevgili, 2004). *Arabis albina* L., an alpine plant species, is a cold-loving plant that lives in the alpine mountain belts of regions such as America, Europe, Africa, the Caucasus and the Levant. It has been reported that this species evolved in the Taurus line and that the gene flow spread from Anatolia to Europe during the Pleistocene climate fluctuations through the passage formed on Marmara (Ansell et al., 2011). As a result of the fact that the Taurus Mountains and North Anatolian Mountains, where Anatolian biological diversity is concentrated, extend in the east-west direction, a faunal distribution or migration route between the Arabian Peninsula and Africa was formed using Europe and the Caucasus-Iran line from the north, and the Taurus Mountains and the Levant line from the south. In the inter-glacial periods, alpine high mountains were habitable areas for cold-loving species, while the Central Anatolian steppes and coastal regions hosted these species as they were climatically more suitable areas during the glacial periods (Atalay, 2006). In this context, species adapted to temperate environments in the quaternary period were in refugium during glacial periods, while taxa adapted to live in cold environments were in refugium during interglacial periods (Stewart et al., 2010). This is not the fact for all species. For example, the temperate species *Spermophilus xhantopyrmnus* (Anatolian ground squirrel, Gelengi) further extended its spread across Anatolia during cold and dry periods (Gür, 2013). Due to Earth's climate getting increasingly warmer after the ice ages, the cold-loving species remained high in the mountains and therefore, became isolated. Thousands of plant and insect species evolved in the high mountains of Anatolia and the high steppes of Eastern Anatolia during this climatic fluctuation, in accordance with the theory of plant-insect coevolution. In summary, the Anatolian mountains have played a central role in the biodiversification of mountains on a global scale. Many studies today show that Anatolia is the gene center of many plant and insect taxa scattered in Europe and nearby regions. If we examine the fauna and flora of Anatolia and its neighbors, it is seen that endemic taxa are more common in mountainous areas, and a positive relationship is observed between endemism and an increase in altitude in some taxa (Noroozi et al., 2018). Not only the existence of tall and isolated mountains, but also the diverse topography have had a reinforcing effect on the evolution of endemic species (Çıplak, 2003; Noroozi et al., 2018). The Anatolian Diagonal divides Central Anatolia and Eastern Anatolia geographically by extending southward from Northeast Anatolia to the Amanos Mountains (Hatay), and this division clearly benefited the genetic diversity in Anatolia by fragmenting the populations of many plant and insect groups.

The average altitude is 1141 meters as 75% of Turkey is mountainous (Dal & Gönençgil, 2018). The remaining part consists of medium-altitude plains, plateaus, hills and wide plains slightly above the sea level (e.g. Çukurova). There exist areas with rich biodiversity that are protected by a special law in Turkey. Known as National Parks (44), Nature Conservation Areas (30), Nature Parks (247), Natural Monuments (114), Wildlife development areas (108), and Wildlife (173). The total number of these regions is 716 (Demirel & Pouya, 2020) (Figure 4). By keeping these special areas protected from constructions and agricultural activities, natural habitats with forest, aquatic and steppe ecosystems are preserved (Guclu & Karahan, 2004; Kaya & Raynal, 2001; Küçük & Ertürk, 2013).

As a result of the tectonic evolution of Anatolia, regional changes in both topography and climate characteristics deeply influenced the mobility and distribution of the fauna. Especially the mountain ranges (Anatolian Diagonal), which act as a potent geographical barrier between Central Anatolia and Eastern Anatolia, the Taurus Mountains, the Eastern Black Sea Mountains and some high mountains in Central and Western Anatolia played an important role in the shaping of the Anatolian fauna. In this context, while the Anatolian diagonal sharply divides many taxa and their geographical distribution (Eastern Anatolia / Central Anatolia) (Çıplak, 2003; Demirsoy, 2008; Gür, 2016; Kapli et al., 2013; S Mutun, 2010), for some species, the higher parts of these mountain ranges function as a passage extending from Northeast Anatolia to the Eastern Mediterranean and Taurus mountain ranges (Demirsoy et al., 2002; Sevgili et al., 2018).

The most obvious reasons that reveal the influence of the Anatolian diagonal are the differences in elevation between the regions in the east and west, and the contrasts in internal ecological dynamics and ecological conditions of the diagonal from north to south (Gür, 2016). The influence of the geographical isolation of the Anatolian diagonal is potent and has led to a genetic divide between the east and the west of the diagonal for many insect groups as well as plants (Demirsoy, 2008; Ekim & Güner, 1986; Mutun & Dinç, 2019).

2.2 Paleoclimatic Effects

Since the end of the Cretaceous period (about 65 million years ago), when nearly half of the plants and animals on Earth faced mass extinction, Earth's climate was progressively growing colder (average dropped by 14) (http://www.alpineanalytics.com/Climate/DeepTime.html). Earth enters the Milankovich cycles (100,000-year axis shift cycle) as a result of its daily motion, annual motion, axial inclination and its movement in its elliptical orbit around the Sun. Since the changes in the orbital axis of the Earth and the primary forces driving plate tectonics are also in constant motion, it is not surprising that severe climatic changes occur at certain intervals on Earth (Zachos et al., 2001). Major climatic fluctuations were recurrent at regular intervals during the Quaternary period, and these fluctuations caused part of the Northern hemisphere to remain under glaciers and led to glacial and interglacial periods. So how did insects react to these harsh climatic fluctuations during these glacial and interglacial periods? Coope (1994) reported that insects respond to these major climate changes in three different ways: "*A species responds to these climate oscillations by 1. Evolution; 2. By somehow surviving; 3. or the species go extinct*". However, as stated by (Gür, 2017), these climatic changes recurred for longer periods in the past, while the recurrent periods in the Pleistocene began to be much shorter and more sudden (100 thousand years). As already mentioned, the periodic increases and decreases in temperature of the earth is due to the Milankovitch cycles, solar density and volcanic eruptions (Bennett, 1990). The biological wealth (asset) of Anatolia, has endured major climate fluctuations in the Pleistocene, with habitat differentiation, blooms or extinctions caused by this tectonic process, it has flourished in genetic diversity as a result (Casale & Vigna Taglianti, 1999; Rögl & Steininger, 1983). When the Quaternary period entered (2,588 million years since) Turkey's geomorphology had almost realized its present form. Subsidence in the Aegean Sea, rising of sea levels, and the hot and humid climate has deeply affected the faunal diversity in Western Anatolia and the Mediterranean. However, during the Pleistocene glacial periods, both the water levels of the seas fell and the higher parts of the mountains (Taurus Mountains, Northeast Anatolia Mountains and high volcanic mountains such as Erciyes in the inner parts) were covered with glaciers (Atalay, 2006). Climatic changes that occurred between glacial and inter-glacial periods, the North-South migration of European, Siberian, and African fauna elements through Anatolia, which is a natural passage between the Caucasus, the Balkans and Africa, have been among the most important factors behind the evolution of Anatolian biological diversity.

In the climatic fluctuations that recurred in the Pleistocene period, the dynamic topography of Anatolia served as a "refugium" in terms of faunal mobility (Ansell et al., 2011; Bilgin, 2011; Hewitt, 1999). The origin of the existence of distinct habitats observed in close proximity to one another in Anatolia, and the rapid climate change are based on the tectonic evolution and paleoclimatic history of this geography. Difference in habitats and the climate observed within a narrow scope reveal faunal colonization and populations that have rapidly isolated from each other. Genetic diversity is generally shaped and structured within the range of species distribution (G. Hewitt, 2000). It is widely accepted that if barriers form between local populations the distribution and the exchange of which become broken as a result, the timescale of the differentiation of populations and the evolution of new species becomes shorter. It is also important to consider that the accumulation and the differentiation of local ecological niche gradients that shape the community is crucial for the emergence of genetic lineages (Ricklefs, 2007). Anatolia is one of the paramount biodiversity reserves in the world, with its Caucasus hotspot in the northeast, the Taurus Mountains hotspot along the Mediterranean and the Aegean coasts in the south, and the hotspots in Central and Eastern Anatolia on the Irano-Turanian line (Myers et al., 2000). About 80% of the total surface area of Turkey constitute the richest regions of plant genetic diversity (Noroozi et al., 2019). The richness of the biodiversity of about 35 prominent hot spots in the world is unfortunately at risk of being lost due to the destruction or eradication of original habitats that formed after the Cretaceous (Mittermeier et al., 2011).

Considering that the number of endemic species is one of the most important factors in determining hot spots, Anatolia, which has a heterogeneous topography, can be likened to an organic laboratory space wherein parallel evolution of plants and insects can be studied. When considered in the context of host-plant relationship, it has to be noted that the specialized insect species that feed on a certain pollinator, and the plant species itself, undergo parallel evolutions. This evolutionary process has also led to the differentiation of the chemical defenses of plants against insects (Ehrlich & Raven, 1964; Futuyma & Agrawal, 2009). Intra-species genetic diversity is therefore higher in areas with heterogeneous habitats and climate characteristics.

On the other hand, it can be said that insects undergo an adaptive differentiation by specializing to the host plant, and acceleration of speciation, niche differentiation, and different nourishment strategies are rapidly becoming widespread in insects (Futuyma & Agrawal, 2009). In addition to the rich diversity of plant species in Turkey, the high number of varieties (29.75% of the 852 endemic varieties) (Güner et al., 2012) is a fine example of diversification in the local populations. In some insect groups, similar to the plants, the number of local endemic taxa is very high, and the rate of endemism reaches 90% especially in populations that have limited mobility. For example, while the species *Sunius* of the Coleoptera order has 140 micropterous species in the palearctic (Staphylinidae), 90% of the species that inhabit Turkey (41 species and 37 endemic) are endemic, and a significant number of endemic taxa inhabit the Mediterranean (Sinan Anlaş, 2020; Assing & Schillhammer, 2001)

(Figure 5). For many other insect groups, uninterrupted mountain ranges extending from Eastern Anatolia to the Mediterranean and from there to the Aegean are an important route for the existence of endemic species, and biodiversity in general (Demirsoy 1982; Koçak 1989; Hesselbarth et al. 1995; Çıplak 2003; Kalkman et al. 2003; Kunt et al. 2008, 2010; Şirin et al. 2010; Darılmaz and İncekara 2011; Sevgili et al. 2012, 2018; Gorochov and Ünal 2012; Kıran and Karaman 2012; Ekiz et al. 2013; Hüseyinoğlu 2014; Akünal and Aslan 2017; Bozdoğan 2020). Additionally, the lakes on these mountains are home to many endemic aquatic insect species (Taşdemir et al. 2011).

3. HABITAT AND PLANT DIVERSITY

Fossil pollen analyses obtained from Anatolia belonging to the Early Miocene period revealed that approximately 90% of the flora at that time was comprised of trees (especially Conifer, Cupressaceae and Pinaceae) (Denk et al., 2019). Pollen and leaf fossil analyses conducted in order to understand the biome type of Anatolia in the Miocene period show that Laurel Forest Biome (Temperate Broadleaf and Mixed Forest) was dominant during the Aquitanian and Burdigalian (early stages of the Miocene Epoch) periods (Denk et al., 2019). It has been previously stated that Anatolia has three important floristic areas in its own unique structure, namely Caucasus, Irano-Turanian and Mediterranian regions. While referring to Turkey, Phytogeography of a sub-classification can be made in the following way; a) Mediterranean (Mediterranean-Taurus mountains), b) Mountain forest remnants (Mediterranean, Aegean, Thrace), c) Mixed broad-leaved and needle-leaved woodland resistant to cold (at the West and the North of Central Anatolia, and Amanos mountains and northern Thrace), d) Open tree and shrub vegetation (Black Sea coastal zone), e) Cold-deciduous forests (Eastern Anatolia), f) Subalpine and alpine vegetation (Taurus and Eastern Black Sea-Anatolian High Mountains), g) Dwarf shrubland (steppe) (Central Anatolia, Southeastern Anatolia and Eastern Anatolia) (Kuzucuoğlu, 2019).

The Mediterranean and Black Sea mountains extending parallel to the coasts, and the mountains that form the Anatolian Diagonal across Anatolia have created both a variety/variability in habitat and climate in Anatolia (Figure 6) (Kuzucuoğlu, 2019). Both the diversification in surface morphology under the effect of tectonic activities, and the variability of climatic processes bolstered diversification and speciation of narrowly-dispersed endemic plant populations by separating the habitats and populations in Anatolia (Ekim & Güner, 1986). Endemic plant diversity in Anatolia is more common in medium-high mountain regions (Noroozi et al., 2019). The fact that herbivorous insects (especially pollinators) constitute the raw material for the evolution of plants, and the issue of plant-insect evolution and diversification has been studied and continues to be investigated in many groups. For many plants species distributed in the Mediterranian and the side of Iran, two hot spots in terms of plant diversity along with the Caucasus and Europe Turkey is the region where the most of their ancestral genes are present (Ansell et al., 2011). For example, when the species identified in Europe belonging to the genus *Orchid* are compared, it is seen that a much richer diversity exists in the Mediterranean, Aegean, Black Sea and Eastern Anatolia regions of Turkey (total 191 taxa, 39 taxa endemic to Anatolia) (Kirca et al., 2020). Although the detected centers of endemism occur only in 16% of the land of Turkey, these areas are hosts to 59% of the endemic taxa that are prioritized for conservation. Since the majority of endemic taxa in Turkey are local and narrowly distributed most of them can be efficiently protected in the identified hot spot regions (Noroozi et al., 2018).

4. INSECT DIVERSITY IN GENERAL

In terms of the numerical distribution of insect species recorded from Turkey, the most recent listing is by (Tezcan, 2020). This study shows that approximately 95% of all insects identified in Turkey belong to the six major insect orders (Coleoptera, Hymenoptera, Lepidoptera, Diptera, Hemiptera ve Orthoptera) (Table 1). On the other hand, orders such as Stresiptera (8 species), Mecoptera (13 species), Megaloptera (2 species), Embioptera (1 species) and Phasmatodea (5 species) are represented by very few taxa (Tezcan, 2020). The order Coleoptera also has the highest number of insect species in the whole world (387,000) (Stork, 2018). However, Turkey hosts less butterfly and insect species in the world according to the ranking. Among the first six orders in the world, accounting for 90% of the total insect species, the order Coleoptera has the highest number of species both in the world and in Turkey. Turkey, unlike the world ranking, hosts fewer Lepidoptera and Diptera species. Only 5.68% of bee species in the world inhabits and/or spreads in Turkey.

Dragonflies and Mayflies are two very important groups of aquatic insects that mate in aquatic habitats and whose larval stages pass in water. Mayflies (Ephemeroptera), one of the most important insect groups in aquatic ecosystems, are also used as bioindicators in determining water quality. Ephemeroptera, while represented by 3,000 species around the world, so far around 157 species have been reported from Turkey (Darilmaz et al., 2016; Kazancı & Türkmen, 2012; Salur et al., 2016). It has been reported that endemic species generally spread in the Taurus and Black Sea regions (Kazancı & Türkmen, 2012). Another group, Hydrophiloidae aquatic insects (Coleoptera: Polyphagia) are has represented by more than 3200 species around the world, yet, only 160 taxa- 4% of which are endemic- have been reported from Anatolia (Darilmaz & İncekara, 2011). 96 species of Odonata, a semi-aquatic predator insect order, have been reported from Turkey. However, when the numbers from studies conducted in recent years are also taken into account, it is understood that a greater number of Odonata species exist in Turkey (Hacet, 2017).

Stoneflies (Plecoptera), a semi-aquatic insect group, that has 117 taxa, 44 species of which are endemic (about 3500 in the world), was reported from Turkey (Darilmaz et al., 2016).

Although Mantodea species of Turkey are well known, the taxonomy of some species is still a matter of controversy. Worldwide, with about 2500 species belonging to 15 families, Mantodea group has 23 species from 5 families reported from Turkey and Cyprus (Ehrmann, 2011; Yilmaz & Sevgili, 2023). The fact that this group consists entirely of predatory natural enemies is important in terms of biological control.

One of the well-researched groups in Turkey is Orthoptera. About 29 thousand valid species have been reported from the world (Cigliano et al., 2023). Around 740 species have been identified from Turkey as well, and more than half of the species belonging to the suborder Ensifera (Long-horned grasshoppers) are endemic (Çıplak et al., 2008; Şirin et al., 2018; Ünal, 2021). The endemism rate of the species in the genera *Poecilimon* and *Isophya* from this group reaches 80% (Sevgili, 2004; Sevgili et al., 2006, 2008, 2018).

Table 1. The insect orders listed below live both in Turkey and around the world and are among the 90% of the total insect diversity

	Number of insect species		**Relative Proportions (%)**	
Order	The World*	Turkey**	The World	Turkey
Coleoptera	387000	11910	3.08	35.22
Hymenoptera	117000	6641	5.68	19.64
Lepidoptera	157000	5580	3.55	16.5
Diptera	155000	3740	2.41	11.06
Hemiptera	104000	3424	3.29	10.13
Orthoptera	28860***	738	2.6	2.17
*(Stork, 2018)(	** (Tezcan, 2020)	(***Cigliano et al., 2023)		

One of the Diptera families, Sarcophagidae (Flesh fly species, Diptera) is represented by more than 150 species in Turkey (Verves et al., 2018) and approximately 3000 species of which have been reported from around the world (Verves et al.,

2017). Seven species of Sarcophagidae (4.58%) are endemic to Anatolia (Verves et al., 2018).

Having the richest insect species in terms of numbers among all insect orders in the world, Coleoptera (beetles) is represented in Turkey with a rich biodiversity of insect fauna. For example, the Ceutorhynchinae subfamily (Coleoptera: Curculionidae) (snout beetles) is represented by more than 260 species in Turkey, 35 species of which are endemic (Hacet & Colonnelli, 2019). 54 of the species belonging to this subfamily have been reported from the Thrace region (Hacet & Colonnelli, 2019). Ants (Hymenoptera: Formicidae), one of the important groups of Hymenopterabees, are represented by more than 13,000 species around the world (Bolton, 2021), and by just over 300 species/subspecies in Turkey (Kiran & Karaman, 2012). Around 1,400 species of Welsh bees (Hymenoptera, Cynipidae) are known to exist worldwide, with 300 species in the western Palearctic region. 148 species of 25 genera are listed from Turkey (Azmaz & Katilmiş, 2017). Some of these species have been recorded recently for Turkey or identified in recent years and are associated with around 80 species of Oak trees (Dinc et al., 2014; Mutun & Dinç, 2015; Mutun, 2016). The family Tipulidae, which is morphologically similar to mosquitoes but with a much larger body, is represented by over 4000 species worldwide. More than 160 species of this group, which is common in moist meadows and habitats, were reported from Turkey (Tekinoğlu & Çalıskan, 2019). Carpenter bees (Xylocopinae: Hymenoptera) are an important group of solitary bees that help pollinate many plants. About 40 species belonging to this subfamily, 25% of which are considered rare, have been identified in Turkey (Özbek & Terzo, 2016)

Cicadas (Cicadidae, Hemiptera), the most important group of insects that produce songs, and the insects we predominantly hear in nature on summers along with grasshoppers, are represented by 168 species in the Palearctic region, while about 32 species have been reported from Anatolia (Demirel, 2019).

Prevalent in Turkey, Auchenorrhyncha (Hemiptera) suborder includes subgroups that are harmful to agricultural areas like Cercopoidea, Cicadoidea, Fulgoroidea, and Mebracoide. Around 780 species belonging to this suborder have been reported from Turkey (Demir, 2006). Earwigs are nocturnal and they mostly feed on debris and litter while some are omnivorous or even predators. The Dermaptera has more than 22,000 species belonging to 12 families around the world, with up to 20 species having been reported Turkey (Anlaş & Kočárek, 2012). Although Earwigs are known from the Tropical Regions, it is clear that there is a need for detailed faunistic and taxonomic research in Turkey.

Of the order Neuroptera, which has many important predator species in terms of biological control, about 207 species have been identified in Turkey which constitutes for more than half of the species (108 pieces) belonging to the Chrysopidae and Myrmeleontidae families (Arı, 2014). Many faunistic studies on this group have been conducted at some regions of Anatolia. For example, in a study conducted in Bursa, 20 species belonging to 7 families (Kovanci & Kovanci, 2015), and 26 species from the Taurus Mountains, which have rich habitats and biological diversity (Atalay, 2006), have been reported (Bozdoğan, 2020).

Thrips (Thysanoptera), important agricultural pests, are represented by approximately 193 species in Turkey and in Bulgaria and Iran, neighbors of Turkey, they are represented by 155 and 201 species respectively (Tunç & Hastenpflug-Vesmanis, 2016). The order Trichoptera (caddisflies) which is a group of insects with aquatic larvae and terrestrial adults, is composed of approximately 402 species/subspecies in Turkey, and 20% of these species spreading in the Northern Anatolia region are endemic to Anatolia (Sipahiler, 2008).

It is reported that there are approximately 19,000 species of diurnal butterflies around the world, but it is estimated that this number represents only the 12% of all species discovered on Earth (Wiemers et al., 2018). More than 5000 species of moths (Koçak & Kemal, 2018) and around 415 butterfly species (including Thrace) have been reported from Turkey (Karaçetin & Welch, 2011; Trakel Web Page, 2021) the number of these current daytime butterfly species was reported as 363 according to Hesselbarth et al. (1995), it has been reported as 406 species by Koçak & Kemal (2018). More than half of these species belong to the Nymphalidae and Lycaenidae families (Wagener, 2006). Over 45 butterfly species are endemic, and more than 20 species have conservation priority status (Karaçetin & Welch, 2011). Endemism in diurnal butterflies is increases from west to east in Turkey, and especially across the Black Sea Region-East Anatolia. These are areas where endemism rates are high (Wagener, 2006).

The Reduvidae (Heteroptera) family, a significant part of which being predators (very few are blood-sucking vector species), is a very large group composed of around 7000 species around the world, and 60 species are in Turkey (Dursun & Salur, 2013).

4.1 Insect Diversity in Caves

The heterogeneous mountainous structure of Turkey has led to the formation of many caves, especially along the Taurus Mountains. The number of important caves discovered by local and foreign cave researchers in Turkey is around 800, but it is estimated that there may exist around 20 thousand caves (Anonymous, 2023). These caves are the best known karst formations, and they are of horizontal and subvertical types located mainly on the Mediterranean Region, Toros Mountains (Especially Antalya, Mersin, Kahramanmaraş, Gaziantep and Hatay) (Günay & Şimşek, 2001; Yamaç et al., 2021). Cappadocia is another region rich in caves. Especially the caves located in the Aegean Region on the Taurus Mountains line have an importance in terms of biological diversity as they hostendemic cave crickets and true crickets

(Orthoptera: Gryllomorphinae and Rhaphidophoridae) (Di Russo et al., 2007; Rampini et al., 2012; Taylan & Şirin, 2016; Taylan et al., 2012, 2013, 2015; Taylan et al., 2011) (Figures 7 and 8). *D. noctivaga,* which has a widespread in the Black Sea region, is an endemic species that has evolved to spread in many habitats, like the rainy and humid forest-covered areas that are green year round (Taylan & Şirin, 2016) (Figure 8). In addition to mines and other small caves in the area, this species also lives in dark places such as old houses, cavities, rock crevices in dense forests (HS, personnel observation). However, since *Dolichopoda* species that live in caves on the Taurus mountains cannot survive outside cave areas (xeric maquis habitat), they have a troglobiont characteristic and a narrow distribution area. On the other hand, *D. lycia* (Galvagni, 2006) inhabits south of the mountains, and *D. sbordoni* (Di Russo & Rampini, 2006) is located in the north (Taylan & Şirin, 2016). Therefore, it was inevitable that these nocturnal, and cave species were localized in a certain cave or in a narrow area. During the pleistocene, under variable climate patterns of glacial and interglacial periods, some other terrestrial insect species adapted to more stable, and sheltered caves and show a typical allopatric speciation process (Gnaspini, 1997). Although karst caves are sparse in the Black Sea region, these species in the region have a wider distribution due to the favorable climatic conditions provided by the uninterrupted cover provided by dense forests. The "Central Lake System" (Meijers et al., 2020), which covers a large area in Central Anatolia from the Early Miocene to the Pleistocene, was an important factor causing the spread and speciation of cave and mountain species on a route following the Taurus Mountains, the Aegean Mountains and the Black Sea Mountains. In the Late Miocene, it is estimated that Central Anatolia showed the characteristics of Old-World savannah and had an intermittent lake and swamp structure, and the humidity was very high (mean annual temperature 16.5-21.3°C) (Akgün et al., 2007). As a result, unsuitable habitat features in the geological history of Central Anatolia have been an effective barrier preventing movement between north-south for many terrestrial insect species. Characteristic topography and climatic features of the Karst structures and south-north contrast Taurus Mountains have provided unique habitats suitable for a distribution corridor or shelters in shaping fauna and flora.

In the insects living in a cave environment, morphological differentiation (for example, locusts have very long antennae that are better adapted to tactile input rather than visual in grasshoppers. Also the loss of pigment, long extremities, specialized mouth parts, etc.) compatible with the environmental factors such as a relatively stable temperature, humidity and lack of natural light, and distinct feeding behaviors have developed (Howarth & Moldovan, 2018). Caves in Turkey are important habitats not only for insects but also for many invertebrate groups (Aranea, Pseudoscorpiones, Collembola, Chilopoda etc.) (Aydin & Şen, 2020; Kunt et al., 2010). It has been reported that due to the effects of climate change, butterflies (especially moths) have started to utilize the cooler environments of caves during the day, indicating that the biological diversity in cave-like environments is subject to change due to these effects (Moog et al., 2021).

4.2 Insect Diversity in Steppes

The vegetations of steppes and steppe forests, are extremely important in terms of biological diversity, and they are particularly threatened by swiddening, agricultural pollution, and overgrazing. While 56% of Turkey's steppe area, which approximately adds up to 320,000 km2, is partially natural steppes and steppe forests the remaining 44% is used for agricultural purposes (Ambarlı et al., 2016). For this reason, many natural steppe areas are separated, the consequence of which is fragmentation in populations. While the Anatolian steppes constitute an important part of Central Anatolia, Eastern Anatolia, and Southeastern Anatolia, the climatic and topographic characteristics of these three regions differ. As previously mentioned, the Anatolian Diagonal helped differentiate the fauna and flora of Eastern Anatolia, which has steppe areas on higher altitudes, and the steppe of Central Anatolia, which is on lower altitudes (Ekim & Güner, 1986; Gür, 2016). The Southeastern Anatolian steppes are cut by the Taurus Mountains extending towards Eastern Anatolia, and they consist of ancient lands that form the northern regions of the fertile crescent. This region is typically the steppe where eremial species are more common due to their hot and arid climatic characteristics (Sevgili & Çıplak, 2000). When Southeastern Anatolia and Central Anatolia steppes are compared, the rate of endemism is significantly lower in Southeastern Anatolia. For example, the genus *Isophya* (43 species) and *Poecilimon* (102 species), which have richer biological diversity in Northern and Central Anatolia, are represented only by a single species (*P. syriacus, I. sikorai*) in the Southeastern Anatolian steppes. Habitat fragmentation turns natural habitats into small patches, and many of the populations living on them, along with natural areas are subject to adverse effects, moreover, these populations and natural areas might even disappear over time (Brooks et al., 2002; Fahrig, 2003). Although there hasn't been a detailed study, this situation negatively affects the genetic diversity within the species. However, no study on insects in the Anatolian steppes has been conducted to clarify the subject. Dung beetles (Coleoptera: Scarabeidae), members of the beetle group, are of great importance in terms of the quality and hydrological properties of soil, as they feed on debris, animal remains, fungi, and decaying plants (Brown et al., 2010). Of the species belonging to this family that live in Turkey (about 650 species), 40% are endemic to Anatolia, and they are concentrated in the Eastern Anatolian steppes, Central Anatolia, the Mediterranean, and the Aegean regions (Carpaneto et al., 2000).

Turkey is very rich in diversity of day butterflies (380 species / 45 endemic species) (Karaçetin & Welch, 2011). Some of its families are represented by quite a large number of species. For example, Lycaenidae members are common especially in steppe areas, and are represented by about 160 species (Ambarlı et al., 2016). Some of these species that spread in the steppes are endemic and under protection. New butterflies (such as *Neolycaena soezen, Polyommatus (Neolysandra) nazugum*) and moth (*Chrysoclista ankaraensis*) species are still identified in steppes and mountainous areas (Çalışkan, 2014; Kemal & Koçak, 2013; Seven, 2014) (Figure 9).

4.3 Aquatic Insect Diversity

The lakes and rivers of Turkey, which are very rich in unique aquatic ecosystems, offer unique habitats to many endemic aquatic / semi-aquatic species. While the Black Sea Region has active rivers and mountain lakes during all seasons of the year with its rainy and relatively cooler climate, the rivers in the semi-arid Southeast Anatolia, the Mediterranean, and the Aegean Regions mostly dry up in the summer. Particularly, a number of rivers that used to provide habitats to many aquatic creatures in the Mediterranean Region in the past have been active only during winters or have completely dried up in recent years. Two important reasons for the increasing number of dried up, or vanished streams are; the excessive use of groundwater resources and rivers bolstered by climate change, and increased need for agricultural irrigation. Many studies have shown that the amount of precipitation that directly affects biota and habitats in these regions are decreasing gradually, and approximately 20-30% of plants and animals will disappear due to the increase in global temperatures (1.5 ° C to 2.5 ° C) (Erol & Randhir, 2012). An important part of Turkey is under the Mediterranean climate regime and this region will be devastated by climate change. The most important groups that will be adversely affected in the region, especially in low altitude regions, especially with the decrease of summer precipitation, the increase of dry days, heat waves, and draught extending to occupy a significant part of the year, will undoubtedly be species that live in aquatic habitats, and water-dependent living plants and animals (Erol & Randhir, 2012). Due to climate change and the excessive use of groundwater for irrigation, the water levels of many lakes in Central Anatolia and the Aegean region have fallen, and some lakes (such as Akşehir- in which two endemic fish species have disappeared- and Meke lakes) have completely disappeared (Jeppesen et al., 2015). It is clear that Turkey's current aquatic life is under great threat. In this context, the current status of the populations of aquatic / semi-aquatic insect species is not well documented. However, a lot of research has documented the aquatic insect fauna.

May flies (Ephemeroptera) are an important insect group since the larvae of many species are indicators of the quality of aquatic ecosystems in streams, they are used as bioindicators in monitoring programs, and they also have widespread and endemic species in Turkish streams (Figure 10). 157 species belonging to approximately 33 genera show distribution in Turkey. 24 of them being endemic, most of them are narrowly distributed, and known only either from the type locality, or several different localities (Kazancı & Türkmen, 2012; Salur et al., 2016). Interestingly, some species have also been identified from very distinct water systems (e.g. *Colliarcys van*) (Figure 10). 48 Chrinomidae (Diptera) species have been identified in a study conducted in the wetlands that are known as the lakes region in Turkey (lakes in Antalya, Denizli, Burdur, Isparta, Konya) (Taşdemir & Ustaoğlu, 2005).

Another important aquatic insect group, Trichoptera (Caddishflies), which spreads in Turkish waters and has a high level of endemism, is represented by 481 species / subspecies, approximately 199 species and two genera of which are endemic to Anatolia (Darılmaz & Salur, 2015; Füsun Sipahiler, 2005). While a significant portion of Trichoptera species in Turkey is of European descent, the main root of the rich diversification in Northeast Anatolia is the Caucasian fauna (Sipahiler, 2008). The European descendant species of this group, approximately 40% of which are endemic to Anatolia, also diversify in the Aegean and the Mediterranean regions (Darılmaz & Salur, 2015; Sipahiler, 2008; F Sipahiler, 1996; Sipahiler, 2005). It is also notable that the Caddishflies diversify in the waters and the wetlands of the inward regions of Anatolia, and that many endemic species with local distribution exist in these regions. The distribution and diversification of aquatic species in these regions have been restricted for reasons such as the scarcity of streams and wetlands in the Southeastern Anatolia region, which has a hot and dry climate, and the drying of existing streams during springs. The dominance of the European descendant species is the characteristic feature of the fauna (up to 42%). While 91 species are also widely distributed in Europe (22.6%), 62 species (15.5%) show distribution in southeastern Mediterranean. 15 of the species are common Caddisflies that live in the Holarctic or the Palearctic. One of the important characteristics is the dominance of Caucasian fauna in northeastern Turkey. 15% of the species found are all Caucasian. Sixteen of the known 402 species/subspecies are found only in Turkey and Iran. While 6% of the species has a widespread distribution (Palearctic or Holarctic), 35% of the known species are endemic, which is the highest rate of endemism among the Mediterranean countries.

Coleoptera, the largest insect group, also has aquatic families. While the entire life cycle of some of these families (such as Dysticidae, Gyrinidae) passes in freshwater habitats (egg-larva-adult), in some family species (such as Hydrophilidae, Lutrochidae) the adulthood period is not dependent on water. Hydrophiloidea (Coleoptera: Polyphaga) superfamily of Turkey is a very crowded group of aquatic beetles represented by 23 genera and 161 species / subspecies (Darilmaz & İncekara, 2011). Six species are endemic to Anatolia. Ongoing studies on aquatic insects in

Anatolia provide records of, and data on new species (Darılmaz et al., 2019), and as a result faunistic lists need to be regularly updated.

Dragonflies are a common group of semi-aquatic predators that diversified in Turkish waters. They are categorized intor two large groups: dragonflies and damselflies. A total of 101 species / subspecies from Turkey have been listed, and 56 of them were reported from Thrace (Boudot et al., 2009; Demirsoy, 1982; Dumont, 1977; Hacet, 2017; Kalkman et al., 2003; Miroğlu et al., 2011). It is estimated that this group, which is a predator at both the larvae and the adult stages, is represented by nearly 6,000 species around the world, and this number might be around 7,000 species (Kalkman et al., 2008).

4.4 Diversity of Insects Effective in Biological Control

In the insect fauna of Turkey, there are many agriculturally harmful species, as well as carnivorous insect species that are important natural enemies in terms of biological control. For example, one species of grasshoppers, *Saga* Charpentier, 1825 (Orthoptera: Tettigoniidae: Saginae, King Crickets = The Saw-legged Bush-crickets) is represented by 14 species in the Palearctic region, while 8 species have been identified from Turkey (Şirin et al., 2019). These predatory katydids, which are quite large and carnivorous, feed by hunting other insect species in both bushes and tall grass in steppe areas (Figure 11). They are a more active species, especially at night. At the same time, they serve as an important food source for many reptiles, small mammals, and bird species due to their large build. Considering the distribution of *Saga* species in our country, it is seen that they have a wide distribution in forests and steppe areas. Four of the species of the genus that spread in Turkey are endemic, and they are seen in Central Anatolia, Eastern Mediterranean, and Hakkari. It is seen that the distribution of most of this particular species in the Eastern Mediterranean (Hatay, Osmaniye, Adana, Mersin) region intersects (Figure 12). On the other hand, five species have been reported to show spread in the Balkans (Kolics et al., 2012).

The most phenomenal predator group in the Insect World is undoubtedly the praying mantis. There are 20 species of these beautiful predators, of which 2400 species belonging to 29 families have been identified all over the world (Otte et al., 2022), two of which are endemic (*Eremiaphila dagi* Doğanlar, 2007; *Holaptilon yagmur* Yılmaz and Sevgili, 2023) from Turkey (Ehrmann, 2011; Yilmaz & Sevgili, 2023).

Another important predator group, Asilidae (Robber flies) family (Diptera) is represented by around 7104 species around the world (Geller-Grimm et al., 2021). Around 240 species belonging to 63 genera have been reported from Turkey (Çalışkan et al., 2020). 309 species have been reported from the neighboring country Iran (Ghahari et al., 2014; Mohammadi et al., 2022). This family, seen almost all over the world, is a well-studied group in Turkey. However, despite this, new taxon definitions are continuously being reported from Anatolia (Hasbenli et al., 2020). Robber flies hunt a lot of species belonging to many insect orders such as Hymenoptera, Orthoptera, Odonata, Coleoptera, Neuroptera, Diptera (Figure 13) (Özbek, 2016; Saghaei et al., 2010). It has been reported that the females hunt more than the males in robber flies that hunt Hemiptera species (Dennis et al. 2010). In a study conducted in Turkey, it was determined that flies mostly feed on Hymenoptera (mostly *Apis mellifera*), Diptera, Coleoptera, Heteroptera and Lepidoptera species respectively (Hayat, 1997).

The Tachinidae family (Diptera), which is an important group in terms of biological control (on pest species in both agricultural and forested areas), contains parasitoid species that are especially effective against Lepidoptera, Coleoptera, Hemiptera, Hymenoptera, Orthoptera and some other Diptera pests (Aytar et al., 2020; Kara & Alaoglu, 2001; Tunçyürek-Soydanbay, 1978). Approximately 340 species have been identified in Turkey belonging to the Tachinidae family (O'Hara et al., 2020), of which, around 8592 species are listed in the world, and 2112 species in the Palearctic region (Kara et al., 2020; Lutovinovas et al., 2018; O'Hara et al., 2020).

There are also many predatory insect species within the Coleoptera families. In a study conducted in Turkey, 40 species belonging to 3 orders (Coleoptera, Hemiptera, Neuroptera) that feed on Bark Beetle (Coleoptera: Curculionidae, Scolytinae, which has about 130 species in Turkey), which damage forested areas, were identified (Cebeci & Baydemir, 2018; Sarikaya & Ibis, 2016). Predatory species are predominant in Coleoptera families such as the Carabidae, Histeridae, Tenebrionidae, Staphylinidae, Cerylonidae and the Elateridae. Members of the Carabidae family (Coleoptera, especially the subfamily Carabinae) are an important group for biological control, and around 360 species of the Carabinae subfamily have been identified in Turkey (Avgin & Cavazzuti, 2011). Although the list of species belonging to the Carabidae family in Turkey is not complete yet, it is estimated that this geography hosts more than 1100 species (Casale & Vigna Taglianti, 1999). It is also noteworthy that this group, many of which are predators, has a rich biological diversity in Anatolia. Of the Tiger Beetle (Coleoptera: Cicindelidae) group which is predatory at both the larva and the adult stages, 46 species belonging to 8 genera have been reported from Anatolia (Avgın & Özdikmen, 2007).

It is necessary to mention that both the parasitoids and the Braconidae (Hymenoptera) family (wasps), which are frequently used commercially and have a rich biological diversity in Anatolia, function on many agricultural pests as natural enemies for biological control. Braconidae is the second largest family in the order Hymenoptera with the number of defined species being around 17 thousand, and it is estimated that there might exist up to 40 thousand species (Jones et al., 2009). A. Beyarslan, who has been studying the Braconidae family in Turkey for many years, listed 195 species belonging to 9 genera from the Braconinae subfamily and reported that more than half of these species spread only in Anatolia (Beyarslan, 2014). The same researcher has listed more than 180 species from the Opiinae subfamily of this family. While the number of species belonging to this subfamily is around 100 in the neighboring country Iran (Gadallah et al., 2016) and about 200 species from the Braconidae family are listed (Fallahzadeh & Saghaei, 2010). The fact that new species from the Braconidae family from Turkey and neighboring countries are still being detected shows that Asia Minor has a very rich biological diversity in species related to agricultural biological control.

Another important parasitoid bee group that is also effective on agricultural pests such as the microlepidoptera pupae, crop pest beetles, butterflies and some other bee species, is Ichneumonidae (Tschopp et al., 2013). About 25 thousand species of this family have been identified around the world, but it is estimated that there may be 60 thousand species with new species waiting to be discovered (Yu et al., 2016). While the Tryphoninae subfamily (Ichneumonidae) has 95 species belonging to 26 genera in Turkey (Çoruh, 2019b) and there are 61 genera and 187 species belonging to the Cryptinae subfamily (Çoruh & Özbek, 2005; Çoruh, 2019a). The Ichneumoninae subfamily is the most common group in Turkey, and approximately 237 species have been identified in total (Riedel et al., 2018). 100 species belonging to 30 genera have been reported from the Pimplinae subfamily (Çoruh, 2016) (Çoruh 2016). The Ichneumonidae family is rich in species diversity in Turkey, and data on new species is continuously being collected from Anatolia (Özdan & Gürbüz, 2019; Riedel et al., 2014).

One of the most important predator groups of Heteroptera, which especially stands out for their potential in biological control, is the species belonging to the Reduviidae (assassin bugs) family (Ambrose, 2003). Apart from this family, the families Anthocoridae, Lygaeidae, Miridae, Nabidae, Tingidae and Pentatomidae in Heteroptera contain other important predatory species (Kıyak, 2019). While more than 40 thousand species of Heteroptera have been identified around the world in general, this number is 8350 in the Palearctic region, and 1526 in Turkey (Kıyak, 2020). It has been reported so far that approximately 60 species belonging to the Reduviidae family are distributed in Turkey (Dursun & Salur, 2013).

4.5 Montane Species Diversity of Anatolia

The formation of great mountain ranges by tectonic uplifts has bolstered plant and insect diversity in many parts of the world. Subalpine and alpine zones in Anatolia, which has a mostly mountainous topography, contain many endemic montane species belonging to numerous insect orders. The "hot spot" areas with the richest endemic plant diversity in Turkey are mountainous regions (the Taurus line, the Black Sea Mountains, the Anatolian Diagonal mountain range, and mountainous Hakkari), similar to the examples of other alpine plant diversity around the world (Atalay, 2006; Ekim & Güner, 1986; Hörandl & Emadzade, 2011; Noroozi et al., 2018, 2019). Although endemism hot spots cover only around 16% of Turkey's surface area, they contain 59% of endemic taxa that are a part of protection priority (Noroozi et al., 2018). Population movements from lower altitude areas onto high mountains, or in the opposite direction (such as narrowing/expanding their distribution areas) in Anatolian interglacial and glacial periods have led to differentiation away from ancestral populations (Some speciation mechanisms by revealing the existence of rich endemic species in the Anatolian geography are illustrated in Figure 14). Especially in increasingly warmer climatic conditions, cold-loving montane species with limited mobility have speciated on high mountains separated by deep valleys. Herbivores and more temperate populations in steppe areas have undergone sympatric speciation by specializing to different plant species, paralleling the rich plant diversity of Anatolia. Some of the populations with limited mobility that live on high mountains have also undergone ethological isolation before mating, as well as sympatric speciation (for example, the male call sounds and reproductive behaviors of the three *Isophya* species; *I. ilkazi, I. obenbergeri* and *I. staneki*, which are distributed in the Ilgaz Mountains, are very different from each other) (Heller, 1988; Sevgili, 2004 unpublished data PhD thesis). Many studies have shown that insect populations living at higher altitudes (for example, aphids, Carabidae bettles, some Diptera, Orthoptera and Neuroptera) have shorter wing forms compared to populations of the same species at lower altitudes. This is result of less flight activity happening in harsh cold environments (Hodkinson, 2005). In many locust groups, ancestral populations have diverged, due to barriers such as deep valleys, and with the internal gene exchange, they have managed to adapt to high alpine/subalpine zones spread in forms resembling islands. With the adaptive radiation process, young species have diversified rapidly from their ancestral species, into a multitude of new forms (Chobanov et al., 2017; Çıplak, 2004; Demirsoy, 2008; Sevgili, 2004; Sevgili et al., 2006; Ünal, 2016).

The *Poecilimon zonatus* species group is one of the best examples of the evolution of populations on isolated lands on mountains. This group of species has vestigial short-wings (Figure 16), is cold-loving, and montane species that are distributed in isolated populations in high mountains. Their distribution in Turkey shows a pattern of extending from the Northeastern Anatolia region across the Anatolian Diagonal along the Taurus Mountains to the Western Mediterranean coast, and across the Amanos Mountains, to Lebanon in the south (Figure 16) (Sevgili et al., 2018). Coastal populations of this group hatch during early spring (possibly in mid-February on Datça peninsula; HS, personal observation) then mate and disappear in May, while populations on tall mountains can be encountered until August. Due to the poikilotherm characteristics of insects, environmental conditions determine the distribution limits of *P. zonatus* species' group populations, much like similar insect groups (Orthoptera: Tettigoniidae: *Parapholidoptera, Psorodonotus*) that spread on sky islands. In the peripatric model, that traces the origins of the distribution process of local populations diverging from ancestral population to staying on different mountain systems during the interglacial periods, showing that habitats with different ecological characteristics will push each population into the allopatric speciation process (Coyne & Orr, 2004). In this context, the speciation process called "budding speciation" takes place by partial or complete interruption of the gene flow between populations, mostly as a result of a combination of geographical isolation, microclimatic features, vegetation, and other ecological differences (Figure 15) (Fitzpatrick et al., 2009; Sevgili et al., 2018; Toussaint et al., 2013). As such, *P. varicornis* is an example of the distribution pattern at the southernmost margin of the *P. zonatus* species group (Lebanon Mountains), and most likely constitutes a "budding speciation" type distribution pattern (Sevgili et al., 2018). Considering the current distribution pattern of *P. tauricola*, a relict species of the same species group isolated on the Bolkar mountains (Eastern Mediterranean Taurus), and its sister species *P. azizsancar* has a disjunct distribution in Northeast Anatolia (Artvin-Erzurum-Bayburt), therefore it can be said that these species probably speciated allopatrically/peripatrically from a widely distributed ancestral population.

Morphological and behavioral differentiation in speciation resulting from allopatric isolation in insects may be weaker than in species with sympatric distribution. Similarly, the speciation processes of plant populations in Anatolia accelerated during the Pleistocene glaciations, which emerged after the topographic positioning process in mountainous areas was nearly complete. The speciation processes of many vicariant endemic species have emerged on sky islands (Figure 15) (Smith & Farrell, 2005; Toussaint et al., 2013) that include both subalpine and alpine zones in the Irano-Turanian and the Mediterranian regions (Ekim & Güner, 1986; Güner et al., 2012; Parolly, 2020). As such, Anatolia offers an ideal system for molecular ecology studies that aim to collect phylogeographic data on the Pleistocene rupture of both plant and insect species with restricted mobility, to show that they have evolved on these sky islands. The Nocarodeini tribus (Orthoptera: Pamphagidae), represented by many plump and endemic grasshopper species in Anatolia, might exhibit atrophied wings (Figure 17), and has limited mobility, and this species diversified in alpine regions on sky islands of tall mountains (Demirsoy, 2008; Ünal, 2016). The males of the species are considerably smaller than the females and exhibit marked sexual dimorphism in body size. The majority of the species belonging to this group have diversified as a separate species or subspecies on almost every mountaintop and have evolved isolated as a result of the disconnected gene exchange between populations due to effective barriers created by mountain ranges, and deep valleys. These species usually overwinter as nymphs or in populations that expand onto the warmer and lower parts of the mountains.

The fact that a significant part of the endemic insect species in Turkey has been discovered on mountainous areas is consistent with the concentration of plant and insect diversity in these regions. For example, the genus *Carabus* (Coleoptera: Carabidae) is a highly effective predator group especially in terms of biological control, that diversified on mountainous regions of Turkey. A significant portion of the 35 endemic species detected in Turkey have been identified in the subalpine and alpine zones of the mountains, or in the forested areas of the cooler and wetter North Anatolia (Figure 18) (Avgin & Prunier, 2015). This large group, which is composed of around 850 species in the world, is represented by around 340 species/subspecies across Turkey (Avgin & Prunier, 2015). This group of predatory beetles, which is very limited in terms of mobility, and generally lives on the ground, shows a wide color variation between its species, and therefore, some of the species' definitions based on their morphology have been invalidated. It is very difficult to estimate their overall distribution, as endemic species are distributed along mountain ranges, and some have been identified within a single locality (Figure 18). Some species of this group have adapted to hotter and more arid environments and have been identified from several localities in Southern and Southeastern Anatolia. However, a significant part of them known from the south are montane species and have been identified from areas on altitudes above 1000 m. For example, a species (*C. (Procrustes) anatolicus*) in Cyprus has been reported from the summits of Troodos Mountain (1600 m) (Avgin & Prunier, 2015). On the other hand, when the distribution pattern of *Carabus* species is examined; a significant portion of the 35 endemic species detected in Turkey are located in the subalpine and alpine zones of mountains or in the forested areas of Northern Anatolia (Figure 18) (Avgin & Prunier, 2015).

Ecological speciation occurs when reproductive isolation develops in populations with natural selection as a result of differences in the environments in which these insects live. It has been revealed that the use of different hosts promotes speciation among gall wasps when the mating frequency among gall-inducing (Cynipidae, Hymenoptera) wasp populations on alternative host plants is higher than between allopatric populations on the same host plant (Egan et al., 2012). Factors such as host change and protection from natural enemies are key factors in the diversity and the evolution of gall wasp populations (Hayward & Stone, 2005). The rich plant diversity of Anatolia, wherein 24 Quercus taxa are distributed (Uslu & Bakış, 2012), has also been effective in the diversification of populations across Anatolia by increasing the host diversity of gall wasps (Mutun, 2016). Gall midges (Diptera: Cecidomyiidae) that induce gall in plants are represented in Turkey by 118 species belonging to 53 genera (Skuhravá, 2016).

4.6. Invasive Insect Species

Comparing populations with high biodiversity to the areas where there is less, the general assumption is that the functioning of the ecosystem increases alongside the diversity, and vice versa (Cardinale et al., 2012). Natural ecosystems are becoming increasingly unsuitable due to these areas being opened for agriculture, as well as deforestation, climate change, and environmental pollution, with some of them disappearing completely (Cardinale et al., 2012). Another threat to natural biodiversity is the prominence of invasive species. Invasive species may adversely affect the dynamics of the original community through various mechanisms, such as attacking native plant species, natural predators, or their hosts, intermating with a native species, competing for food with local species, and carrying or causing a disease that will disrupt the normal functioning of the natural ecosystem (Kenis et al., 2009; Snyder & Evans, 2006). For example, the endemic *Pieris brassicae wollastoni* found in the Madeira Islands of the Atlantic Ocean became extinct sometime after the invasion of the island by *Pieris rapae*, which is now the dominant species on the island. It has been reported that the reason for this extinction was that the native species had no resistance to the virus transmitted by the invading species (Kenis et al., 2009). Extensive research is required to understand how the existing ecosystem will respond to invading species and how native interspecies relationships may change. Invasive species may suppress the development and reproduction of plants, causing them to weaken, and indirectly accelerate the destruction of the ecosystem, making it vulnerable to attacks by other pest species.

Many invasive insect species have been reported from Turkey (Table 2), but more studies have been conducted on the distribution and control of agricultural and forest pests since then. Unfortunately, no study that investigates how invasive species affect other elements of the ecosystem and their possible relationships with natural species has been conducted yet. It is also becoming apparent that the transfer of honeybees (*Apis mellifera*) that produce the irreplaceable honey for humans, between countries and continents may be harming the local pollinator species (Moritz et al., 2005). Anatolia, located in the middle of three continents, is under a great threat since it is on the route of travel of many invading species that may come from Africa, Asia and Europe, and thus, their relations with the natural fauna and flora ought to be investigated, and models have to be put forward.

Due to international trade (Battiston et al., 2020; Pintilioaie et al., 2021; Westphal et al., 2008), climate change, human migration, tourism, and some other factors, many species can be intentionally or unintentionally transferred to areas away from their natural range, and resilient ones may become invasive species by adapting to new habitats/niches. Climate change both directly and indirectly affects the fight against invasive species by altering their geographic spread rate and strength, their ability to settle in habitats and their adaptation possibilities (Hulme, 2017).

One of the most important agricultural pest groups in Turkey is the species of the Aphididae family (aphidofauna) (Görür et al., 2020). With studies conducted, 570 species belonging to 141 genera have been identified, and 2.4% of them have only been reported from Turkey (Görür et al., 2020; Şenol et al., 2015, 2017). The ability of aphids to reproduce both sexually and asexually increases their tolerance to cover wide ranges of environmental variability, increasing their survival success and facilitating them to become harmful and invasive species. Due to the rich, natural and agricultural habitat diversity of Anatolia, the possible ecological interactions of these invasive species with the populations of native species are an important research area.

Another important agricultural pest group is Fulgoromorpha species, and there are about 150 species that cause damage to 99 different cultivated plants, and they cause very serious crop losses, resulting in economic harm as well (Bourgoin, 2020). Fulgoromorpha families contain around 220 species in Turkey, including invasive species that attack many plants (Demir, 2018). *Orosanga japonica* entered Northeast Anatolia through Georgia in 2007. After a short time period *O. japonica* had also been reported at the coast of Southeast Bulgaria in 2011 (Akıner et al., 2019) (Figure 19). It has also been documented that this invasive species, which causes serious damage to the fresh branches of almost all kinds of plants within a 500 m altitude range from the sea, is the natural prey of some bird species and dragonflies (Karataş et al., 2020).

A significant portion of the invasive insect species seen in Turkey cause damage to vineyards, vegetable gardens, and forests. For example, the Citrus Long-horned Beetle (*Anoplophore chinensis*) is an invasive species that can damage not only citrus fruits but also park and garden plants (Bozkurt, 2018). Therefore, invasive species can change their known natural diet and mating habits to adapt to new areas they enter and put pressure on many native species.

In addition to the spread of invasive species that can only harm agriculture, predatory invasive species that can affect the natural fauna composition and food chain have also started to appear in Turkey (Wainright et al., 2021). For example, in recent years, Giant Asian Mantes and *Hierodula tenuidendata* Saussure, 1869 have started to spread from Thrace to Anatolia (Sevgili & Yılmaz, 2022).

Table 2. The number of invasive species belonging to major insect orders that have been reported in recent years from Turkey

Order	Number of Invasive Species	Literature
Lepidoptera	**20**	(Bayram & Tonğa, 2016; Benyamini et al., 2007; Can et al., 2022; Cengiz et al., 2016; Elekcioglu & Olculu, 2017; A. Ö. Koçak & Kemal, 2018; A. et al Koçak, 2006; Öztemiz & Doğanlar, 2015; Soyhan et al., 2013)
Coleoptera	**31**	(Arslangündoğdu & Hızal, 2010; Aysal & Kıvan, 2014; Bukejs & Telnov, 2014; Jelínek et al., 2016; Kalaentzis et al., 2019; Oğuzoğlu et al., 2017; Özdikmen, 2017; Öztemiz & Doğanlar, 2015; Tuncer et al., 2020; Yalçın et al., 2019)
Hemiptera	**80 (most of them belong to aphidofauna)**	(Ak et al., 2019; Akyıldırım, 2013; Arslangündoğdu & Hızal, 2010; Elekcioglu & Olculu, 2017; Görür et al., 2020; Güncan & Gümüş, 2019; Hizal et al., 2019; Karaca et al., 2015; Küçükbasmacı, 2014; Şenol et al., 2015; Tuncer et al., 2020; Ülgentürk, 2016)
Diptera	**14**	(Öztemiz & Doğanlar, 2015)
Thysanoptera	**20**	(Atakan & Pehlivan, 2020; Öztemiz & Doğanlar, 2015)
Hymenoptera	**4**	(Öztemiz & Doğanlar, 2015)
Mantodea	**1**	(Sevgili & Yılmaz, 2022)

5. THREATS TO BIODIVERSITY

Intensive use of habitats for agricultural purposes has a wide range of adverse effects on biodiversity as it reduces habitat diversity and landscape heterogeneity in natural areas (Hendrickx et al., 2007). The decrease in total species diversity is due to the decrease in species diversity among local populations. For example, Arthropod species diversity has been found to decrease in habitats disturbed by intensive agricultural activities (Hendrickx et al., 2007). Patchy habitats fragmented by agricultural activities accelerate the process of extinction of biodiversity and for species with limited mobility, biological contact between local populations is eliminated. This creates negative effects on the general well-being of the population, causes a decrease in genetic diversity and an increase in mating (homozygosity) in fragmented, small populations. Although insect groups with superior mobility (such as flies, butterflies, bees, dragonflies) can find a more comfortable distribution area in the heterogeneous topographic structure of Anatolia, the degree of variation within the population is quite wide due to different climatic, vegetation and harsh topographic changes in short distances. In populations of the same species in low and high-altitude areas, the differences in color and pattern, and body size parameters make it difficult to distinguish species and cause taxonomic errors in the identification of species. Specializing in certain insect groups in heterogeneous geographies such as Anatolia requires a thorough understanding of local populations and sufficient knowledge of the ecological pattern of the population. Although there are many endemic species with a very fragile distribution in Anatolia makes this geography important in terms of biodiversity, it also imposes an important responsibility on the inhabitants of this land to prevent the extinction of these populations.

Biodiversity in Turkey is partly composed of species belonging to three main hotspots (Caucasus, Mediterranian, Irano-Anatolien) (Hanson et al., 2009). Loss of biodiversity in a particular region means not only the extinction of species, but also the disruption of the functionality of the ecosystem as a result of the loss of genes, individuals, and populations. It would be accurate to state that the main reason for all these extinctions is the rapid growth in human population, and the damage to natural areas that the increase in the demand for natural resources excessive consumption creates. Protecting the well-being of biological diversity is crucial both for the internal dynamics of biological diversity and for the sustainability of species, ecosystems, and ecosystems' utility for the quality and the future of human life (Winn & Pogutz, 2013). Briefly, the factors negatively effect on biological diversity can be summarized as:

1. **Loss of specific or general habitats** (e.g., deterioration and loss of woodlands, steppes, freshwater ecosystems, Figure 20) (Brooks et al., 2002; Hendrickx et al., 2007; Jactel et al., 2021; Tilman et al., 2017): Loss of habitat can change the distribution areas of living things: drought and decrease in precipitation and increase in temperature negatively affect life in these regions and the species here will have to migrate towards cool/cooler, and livable areas where they can mate. Some species will have gone extinct completely during the process. Habitats for more northern, and cold-loving species will shrink and many local populations will be threatened with extinction. However, some resilient species that can rapidly spread in degraded areas will become dominant, and prominent populations becoming new threats to agricultural areas.
2. **Environmental pollution** (Noxious waste for the environment and living beings resulting from industrial or agricultural production) (Agathokleous et al., 2020; Cardoso et al., 2020; Jactel et al., 2021; Tilman et al., 2017). Global or local environmental factors such as pollution, and land transformation cause concerning loss in insect diversity (Cardoso et al., 2020). Conservation of habitats of all sizes is of the utmost importance for all nations.
3. **Influence of invasive species** (Jactel et al., 2021; Kenis et al., 2009; Mainka & Howard, 2010): It has been reported that there are around 14,000 invasive species in Europe, more than half of which have come from regions outside

of Europe, and their numbers continue to rapidly increase (Roy et al., 2019). For example, 30 invasive mammal species have caused the extinction of 738 species (birds, reptiles, other mammals) (Doherty et al., 2016).

4. **Excessive demand for species that have been commodified for leisure or human consumption** (Cardoso et al., 2020; Tilman et al., 2017): Although it is not a common situation in Turkey, insects are consumed as food in many parts of the world. Insects are utilized as traditional medicine, and even some are used as pets or decorations (Cardoso et al., 2020). Butterflies and beetles are used as ornaments and souvenirs. Praying mantises are the most preferred insect pets in homes. In the future, the rate of consuming insects for nutrition may increase gradually for reasons such as population growth, and a probable global protein crisis.
5. **Consequences of climate change (Adverse effects of climate change (negative changes in the food chain/trophic hierarchy due to loss of suitable habitats and changes in species phenologies, loss of natural corridors, inappropriate fire regimes etc.))** (Cardoso et al., 2020; Harris et al., 2019; Jactel et al., 2021; Mitton & Ferrenberg, 2012; Raven & Wagner, 2021; Sands, 2018; Wilson & Maclean, 2011):
    a. Living organisms are only able to continue their vital activities between certain temperatures (air and water temperature). While direct deaths may increase due to thermal stress, subthermal temperatures might not cause the death of the species, in fact, it may cause infertility, ending the reproductive activity and making the populations disappear in time (Parratt et al., 2021).
    b. The reproductive periods of living organisms and the growth periods of plants will vary immensely: A species is born/hatched/germinated from seed within a certain period. Due to the temperature change, these vital cycles will be altered and there will be a lack of target organisms in activities such as finding food. On the other hand, if the synchronization between pollinator insects and host plants is lost, serious damage may befall the ecosystem.
    c. The number of invasive and alien species will increase: While climate change will put stress on the activities of many species in their current range, it may provide better mating and feeding conditions for other species. In particular, populations of some invasive and alien species in forests and agricultural areas may become immensely denser and more harmful. By nature, the adaptation abilities of the invasive species are also high. Under the stress of climate change, the current activities of natural enemies, which to some extent limit the spread of invasive species and their population density, may be disrupted. There are many studies

demonstrating that invasive species are also a threat to native species in the community by changing the process and function of the ecosystem (Dueñas et al., 2018).

d. Possible breaks in the food chain/web: Damage to a population with trophic levels of terrestrial ecosystems (for example, phytoplankton, insect population, disappearance of a key species, etc.) may cause disruptions in the food chain, resulting in negative effects on many other species in the network. For example, insect parasitoids are organisms that can exist in an ecosystem with their hosts. With the effects of climate change, alterations in the seasonal ecology of the parasitoid will disrupt the host-parasitoid synchronization, the phenology of other predators in that network, and the food web relationship between the host and the plant. The disruption may also cause the species to evolve according to the malleability of the species to the new environmental conditions (Tougeron et al., 2020).

e. Naturally occurring and inappropriate fire regimes in the Anthropocene: A positive relationship is be expected between rich habitat diversity and insect diversity. As it has been repeatedly stated, sudden changes in the climate, widespread misuse of pesticides, habitat fragmentation, large fires (because of wildfires such as deliberate fires) and the loss of natural corridors between local populations threaten the habitats of insects and lead them to their ultimate disappearance. In many ecosystems worldwide, wildfires can be beneficial as it promotes distribution, biodiversity, floristic diversity and seed germination, resulting in more fire-resistant plant generations (Bond et al., 2005; Kelly et al., 2020; Pausas & Keeley, 2019; Tavşanoğlu & Serter Çatav, 2012). Insects with limited mobility (such as wingless insects) and other groups of invertebrates (such as ants, spiders) living under tree bark, on leaves and stems, and in forest leaf litter are adversely affected by inappropriate fire regimes (but see Sileshi & Mafongoya, 2006; Underwood & Quinn, 2010; Zaitsev et al., 2016). However, research on some invertebrate groups in the fire area also revealed that the populations could be recovered after a while. For example, a few studies conducted on *Pinus brutia* forests of different ages after fire in Marmaris (Turkey), revealed that different surface-active beetle and ant communities showed different successional gradients in long term (Kaynaş, 2017; Kaynaş et al., 2018). However, as the Earth gets hotter and drier, natural, and ecologically sustainable fire regimes in fire-prone ecosystems are becoming unsustainable. Local high temperatures that are the results of deliberately set fires and human-induced global warming negatively affect other habitats outside of fire-adapted ecosystems (Kelly et al., 2020). On the other hand, many species may be under threat due to the frequency of fires, and their effects effect on large areas. Loss of natural corridors between forest interiors, forest edges, shrubberies, water edges and open habitats with grassy vegetation renders metapopulations unhealthy and can lead to self-infestation in isolated local populations (Sands, 2018). In addition to large and unnatural fires, fires in non-fire-adapted ecosystems cause serious damage to insect habitats due to global warming and other anthropogenic factors (such as plowing, deforestation, overgrazing). Issues such as the dynamics of the ecosystem where a wildfire take place and whether anthropogenic intervention (such as afforestation) should be made to the burnt areas after the fire will also affect the resettlement dynamics of insects in these areas. In particular, the prevention of excessive unnatural fires and the control of fires in forest areas that are not adapted to fire require thorough planning in terms of local fire management policies (Kelly et al., 2020; Moreira et al., 2020). Prevention of uncontrolled fires which lead to the loss of biodiversity is of the utmost importance.

Turkey is at risk of losing its biological and genetic diversity (Gross, 2012). Incorrect agricultural policies (such as over irrigation, excessive chemical insecticide use, excessive fertilization, intensive agriculture), especially in the Mediterranean, Central Anatolian, Aegean and Southeastern Anatolian regions, urbanization and construction expanding towards agricultural and natural areas, failure to protect forest and steppe ecosystems, overgrazing (Oztas et al., 2003) devastate biological diversity at both the genetic, population, and species level. The tendency towards monoculture agriculture, driven by economic interests, causes a rapid increase in invasive or native insect populations and alters the faunal composition of local species. In addition, serious losses are observed in the genetic diversity of agricultural plants due to harmful agricultural policies (Uzun & Bayır, 2009).

6. FUTURE PERSPECTIVES

In Turkey, the insect orders such as Lepidoptera, Mantodea, Orthoptera, Neuroptera, Odonata, Trichoptera, Ephemeroptera have been largely studied, and some families of Diptera, Hymenoptera, Coleoptera and Hemiptera have been studied well, both faunistically and systematically. Some families, which are harmful to agriculture and have important natural enemies, and the biology/ecology of these have been discussed in detail. There are also studies available that evaluate biogeographic approaches related to some insect orders. In particular, it is seen that revision, zoogeography and phylogeography publications based on data obtained with morphological and molecular tools, with an evolutionary and ecological approach, are mainly focused on Orthoptera, and some Coleoptera, Diptera, Lepidoptera, Hymenoptera families, and Welsh wasps. There is an extensive literature available on mosquitoes and sandflies within the context of vector ecology. The subject of vector ecology is not mentioned in this review, as it includes a separate, and detailed process. Sand flies (Diptera: Psychodidae), 28 species of which have been identified in Anatolia, are an important group in terms of transmission of diseases(leishmaniasis) and are a thoroughly studied family, both ecologically and molecularly (Erisoz Kasap et al., 2019). The Culicidae family (Diptera), which is composed of millions of different disease vector species carrying malaria, yellow fever, zika, and West Nile virus, also includes other well-known species, and 62 species belonging to seven genera of this family have been listed from Turkey (Kuçlu & Dik, 2018).

Studies are still being conducted on faunistic listing and systematics of the species that spread locally, regionally and all across, and around Turkey, and new species belonging to almost every large insect order are still being discovered. However, studies on the well-being and protection priorities of these populations due to factors such as destruction of habitats of a large number of endemic species, global climate change, environmental pollution, extinctions caused by invading species and overgrazing, are almost nonexistent. Studies on the conversations of insects, endemic species, and genetic diversity of local populations in Turkey should be the priority, especially when we consider the aforementioned factors, in order to determine conservation statuses. DNA barcoding of species is also a requirement. Unfortunately, there is an absence of a full-fledged Natural History Museum in Turkey since samples (dry material, alcohol storage or preparation) are stored in personal collections.

One of the other important shortcomings is the inadequacy of ecological and behavioral studies on the species. A significant part of the existing studies is made up entirely of descriptive data. Plant-insect relationships and coevolution, metapopulation and molecular ecological studies are scarce. Studies on the ecological and conservation status of pollinator species and their population densities also seem insufficient.

It is clear that one of the most important factors that will affect the population density, distribution and behavior of insects in the coming years will be climate change. This includes the rapid increase in residential areas, the expansion of agricultural areas due to food demands, the destruction of habitats due to the increase in transportation, production and energy demands, with new roads, factories and dams being constructed, and environmental pollution. Therefore, action plans are needed to fight against the risk of extinction of many endemic species.

On the other hand, with the effects of immigration, international trade, and climate change, it is seen that invasive species are entering Turkey growingly, and settling in suitable habitats around the country faster than predicted. It is vital to investigate the points of contact for these invasive species, most of which are agricultural pests, not only in agriculture and forest areas, but also the existing natural fauna (niche competition, hybridization, increased competition for food, hunting of predatory species). Ecological interactions with biological control elements should also be evaluated in this context.

Invasive species from many disease-carrying vector insect groups have also been detected in Turkey in recent years, and these species are rapidly expanding their distribution area and consequently becoming established populations in suitable habitats. For example, Mosquito *Aedes albopinctus* has been sighted in Thrace in 2011, and since then, the species has now spread to and reproduced almost all over the Black Sea region (Ordu province observation, HS) (Akıner et al., 2016, 2018). Similarly, the presence of resident populations of Aedes aegypti in the Eastern Black Sea region was detected (Akıner et al., 2018). The fact that the Black Sea region is a rainy and humid region, and the annual average temperature is rising creates the fundamentals for these invasive species to settle in the habitats easily. In this context, due to climate change, it is of great importance in terms of public health that the population dynamics and ecology of vector insects are studied well.

Populations with lower ecological tolerances that live in undisturbed habitats, especially alpine and steppe endemic species, face a great threat because of climate change and anthropological impacts. Overgrazing, clearing of agricultural areas and excessive pesticide use, construction and other factors put these already narrowly distributed local populations under pressure. Insect groups with limited mobility either have their genetic diversity reduced or unfortunately disappear completely in habitats where gene exchange is completely interrupted and fragmented (Figure 20). It is urgent to investigate the populations remaining in cities and areas that are being opened to urbanization, and to prepare the necessary protection measures with action plans. The National Parks, where the biological diversity is very high in Turkey, are areas where natural habitats are protected to a large extent. Therefore, it is necessary to expand these genetic diversity reserves and to continue to protect the well-being of the habitats of these areas in a sustainable way.

ACKNOWLEDGMENT

I thank acknowledge M. Sait Taylan for providing some data about cave crickets. I also thank Hilal Özdağ Sevgili whose comments have improved the manuscript. The language edit of the manuscript made by Onur Sevgili.

REFERENCES

Abrol, D. P., Gorka, A. K., Ansari, M. J., Al-Ghamdi, A., & Al-Kahtani, S. (2019). Impact of insect pollinators on yield and fruit quality of strawberry. *Saudi Journal of Biological Sciences*, *26*(3), 524–530. doi:10.1016/j.sjbs.2017.08.003 PMID:30899167

Agathokleous, E., Feng, Z., Oksanen, E., Sicard, P., Wang, Q., Saitanis, C. J., Araminiene, V., Blande, J. D., Hayes, F., Calatayud, V., Domingos, M., Veresoglou, S. D., Peñuelas, J., Wardle, D. A., De Marco, A., Li, Z., Harmens, H., Yuan, X., Vitale, M., & Paoletti, E. (2020). Ozone affects plant, insect, and soil microbial communities: A threat to terrestrial ecosystems and biodiversity. *Science Advances*, *6*(33), eabc1176. doi:10.1126ciadv.abc1176 PMID:32851188

Ak, K., Uluca, M., Aydin, Ö., & Gokturk, T. (2019). Important invasive species and its pest status in Turkey: *Halyomorpha halys* (Stål) (Heteroptera: Pentatomidae). *Journal of Plant Diseases and Protection*, *126*(5), 401–408. doi:10.100741348-019-00243-7

Akgün, F., Kayseri, M. S., & Akkiraz, M. S. (2007). Palaeoclimatic evolution and vegetational changes during the Late Oligocene–Miocene period in Western and Central Anatolia (Turkey). *Palaeogeography, Palaeoclimatology, Palaeoecology*, *253*(1–2), 56–90. doi:10.1016/j.palaeo.2007.03.034

Akıner, M. M., Beriş, F. Ş., Seyis, F., Öztürk, M., Sevgili, H., & Demir, E. (2019). Annual variation of the *Orosanga japonica* Melichar 1898 (Hemiptera: Ricaniidae) populations in the eastern Black Sea region of Turkey and possible molecular separation with based on 28S rDNA sequences from other Ricaniidae groups. *Bitki Koruma Bulteni*, *59*(4), 11–19. doi:10.16955/bitkorb.546120

Akıner, M. M., Demirci, B., Babuadze, G., Robert, V., & Schaffner, F. (2016). Spread of the invasive mosquitoes *Aedes aegypti* and *Aedes albopictus* in the Black Sea Region increases risk of Chikungunya, Dengue, and Zika Outbreaks in Europe. *PLoS Neglected Tropical Diseases*, *10*(4), e0004664. doi:10.1371/journal.pntd.0004664 PMID:27115737

Akıner, M. M., Demirci, B., Bedir, H., Doğan, A. F., Gökdemir, A., Topluoğlu, S., Altuğ, Ü., Kurtcebe, Z. Ö., & Irmak, H. (2018). Surveillance and control of invasive *Aedes* species in the eastern black sea area of Turkey. *Turk Hijiyen ve Deneysel Biyoloji Dergisi*, *75*(3), 225–238. doi:10.5505/TurkHijyen.2018.68736

Akyıldırım, H. (2013). Evaluation of the zoogeographical contents of Turkey aphid (Hemiptera, Aphidoidea) fauna and invasive components. *Biyoloji Bilimleri Araştırma Dergisi*, *6*(1), 44–48.

Ambarlı, D., Zeydanlı, U. S., Balkız, Ö., Aslan, S., Karaçetin, E., Sözen, M., Ilgaz, Ç., Gürsoy Ergen, A., Lise, Y., Demirbaş Çağlayan, S., Welch, H. J., Welch, G., Turak, A. S., Bilgin, C. C., Özkil, A., & Vural, M. (2016). An overview of biodiversity and conservation status of steppes of the Anatolian Biogeographical Region. *Biodiversity and Conservation*, *25*(12), 2491–2519. doi:10.100710531-016-1172-0

Ambrose, D. P. (2003). Biocontrol potential of assassin bugs (Hemiptera: Reduviidae). *Journal of Experimental Zoology India*, *6*(1), 1–44.

Anlaş, Sinan, & Kočárek, P. (2012). Current status of Dermaptera (Insecta) fauna of Turkey and Cyprus. *Turkiye Entomoloji Dergisi*, *36*, 43–58.

Anlaş, S. (2020). A new micropterous species of the genus *Sunius* Stephens, 1829 from Central Anatolia (Coleoptera: Staphylinidae). *Zoology in the Middle East*, *66*(4), 342–346. doi:10.1080/09397140.2020.1826676

Anonymous. (2023). *Mağara Turizmi (Webpage of Republic of Turkey Ministry of Culture and Tourism)*. KTB. https://yigm.ktb.gov.tr/ TR-10335/ magara-turizmi.html

Ansell, S. W., Stenøien, H. K., Grundmann, M., Russell, S. J., Koch, M. A., Schneider, H., & Vogel, J. C. (2011). The importance of Anatolian mountains as the cradle of global diversity in *Arabis alpina*, a key arctic-alpine species. *Annals of Botany*, *108*(2), 241–252. doi:10.1093/aob/mcr134 PMID:21712298

Arı, İ. (2014). A checklist of Turkish Neuropteran species (Insecta: Neuroptera) with remarks on provincial distribution in Turkey. *Acta Entomologica Slovenicaa*, *22*(1), 59–86.

Arslangündoğdu, Z., & Hızal, E. (2010). The Western conifer seed bug, *Leptoglossus occidentalis* (Heidemann, 1910), recorded in Turkey (Heteroptera: Coreidae). *Zoology in the Middle East*, *50*(1), 138–139. doi:10.1080/09397140.2010.10638426

Assing, V., & Schillhammer, H. (2001). On theTurkish species of *Sunius* Curtis 1829 (Coleoptera : Staphylinidae, Paederinae). *Linzer Biologische Beitrage*, *33*, 195–210.

Atakan, E., & Pehlivan, S. (2020). A harmful *Thrips* species on Lemon in the Eastern Mediterranean Region of Turkey: *Thrips hawaiiensis* (Morgan) (Thysanoptera: Thripidae). *Kahramanmaraş Sütçü İmam Üniversitesi Tarım ve Doğa Dergisi*, *23*(1), 19–25. doi:10.18016/ksutarimdoga.vi.578082

Atalay, I. (2006). The effects of mountainous areas on biodiversity: a case study from the northern Anatolian Mountains and the Taurus Mountains. *Proceedings of the 8th International Symposium on High Mountain Remote Sensing Cartography,* (*vol. 41*, pp. 17–26). Uni Graz..

Atalay, I., Efe, R., & Öztürk, M. (2014). Effects of topography and climate on the ecology of taurus mountains in the mediterranean region of Turkey. *Procedia: Social and Behavioral Sciences*, *120*, 142–156. doi:10.1016/j.sbspro.2014.02.091

Avcı, M. (2005). Diversity and endemism in Turkey's vegetation. *Journal of Geography*, *13*, 27–55.

Avgın, S., & Özdikmen, H. (2007). Check list of the Tiger Beetles of Turkey with a review of distribution and biogeography Coleoptera Cicindelidae. *Munis Entomology & Zoology*, *2*, 87–102.

Avgin, S. S., & Cavazzuti, P. (2011). The studies made on Turkish Carabinae with checklist and bibliography (Coleoptera: Carabidae). *Turkish Journal of Zoology*, *35*, 403–432. doi:10.3906/zoo-0905-13

Avgin, S. S., & Prunier, D. (2015). Endemic *Carabus* species of Turkey: Some researches with biogeographic and ecological notes (Coleoptera, Carabidae, Carabinae). *Ekoloji Çevre Dergisi*, *24*, 26–40. doi:10.5053/ekoloji.2015.19

Aydin, G., & Şen, İ. (2020). Determination of arthropod biodiversity and some ecological parameters of Erdal Şekeroğlu (Isparta, Turkey) and Kadıini (Antalya, Turkey) cave ecosystems with evaluation of usability of insects in cave mapping. *Turkiye Entomoloji Dergisi*, *44*, 539–557. doi:10.16970/entoted.770018

Aysal, T., & Kıvan, M. (2014). Occurrence of an invasive alien species *Harmonia axyridis* (Pallas) (Coleoptera: Coccinellidae) in Turkey. *Türkiye Entomoloji Bülteni*, *4*(3), 141. doi:10.16969/teb.17261

Aytar, F., Kara, K., & Atay, T. (2020). Tachinid (Diptera: Tachinidae) parasitoids reared from lepidopterous and hymenopterous hosts in southern forests of Turkey. *Turkiye Entomoloji Dergisi*, *45*, 3–7. doi:10.16970/entoted.779465

Azmaz, M., & Katilmiş, Y. (2017). Updated species list of Cynipidae (Hymenoptera) from Turkey. *Zootaxa*, *4303*(3), 361–378. doi:10.11646/zootaxa.4303.3.3

Barkan, N. P., & Aytekin, A. M. (2013). Systematical studies on the species of the subgenus *Bombus* (*Thoracobombus*) (Hymenoptera: Apidae, *Bombus* Latreille) in Turkey. *Zootaxa*, *3737*(2), 167–183. doi:10.11646/zootaxa.3737.2.5 PMID:25112746

Battiston, R., Amerini, R., Di Pietro, W., Guariento, L. A., Bolognin, L., & Moretto, E. (2020). A new alien mantis in Italy: Is the Indochina mantis *Hierodula patellifera* chasing the train for Europe? *Biodiversity Data Journal*, *8*, e50779. doi:10.3897/BDJ.8.e50779 PMID:32194321

Bayram, A., & Tonğa, A. (2016). First report of *Chilo partellus* in Turkey, a new invasive maize pest for Europe. *Journal of Applied Entomology*, *140*(3), 236–240. doi:10.1111/jen.12232

Bennett, K. D. (1990). Milankovitch cycles and their effects on species in ecological and evolutionary time. *Paleobiology*, *16*(01), 11–21. doi:10.1017/S0094837300009684

Benyamini, D., Bruschini, C., Serra, G., & John, E. (2007). First confirmed records of *Papilio demoleus* (Linnaeus 1758) in Syria, wih comments on the species' appearance in Turkey and the Dominican Republic. *News of the Israeli Lepidopterists Society*, *24*(2), 4–11.

Beyarslan, A. (2014). Checklist of Braconinae species of Turkey (Hymenoptera: Braconidae). *Zootaxa*, *3790*(2), 201–242. doi:10.11646/zootaxa.3790.2.1 PMID:24869868

Bilgin, R. (2011). Back to the suture: The distribution of intraspecific genetic diversity in and around anatolia. *International Journal of Molecular Sciences*, *12*(6), 4080–4103. doi:10.3390/ijms12064080 PMID:21747726

Blüthgen, N., & Klein, A.-M. (2011). Functional complementarity and specialisation: The role of biodiversity in plant–pollinator interactions. *Basic and Applied Ecology*, *12*(4), 282–291. doi:10.1016/j.baae.2010.11.001

Bolton, B. (2021). *AntCat: An Online Catalog of the Ants of the World.*

Bond, W. J., Woodward, F. I., & Midgley, G. F. (2005). The global distribution of ecosystems in a world without fire. *The New Phytologist*, *165*(2), 525–537. doi:10.1111/j.1469-8137.2004.01252.x PMID:15720663

Boudot, J.-P., Kalkman, V. J., & Azpilicueta AMorin, M. (2009). Atlas of the Odonata of the Mediterranean and North Africa | IUCN Library System. Libellula Supplement, 9, 1–256.

BourgoinT. (2020). Fulgoromorpha Lists On the Web. *The Catalogue of Life Partnership*. doi:10.15468/57cc67

Bozdoğan, H. (2020). Species richness and composition of Neuroptera in the forests fregments of the Taurus Mountains Range, Turkey. *Saudi Journal of Biological Sciences*, *27*(5), 1201–1207. doi:10.1016/j.sjbs.2020.02.018 PMID:32346325

Bozkurt, V. (2018). İstilacı böcek türlerinin mücadelesinin yönetimi: *Anoplophora chinensis* (Forster, 1771) (Coleoptera: Cerambycidae) örneği. *Doğal Afetler ve Çevre Dergisi, 4*, 25–31. doi:10.21324/dacd.441390

Brooks, T. M., Mittermeier, R. A., Mittermeier, C. G., da Fonseca, G. A. B., Rylands, A. B., Konstant, W. R., Flick, P., Pilgrim, J., Oldfield, S., Magin, G., & Hilton-Taylor, C. (2002). Habitat loss and extinction in the hotspots of biodiversity. *Conservation Biology, 16*(4), 909–923. doi:10.1046/j.1523-1739.2002.00530.x

Brown, J., Scholtz, C. H., Janeau, J.-L., Grellier, S., & Podwojewski, P. (2010). Dung beetles (Coleoptera: Scarabaeidae) can improve soil hydrological properties. *Applied Soil Ecology, 46*(1), 9–16. doi:10.1016/j.apsoil.2010.05.010

Bukejs, A., & Telnov, D. (2014). The first record of the invasive lady beetle *Harmonia axyridis* (Pallas, 1773) (Coleoptera: Coccinellidae) in Turkey. *Zoology and Ecology, 1–4*, 1–4. Advance online publication. doi:10.1080/21658005.2014.984490

Bush, G. L., & Butlin, R. K. (2004). Sympatric Speciation in Insects. I: U. Dieckmann, M. Doebeli, J. A. J. Metz, & D. Tautz (Eds.), Adaptive Speciation (pp. 229–248). Cambridge University Press.

Çalışkan, H., Şirin, D. Ü., Aslan, A., Fidan, E. C., & Şahin, Y. (2020). Five New Records for Turkish Robber Flies (Diptera: Asilidae) Fauna. *Journal of Entomological Research Society, 22*(3), 239–254.

Çalışkan, S. S. (2014). Review of the the Genus *Chrysoclista* Stainton, 1854 (Lepidoptera: Agonoxenidae) in Turkey with a New Species. *Journal of Entomological Research Society, 16*(3), 67–73.

Cameron, S. A., Hines, H. M., & Williams, P. H. (2007). A comprehensive phylogeny of the bumble bees (*Bombus*). *Biological Journal of the Linnean Society. 91*(1), 161–188. doi:10.1111/j.1095-8312.2007.00784.x

Can, F., Ercan, F., & Ulasli, B. (2022). First discovery of the natural egg parasitoid of *Cydalima perspectalis* (Walker, 1859) in Turkey with molecular methods (Lepidoptera: Crambidae). *SHILAP Revista de Lepidopterologia, 50*(200), 607–616. doi:10.57065hilap.253

Canpolat, E. (2005). *Karacadağ (Diyarbakır) volkanı jeomorfolojisi/Geomorphology of Karacadağ (Diyarbakır) volcano* [Master thesis, İstanbul Üniversitesi].

Cardinale, B. J., Duffy, J. E., Gonzalez, A., Hooper, D. U., Perrings, C., Venail, P., Narwani, A., Mace, G. M., Tilman, D., Wardle, D. A., Kinzig, A. P., Daily, G. C., Loreau, M., Grace, J. B., Larigauderie, A., Srivastava, D. S., & Naeem, S. (2012). Biodiversity loss and its impact on humanity. *Nature*, *486*(7401), 59–67. doi:10.1038/nature11148 PMID:22678280

Cardinale, B. J., Wright, J. P., Cadotte, M. W., Carroll, I. T., Hector, A., Srivastava, D. S., Loreau, M., & Weis, J. J. (2007). Impacts of plant diversity on biomass production increase through time because of species complementarity. *Proceedings of the National Academy of Sciences of the United States of America*, *104*(46), 18123–18128. doi:10.1073/pnas.0709069104 PMID:17991772

Cardoso, P., Barton, P. S., Birkhofer, K., Chichorro, F., Deacon, C., Fartmann, T., Fukushima, C. S., Gaigher, R., Habel, J. C., Hallmann, C. A., Hill, M. J., Hochkirch, A., Kwak, M. L., Mammola, S., Ari Noriega, J., Orfinger, A. B., Pedraza, F., Pryke, J. S., Roque, F. O., & Samways, M. J. (2020). Scientists' warning to humanity on insect extinctions. *Biological Conservation*, *108426*, 108426. Advance online publication. doi:10.1016/j.biocon.2020.108426

Carpaneto, G., M., Piattella, E., & Pittino, R. (2000). The scarab beetles of Turkey: an updated checklist and chorotype analysis (Coleoptera, Scarabaeoidea). *Biogeographia – The Journal of Integrative Biogeography*, *21*. doi:10.21426/B6110170

Casale, A., & Vigna Taglianti, A. (1999). Caraboid beetles (excl. Cicindelidae) of Anatolia, and their biogeographical significance (Coleoptera, Caraboidea). *Biogeographia – The Journal of Integrative Biogeography*, *20*. doi:10.21426/B620110002

Cebeci, H. H., & Baydemir, M. (2018). Depredadores de escarabajos de la corteza (Coleoptera) en la región de Balikesir de Turquía. *Revista Colombiana de Entomologia*, *44*(2), 283–287. doi:10.25100ocolen.v44i2.7326

Cengiz, F. C., Kaya, K., Ulaşlı, B., & Morinière, J. (2016). First record of the egg parasitoids of *Chilo partellus* (Swinhoe, 1885) (Lepidoptera: Crambidae) in Turkey unveiled by DNA Barcoding. *Turkiye Entomoloji Dergisi*, *40*(2). doi:10.16970/ted.80849

Chobanov, D. P., Kaya, S., Grzywacz, B., Warchałowska-Śliwa, E., & Çıplak, B. (2017). The Anatolio-Balkan phylogeographic fault: A snapshot from the genus *Isophya* (Orthoptera, Tettigoniidae). *Zoologica Scripta*, *46*(2), 165–179. doi:10.1111/zsc.12194

Cigliano, M. M., Braun, H., Eades, D. C., & Otte, D. (2023). *Orthoptera Species File. Version 5.0/5.0.* Orthoptera. http://orthoptera.speciesfile.org/ HomePage/ Orthoptera/ HomePage.aspx

Çıplak, B. (2003). Distribution of Tettigoniinae (Orthoptera, Tettigoniidae) bush-crickets in Turkey: The importance of the Anatolian Taurus Mountains in biodiversity and implications for conservation. *Biodiversity and Conservation, 12*(1), 47–64. doi:10.1023/A:1021206732679

Çıplak, B. (2004). Biogeography of Anatolia: The marker group Orthoptera. *Memorie della Societa Entomologica Italiana, 82*(2), 357–372.

Çıplak, B., Demirsoy, A., Yalım, B., & Sevgili, H. (2008). Türkiye Orthoptera (Düzkanatlılar, Çekirgeler) Faunası. I: A. Demirsoy (Ed.), Genel Zoocoğrafya ve Türkiye Zoocoğrafyası "Hayvan Coğrafyası" (8. Basım, pp. 681–707). Meteksan.

Coope, G. R. (1994). The response of insect faunas to glacial-interglacial climatic fluctuations. *Philosophical Transactions of the Royal Society of London. Series B, Biological Sciences, 344*(1307), 19–26. doi:10.1098/rstb.1994.0046

Çoruh, S. (2016). Biogeography and host evaluation of the subfamily Pimplinae (Hymenoptera: Ichneumonidae) in Turkey. *Journal of Entomological Research Society, 18*(2), 33–60.

Çoruh, S. (2019a). Taxonomic and biogeographic evaluations of the subfamily Cryptinae (Hymenoptera: Ichneumonidae). *Turkiye Entomoloji Dergisi*, 313–337. doi:10.16970/entoted.520717

Çoruh, S. (2019b). Taxonomical and biogeographical evaluation of the subfamily Tryphoninae (Hymenoptera: Ichneumonidae) in Turkey. *Journal of Entomological Research Society, 21*(3), 301–321.

Çoruh, S., & Özbek, H. (2005). New records of Cryptinae (Hymenoptera: Ichneumonidae) from Turkey with some hosts. *Turkiye Entomoloji Dergisi, 29*(3), 183–186.

Coyne, J. A., & Orr, H. A. (2004). *Speciation*. Sinauer Associates.

Dal, N., & Gönençgil, B. (2018). *Türkiye'de dağ ve dağlık alan sınırlandırması için bir yaklaşım/An approach for the limitation of mountain and mountain areas in Turkey.*, International Geography Symposium on the 30th Anniversary of TUCAUM. 907–913.

Darilmaz, M. C., Salur, A., Murányi, D., & Vinçon, G. (2016). Contribution to the knowledge of Turkish stoneflies with annotated catalogue (Insecta: Plecoptera). *Zootaxa*, *4074*(1), 1–74. doi:10.11646/zootaxa.4074.1.1 PMID:27395953

Darilmaz, M. C., & İncekara, Ü. (2011). Checklist of hydrophiloidea of Turkey (Coleoptera: Polyphaga). *Journal of Natural History*, *45*(11–12), 685–735. doi:10.1080/00222933.2010.535916

Darılmaz, M. C., Polat, A., & Incekara, U. (2019). Faunistic study on aquatic coleoptera of the eastern mediterranean region of Turkey. *Turkish Journal of Fisheries and Aquatic Sciences*, *19*(5), 409–421. doi:10.4194/1303-2712-v19_5_06

Darılmaz, M. C., & Salur, A. (2015). Annotated Catalogue of the Turkish Caddisflies (Insecta: Trichoptera). *Munis Entomology & Zoology*, *10*(10), 521–734.

Demir, E. (2006). Preliminary report on the Auchenorrhyncha (Hemiptera) fauna of Kazdağı National Park with two new records for Turkey. *Acta Entomologica Slovenica (Ljubljana)*, *14*(1), 89–102.

Demir, E. (2018). The economically important alien invasive planthoppers in Turkey (Hemiptera: Fulgoromorpha). *Acta Entomologica Slovenica (Ljubljana)*, *2*(2), 73–78.

Demirel, E. (2019). Cicadidae (Hemiptera: Auchenorrhyncha) from Mt.Musa, Turkey. *Monographs of the Upper Silesian Museum*, *10*, 29–45.

Demirel, Ö., & Pouya, S. (2020). Sürdürülebilir dağ yönetimi koruma boyutu ve kaynak rezervi olarak Ağrı Dağı/Sustainable Mountain Management, Ağrı Mount example. *International Journal of Mountaineering and Climbing*, *3*(1), 1–14. doi:10.36415/dagcilik.686665

Demirsoy, A. (1982). [). TÜBİTAK.]. *Türkiye Faunası Odonata*, *8/14*, 156–156.

Demirsoy, A. (2008). Genel Zoocoğrafya ve Türkiye Zoocoğrafyası (Hayvan Coğrafyası) (pp. 1007–1007). Palme yayıncılık, Ankara.

Demirsoy, A., Salman, S., & Sevgili, H. (2002). Novadrymadusa, a new genus of bushcricket with a new species and notes on related genera (Orthoptera: Tettigoniidae). *Journal of Orthoptera Research*, *11*(2), 175–183. doi:10.1665/1082-6467(2002)011[0175:NANGOB]2.0.CO;2

Denk, T., Güner, H. T., & Bouchal, J. M. (2019). Early Miocene climate and biomes of Turkey: Evidence from leaf fossils, dispersed pollen, and petrified wood. *Palaeogeography, Palaeoclimatology, Palaeoecology*, *530*, 236–248. doi:10.1016/j.palaeo.2019.05.042

Di Russo, C., Rampini, M., & Landeck, I. (2007). The cave crickets of northeast Turkey and transCaucasian regions, with descriptions of two new species of the genera *Dolichopoda* and *Troglophilus* (Orthoptera, Rhaphidophoridae). *Journal of Orthoptera Research, 16*(1), 67–76. doi:10.1665/1082-6467(2007)16[67:TCCONT]2.0.CO;2

Dinc, S., Mutun, S., & Melika, G. (2014). A new species of Andricus Hartig oak gall wasp from Turkey (Hymenoptera: Cynipidae, Cynipini). *North-Western Journal of Zoology, 10*(1), 122–127.

Doğan, M., Akaydın, G., & Erdal, J. (2020). Phytogeography, ecology and conservation of the genus *Limonium* (Plumbaginaceae) in Turkey and a taxonomic revision. *Plant Systematics and Evolution, 306*(6), 89. doi:10.100700606-020-01706-3

Doherty, T. S., Glen, A. S., Nimmo, D. G., Ritchie, E. G., & Dickman, C. R. (2016). Invasive predators and global biodiversity loss. *Proceedings of the National Academy of Sciences of the United States of America, 113*(40), 11261–11265. doi:10.1073/pnas.1602480113 PMID:27638204

Dueñas, M.-A., Ruffhead, H. J., Wakefield, N. H., Roberts, P. D., Hemming, D. J., & Diaz-Soltero, H. (2018). The role played by invasive species in interactions with endangered and threatened species in the United States: A systematic review. *Biodiversity and Conservation, 27*(12), 3171–3183. doi:10.100710531-018-1595-x

Dumont, H. J. (1977). A review of the dragonfly fauna of Turkey and adjacent Mediterranean islands (Insecta: Odonata). *Bulletin et Annales de La Société Royale Belge d'entomologie, 113*, 119–171.

Dursun, A., & Salur, A. (2013). Presence of Sphedanolestes sanguineus (Fabricius, 1794) in Turkey, followed by an annotated checklist of Reduviidae (Hemiptera: Heteroptera). *Turkish Journal of Zoology, 37*, 610–620. doi:10.3906/zoo-1203-17

Egan, S. P., Hood, G. R., Feder, J. L., & Ott, J. R. (2012). Divergent host-plant use promotes reproductive isolation among cynipid gall wasp populations. *Biology Letters, 8*(4), 605–608. doi:10.1098/rsbl.2011.1205 PMID:22337505

Ehrlich, P. R., & Raven, P. H. (1964). Butterflies and plants: A study in coevolution. *Evolution; International Journal of Organic Evolution, 18*(4), 586–608. doi:10.2307/2406212

Ehrmann, R. (2011). Mantodea from Turkey and Cyprus (Dictyoptera: Mantodea). *Articulata, 26*(1), 1–42.

Ekim, T., & Güner, A. (1986). The Anatolian Diagonal: Fact or fiction? *Proceedings of the Royal Society of Edinburgh. Section B: Biological Sciences*, *89*, 69–77. doi:10.1017/S0269727000008915

Elekcioglu, N. Z., & Olculu, M. (2017). Pest, predator and parasitoid species in persimmon orchards in the eastern Mediterranean region of Turkey, with new records. *Fresenius Environmental Bulletin*, *26*(8), 5170–5176.

Erisoz Kasap, O., Linton, Y.-M., Karakus, M., Ozbel, Y., & Alten, B. (2019). Revision of the species composition and distribution of Turkish sand flies using DNA barcodes. *Parasites & Vectors*, *12*(1), 410. doi:10.118613071-019-3669-3 PMID:31439012

Erol, A., & Randhir, T. O. (2012). Climatic change impacts on the ecohydrology of Mediterranean watersheds. *Climatic Change*, *114*(2), 319–341. doi:10.100710584-012-0406-8

Esin, N. V., Yanko-Hombach, V. V., & Esin, N. I. (2018). Evolutionary mechanisms of the Paratethys Sea and its separation into the Black Sea and Caspian Sea. *Quaternary International*, *465*, 46–53. doi:10.1016/j.quaint.2016.06.019

Fahrig, L. (2003). Effects of habitat fragmentation on biodiversity. *Annual Review of Ecology, Evolution, and Systematics*, *34*(1), 487–515. doi:10.1146/annurev.ecolsys.34.011802.132419

Fallahzadeh, M., & Saghaei, N. (2010). Checklist of Braconidae (Insecta: Hymenoptera) from Iran. *Munis Entomology & Zoology*, *5*(1), 170–186.

FAO. (2021). *Global review of forest pests and diseases*. FAO.

Fitzpatrick, B. M., Fordyce, J. A., & Gavrilets, S. (2009). Pattern, process and geographic modes of speciation. *Journal of Evolutionary Biology*, *22*(11), 2342–2347. doi:10.1111/j.1420-9101.2009.01833.x PMID:19732257

Futuyma, D. J., & Agrawal, A. A. (2009). Macroevolution and the biological diversity of plants and herbivores. *Proceedings of the National Academy of Sciences of the United States of America*, *106*(43), 18054–18061. doi:10.1073/pnas.0904106106 PMID:19815508

Gadallah, N. S., Ghahari, H., Peris-Felipo, F. J., & Fischer, M. (2016). Updated checklist of Iranian Opiinae (Hymenoptera: Braconidae). *Zootaxa*, *4066*(1), 1–40. doi:10.11646/zootaxa.4066.1.1 PMID:27395527

Geller-Grimm, F., Dikow, T., & Lavigne, R. J. (2021). Asilidae Homepage. *Information (Basel)*.

Ghahari, H., Hayat, R., Lavigne, R. J., & Ostovan, H. (2014). An annotated checklist of Iranian Asilidae (Insecta: Diptera: Brachycera: Asiloidea). *Linzer Biologische Beitrage*, *46*(2), 1379–1446.

Gnaspini, P. (1997). Climatic fluctuations and tropical troglobitic evolution. *International Journal of Speleology*, *26*(1/2), 33–36. doi:10.5038/1827-806X.26.1.3

Gorochov, A. V., & Ünal, M. (2012). Review of Turkish crickets from the subfamily Gryllomorphinae (Orthoptera: Gryllidae). *Zoosystematica Rossica*, *21*(1), 28–59. doi:10.31610/zsr/2012.21.1.28

Görür, G., Şenol, Ö., Akyildirim Beğen, H., & Akyürek, B. (2020). Son zamanlarda yapılan çalışmalara göre Türkiye afit faunasıyla ilgili öngörüler. *Atatürk Üniversitesi Ziraat Fakültesi Dergisi*, *51*, 63–68. doi:10.17097/ataunizfd.597925

Gross, M. (2012). Turkey's biodiversity at the crossroads. *Current Biology*, *22*(13), R503–R505. doi:10.1016/j.cub.2012.06.051 PMID:22970449

Guclu, K., & Karahan, F. (2004). A review: The history of conservation programs and development of the national parks concept in Turkey. *Biodiversity and Conservation*, *13*(7), 1373–1390. doi:10.1023/B:BIOC.0000019399.35427.64

Günay, G., & Şimşek, Ş. (2001). *Karst hydrogeology in hydrothermal system*, 1–13.

Güncan, A., & Gümüş, E. (2019). Brown Marmorated Stink Bug, *Halyomorpha halys* (StåL, 1855) (Hemiptera: Heteroptera, Pentatomidae), a new and important pest in Turkey. *Entomological News*, *128*(2), 204. doi:10.3157/021.128.0208

Güner, A., Aslan, S., Ekim, T., Vural, M., & Babaç, M. T. (2012). Türkiye Bitkileri Listesi (Damarlı Bitkiler) (pp. 1290–1290). Nezahat Gökyiğit Botanik bahçesi ve Flora Araştırmaları Derneği Yayını.

Güngöroğlu, C. (2017). Natura 2000: Its status quo, opportunities and challenges for forest biodiversity in Turkey. *Fresenius Environmental Bulletin*, *26*(10), 6217–6224.

Gür, H. (2013). The effects of the Late Quaternary glacial-interglacial cycles on Anatolian ground squirrels: Range expansion during the glacial periods? *Biological Journal of the Linnean Society. Linnean Society of London*, *109*(1), 19–32. doi:10.1111/bij.12026

Gür, H. (2016). The Anatolian diagonal revisited: Testing the ecological basis of a biogeographic boundary. *Zoology in the Middle East*, *62*(3), 189–199. doi:10.1080/09397140.2016.1226544

Gür, H. (2017). Geç kuvaterner buzul buzullararası döngülerinin anadolu'nun biyolojik çeşitliliği üzerine etkileri. *Türkiye Jeoloji Bülteni / Geological Bulletin of Turkey, 60*(4), 507–528. doi:10.25288/tjb.363813

Hacet, N. (2017). Updated checklist of Odonata fauna in the Turkish Thrace Region, with additional records of new, rare, and threatened taxa. *Turkish Journal of Zoology, 41*, 33–42. doi:10.3906/zoo-1601-65

Hacet, N., & Colonnelli, E. (2019). On the Ceutorhynchinae (Coleoptera: Curculionidae) Fauna of Turkish Thrace, with Additional Records for Turkey. *Journal of Entomological Research Society, 21*(2), 175–183.

Hanson, T., Brooks, T. M., Da Fonseca, G. A. B., Hoffmann, M., Lamoreux, J. F., Machlis, G., Mittermeier, C. G., Mittermeier, R. A., & Pilgrim, J. D. (2009). Warfare in biodiversity hotspots. *Conservation Biology, 23*(3), 578–587. doi:10.1111/j.1523-1739.2009.01166.x PMID:19236450

Harris, J. E., Rodenhouse, N. L., & Holmes, R. T. (2019). Decline in beetle abundance and diversity in an intact temperate forest linked to climate warming. *Biological Conservation, 240*, 108219. doi:10.1016/j.biocon.2019.108219

Hasbenli, A., Çiftçi, D., & Çağlar, Ü. (2020). Tanap, a new robber fly genus from Turkey (Diptera: Asilidae: Dioctriinae). *Zootaxa, 4822*(1). doi:10.11646/zootaxa.4822.1.6

Hayat, R. (1997). Prey of some Robber Flies (Diptera: Asilidae) in Turkey. *Zoology in the Middle East, 15*(1), 87–94. doi:10.1080/09397140.1997.10637743

Hayward, A., & Stone, G. N. (2005). Oak gall wasp communities: Evolution and ecology. *Basic and Applied Ecology, 6*(5), 435–443. doi:10.1016/j.baae.2005.07.003

Heller, K.-G. (1988). *Bioakustik der europäischen Laubheuschrecken* (pp. 358–358). J. Margraf.

Hendrickx, F., Maelfait, J.-P., Van Wingerden, W., Schweiger, O., Speelmans, M., Aviron, S., Augenstein, I., Billeter, R., Bailey, D., Bukacek, R., Burel, F., Diekötter, T. I. M., Dirksen, J., Herzog, F., Liira, J., Roubalova, M., Vandomme, V., & Bugter, R. O. B. (2007). How landscape structure, land-use intensity and habitat diversity affect components of total arthropod diversity in agricultural landscapes. *Journal of Applied Ecology, 44*(2), 340–351. doi:10.1111/j.1365-2664.2006.01270.x

Hesselbarth, G., Wagener, S., & Van Oorschot, H. (1995). *Die Tagfalter der Türkei : unter Berücksichtigung der angrenzenden Länder*. Verlag Goecke & Evers.

Hewitt, G. (2000). The genetic legacy of the Quaternary ice ages. *Nature*, *405*(6789), 907–913. doi:10.1038/35016000 PMID:10879524

Hewitt, G. M. (1999). Post-glacial re-colonization of European biota. *Biological Journal of the Linnean Society. Linnean Society of London*, *68*(1–2), 87–112. doi:10.1111/j.1095-8312.1999.tb01160.x

Hewitt, G. M. (2011). Mediterranean peninsulas: the evolution of hotspots. I: F. E. Zachos & J. C. Habel (Eds.), Biodiversity Hotspots (pp. 123–147). Springer Berlin Heidelberg. doi:10.1007/978-3-642-20992-5_7

HizalE.OztemizS.GjonovI. (2019). *Ricania shantungensis* Chou & Lu 1977 (Hemiptera: Fulgoromorpha: Ricaniidae) a new invasive insect species in European Turkey. Zenodo. doi:10.5281/zenodo.3547677

Hodkinson, I. D. (2005). Terrestrial insects along elevation gradients: Species and community responses to altitude. *Biological Reviews of the Cambridge Philosophical Society*, *80*(3), 489–513. doi:10.1017/S1464793105006767 PMID:16094810

Hörandl, E., & Emadzade, K. (2011). The evolution and biogeography of alpine species in *Ranunculus* (Ranunculaceae): A global comparison. *Taxon*, *60*(2), 415–426. doi:10.1002/tax.602011

Howarth, F. G., & Moldovan, O. T. (2018). The ecological classification of cave animals and their adaptations. I: O. T. Moldovan, Ľ. Kováč, & S. Halse (Eds.), Cave Ecology (Vol. 235, pp. 41–67). Springer International Publishing. doi:10.1007/978-3-319-98852-8_4

Hulme, P. E. (2017). Climate change and biological invasions: Evidence, expectations, and response options. *Biological Reviews of the Cambridge Philosophical Society*, *92*(3), 1297–1313. doi:10.1111/brv.12282 PMID:27241717

Ilgar, A. (2015). MessiniyenTuzluluk Krizi: Akdeniz'in kurumasına ilişkin bir derleme. *Doğal Kaynaklar ve Ekon Bülteni*, *20*, 73–80.

Jactel, H., Imler, J.-L., Lambrechts, L., Failloux, A.-B., Lebreton, J. D., Maho, Y. L., Duplessy, J.-C., Cossart, P., & Grandcolas, P. (2021). Insect decline: Immediate action is needed. *Comptes Rendus Biologies*, *343*(3), 267–293. doi:10.5802/crbiol.37 PMID:33621456

Jelínek, J., Audisio, P., Hájek, J., Baviera, C., Moncoutier, B., Barnouin, T., Brustel, H., Genç, H., & Leschen, R. A. B. (2016). *Epuraea Imperialis* (Reitter, 1877). New invasive species of Nitidulidae (Coleoptera) in Europe, with a checklist of Sap Beetles introduced to Europe and Mediterranean areas. *AAPP Atti Della Accademia Peloritana Dei Pericolanti, Classe Di Scienze Fisiche. Matematiche e Naturali, 94*(2), 1–24.

Jeppesen, E., Brucet, S., Naselli-Flores, L., Papastergiadou, E., Stefanidis, K., Nõges, T., Nõges, P., Attayde, J. L., Zohary, T., Coppens, J., Bucak, T., Menezes, R. F., Freitas, F. R. S., Kernan, M., Søndergaard, M., & Beklioğlu, M. (2015). Ecological impacts of global warming and water abstraction on lakes and reservoirs due to changes in water level and related changes in salinity. *Hydrobiologia, 750*(1), 201–227. doi:10.100710750-014-2169-x

Johnson, C. N., Balmford, A., Brook, B. W., Buettel, J. C., Galetti, M., Guangchun, L., & Wilmshurst, J. M. (2017). Biodiversity losses and conservation responses in the Anthropocene. *Science, 356*(6335), 270–275. doi:10.1126cience.aam9317 PMID:28428393

Jones, O. R., Purvis, A., Baumgart, E., & Quicke, D. L. J. (2009). Using taxonomic revision data to estimate the geographic and taxonomic distribution of undescribed species richness in the Braconidae (Hymenoptera: Ichneumonoidea). *Insect Conservation and Diversity, 2*(3), 204–212. doi:10.1111/j.1752-4598.2009.00057.x

Kalaentzis, K., Mpamnaras, A., & Kazilas, C. (2019). First record of the alien exotic sap beetle *Phenolia (Lasiodites) picta* (Coleoptera: Nitidulidae) in Greece. *Entomologia Hellenica, 28*(2), 11. doi:10.12681/eh.21323

Kalkman, V. J., Clausnitzer, V., Dijkstra, K.-D. B., Orr, A. G., Paulson, D. R., & van Tol, J. (2008). Global diversity of dragonflies (Odonata) in freshwater. *Hydrobiologia, 595*(1), 351–363. doi:10.100710750-007-9029-x

Kalkman, V. J., Wasscher, M., & Vanpelt, G. J. (2003). An annotated checklist of the Odonata of Turkey. *Odonatologica, 32*(3), 215–236.

Kapli, P., Botoni, D., Ilgaz, C., Kumlutaş, Y., Avcı, A., Rastegar-Pouyani, N., Fathinia, B., Lymberakis, P., Ahmadzadeh, F., & Poulakakis, N. (2013). Molecular phylogeny and historical biogeography of the Anatolian lizard *Apathya* (Squamata, Lacertidae). *Molecular Phylogenetics and Evolution, 66*(3), 992–1001. doi:10.1016/j.ympev.2012.12.002 PMID:23261710

Kara, K., & Alaoglu, Ö. (2001). Some new host records of Tachinidae (Diptera) from Turkey. *Studia Dipterologica, 8*(1), 349–351.

Kara, K., Tschorsnig, H. P., & Atay, T. (2020). Checklist of Turkish tachinidae (insecta, diptera) with new records. *Journal of Entomological Research Society, 22*(2), 163–190.

Karaca, İ., Kayahan, A., Şimşek, B., & Çelikpençe, Y. (2015). First record of *Glycaspis brimblecombei* Moore (Hemiptera: Aphalaridae), in Turkey. *Phytoparasitica, 43*(2), 171–175. doi:10.100712600-015-0457-2

Karataş, A., Karataş, A., Yavuz, N., & Genç, M. (2020). Distribution and activity period of the invasive *Orosanga japonica* (Melichar, 1898) (Hemiptera: Ricaniidae) in Turkey. *Zoology in the Middle East, 66*(3), 246–252. doi:10.1080/09397140.2020.1793502

Kaya, Z., & Raynal, D. J. (2001). Biodiversity and conservation of Turkish forests. *Biological Conservation, 97*(2), 131–141. doi:10.1016/S0006-3207(00)00069-0

Kaynaş, B Y. (2017). Long-term changes in surface-active beetle communities in a post-fire successional gradient in *Pinus brutia* forests. *IForest - Biogeosciences and Forestry, 10*(2), 376–382. doi:10.3832/ifor2140-009

Kaynaş, B. Y. Kıran, K., & Karaman, C. (2018). Long-term effects of fire on ant communities (Hymenoptera: Formicidae) inPinus brutia forests of south-western Turkey. Zoology in the Middle East, 64(2), 160–168. doi:10.1080/09397140.2018.1442302

Kazancı, N., & Türkmen, G. (2012). The checklist of Ephemeroptera (Insecta) species of Turkey. *Review of Hydrobiology, 5*(2), 143–156.

Kelly, L. T., Giljohann, K. M., Duane, A., Aquilué, N., Archibald, S., Batllori, E., Bennett, A. F., Buckland, S. T., Canelles, Q., Clarke, M. F., Fortin, M.-J., Hermoso, V., Herrando, S., Keane, R. E., Lake, F. K., McCarthy, M. A., Morán-Ordóñez, A., Parr, C. L., Pausas, J. G., & Brotons, L. (2020). Fire and biodiversity in the Anthropocene. *Science, 370*(6519), eabb0355. doi:10.1126cience.abb0355 PMID:33214246

Kemal, M., & Koçak, A. Ö. (2013). Notes on the subgenus Polyommatus (Neolysandra) Koçak in East Turkey, with descriptions of new taxa (Lycaenidae, Lepidoptera). *Miscellaneous Papers, 160*, 5–6.

Kenis, M., Auger-Rozenberg, M.-A., Roques, A., Timms, L., Péré, C., Cock, M. J. W., Settele, J., Augustin, S., & Lopez-Vaamonde, C. (2009). Ecological effects of invasive alien insects. *Biological Invasions, 11*(1), 21–45. doi:10.100710530-008-9318-y

Kerey, I. E., Meriç, E., Tunoğlu, C., Kelling, G., Brenner, R. L., & Doğan, A. U. (2004). Black Sea–Marmara Sea Quaternary connections: New data from the Bosphorus, İstanbul, Turkey. *Palaeogeography, Palaeoclimatology, Palaeoecology, 204*(3–4), 277–295. doi:10.1016/S0031-0182(03)00731-4

Kiran, K., & Karaman, C. (2012). First annotated checklist of the ant fauna of Turkey (Hymenoptera: Formicidae). *Zootaxa, 3548*(1), 1. doi:10.11646/zootaxa.3548.1.1 PMID:23847408

Kirca, S., Kreutz, K., & Çolak, A. H. (2020). A biogeographical and ecological classification of orchids in Turkey. *Phytocoenologia, 50*(1), 65–77. doi:10.1127/phyto/2019/0292

Kıyak, S. (2019). Dünya ve Türkiye'de Heteroptera Taksonu Biyo-çeşitliliği ve Çalışmalarına Kısa Bir bakış. *Journal of the Heteroptera of Turkey, Special Is*, 4–5.

Kıyak, S. (2020). Analysis of the insect fauna of Turkey and suggestions for future studies. *Analysis of the Insect Fauna of Turkey and Suggestions for Future Studies, 15*(2), 690–710.

Klein, A. M., Vaissière, B. E., Cane, J. H., Steffan-Dewenter, I., Cunningham, S. A., Kremen, C., & Tscharntke, T. (2007). Importance of pollinators in changing landscapes for world crops. *Proceedings. Biological Sciences, 274*(1608), 303–313. doi:10.1098/rspb.2006.3721 PMID:17164193

Koçak, A. (2006). A new genus and species to the Lepidoptera Fauna of Turkey and Syria (Papilionidae). *Miscellaneous Papers, 97*(97), 39–45.

Koçak, A. Ö., & Kemal, M. (2018). A synonymous and distributional list of the species of the Lepidoptera of Turkey. *Memoirs, 8*, 1–487.

Kolics, B., Ács, Z., Chobanov, D. P., Orci, K. M., Qiang, L. S., Kovács, B., Kondorosy, E., Decsi, K., Taller, J., Specziár, A., Orbán, L., & Müller, T. (2012). Re-visiting phylogenetic and taxonomic relationships in the genus *Saga* (Insecta: Orthoptera). *PLoS One, 7*(8), e42229. doi:10.1371/journal.pone.0042229 PMID:22912691

Kovanci, O. B., & Kovanci, B. (2015). Contribution to the knowledge of Mantispoidea, Osmyloidea, and Myrmeleontoidea with new records for the Neuroptera fauna of northwestern Turkey. *Turkish Journal of Zoology, 39*, 110–118. doi:10.3906/zoo-1401-45

Krijgsman, W., Hilgen, F. J., Raffi, I., Sierro, F. J., & Wilson, D. S. (1999). Chronology, causes and progression of the Messinian salinity crisis. *Nature, 400*(6745), 652–655. doi:10.1038/23231

Kuçlu, Ö., & Dik, B. (2018). Mosquito (Diptera: Culicidae) fauna of western Black Sea region of Turkey. *Turkiye Parazitolojii Dergisi, 42*(2), 138–143. doi:10.5152/tpd.2018.5339

Küçük, M., & Ertürk, E. (2013). Biodiversity and protected areas in Turkey. *Sains Malaysiana, 42*(10), 1455–1460.

Küçükbasmacı, İ. (2014). Two new invasive species recorded in Kastamonu (Turkey): Oak lace bug [*Corythucha arcuata* (Say, 1832)] and sycamore lace bug [*Corythucha ciliata* (Say, 1832)] (Heteroptera: Tingidae). *Journal of Entomology and Nematology, 6*(8), 104–111. doi:10.5897/JEN2014.0102

Kunt, B. K., Yağmur, A. E., Özkütük, S., & Anlaş, S. (2010). Checklist of the cave Dwelling Invertebrates (Animalia) of Turkey. *Biological Diversity and Conservation, 3*(2), 26–41.

Kuzucuoğlu, C. (2019). The Physical Geography of Turkey: An Outline. I: C. Kuzucuoğlu, A. Çiner, & N. Kazancı (Eds.), Landscapes and Landforms of Turkey (pp. 7–15). Springer.

le Roux, E., van Veenhuisen, L. S., Kerley, G. I. H., & Cromsigt, J. P. G. M. (2020). Animal body size distribution influences the ratios of nutrients supplied to plants. *Proceedings of the National Academy of Sciences of the United States of America, 117*(36), 22256–22263. doi:10.1073/pnas.2003269117 PMID:32839336

Lutovinovas, E., Tschorsnig, H.-P., Barták, M., Kubík, Š., Dursun, O., Civelek, H.-S., & Kara, K. (2018). Contribution to the tachinid fauna of southwestern Turkey (Diptera: Tachinidae). *Annales de La Société Entomologique de France (N.S.), 54*(4), 335–366. doi:10.1080/00379271.2018.1468724

Mainka, S. A., & Howard, G. W. (2010). Climate change and invasive species: Double jeopardy. *Integrative Zoology, 5*(2), 102–111. doi:10.1111/j.1749-4877.2010.00193.x PMID:21392328

Meijers, M. J. M., Brocard, G. Y., Whitney, D. L., & Mulch, A. (2020). Paleoenvironmental conditions and drainage evolution of the central Anatolian lake system (Turkey) during late Miocene to Pliocene surface uplift. *Geosphere, 16*(2), 490–509. doi:10.1130/GES02135.1

Meijers, M. J. M., Peynircioğlu, A. A., Cosca, M. A., Brocard, G. Y., Whitney, D. L., Langereis, C. G., & Mulch, A. (2018). Climate stability in central Anatolia during the Messinian Salinity Crisis. *Palaeogeography, Palaeoclimatology, Palaeoecology, 498*, 53–67. doi:10.1016/j.palaeo.2018.03.001

Miroğlu, A., Kartal, V., & Salur, A. (2011). Odonata of the eastern Black Sea region of Turkey, with some taxonomic notes | Request PDF. *Odonatologica, 40*(2), 105–122.

Mittermeier, R. A., Turner, W. R., Larsen, F. W., Brooks, T. M., & Gascon, C. (2011). Global Biodiversity Conservation: The Critical Role of Hotspots. I: F. E. Zachos & J. C. Habel (Eds.), Biodiversity Hotspots (pp. 3–22). Springer Berlin Heidelberg. doi:10.1007/978-3-642-20992-5_1

Mitton, J. B., & Ferrenberg, S. M. (2012). Mountain pine beetle develops an unprecedented summer generation in response to climate warming. *American Naturalist, 179*(5), E163–E171. doi:10.1086/665007 PMID:22504550

Mohammadi, R., Talebi, A. A., Sakhvon, V. V., Khaghaninia, S., Fathipour, Y., & Kazerani, F. (2022). Study of the robber flies (Diptera: Asilidae) in northwestern Iran: new records, description of a new species, and an updated checklist of Iranian fauna. *Biologia, 77*(9), 2487–2496. doi:10.100711756-022-01116-5

Moog, O., Christian, E., & Eis, R. (2021). Increased cave use by butterflies and moths: A response to climate warming? *International Journal of Speleology, 50*(1), 15–24. doi:10.5038/1827-806X.50.1.2361

Moreira, F., Ascoli, D., Safford, H., Adams, M. A., Moreno, J. M., Pereira, J. M. C., Catry, F. X., Armesto, J., Bond, W., González, M. E., Curt, T., Koutsias, N., McCaw, L., Price, O., Pausas, J. G., Rigolot, E., Stephens, S., Tavsanoglu, C., Vallejo, V. R., & Fernandes, P. M. (2020). Wildfire management in Mediterranean-type regions: Paradigm change needed. *Environmental Research Letters, 15*(1), 011001. doi:10.1088/1748-9326/ab541e

Moritz, R. F. A., Härtel, S., & Neumann, P. (2005). Global invasions of the western honeybee (*Apis mellifera*) and the consequences for biodiversity. *Ecoscience, 12*(3), 289–301. doi:10.2980/i1195-6860-12-3-289.1

Mutun, S. (2010). Intraspecific genetic variation and phylogeography of the oak gallwasp andricus caputmedusae (hymenoptera: cynipidae): Effects of the anatolian diagonal. *Acta Zoologica Academiae Scientiarum Hungaricae, 56*(2), 153–172.

Mutun, Serap. (2016). Türkiye'deki mazı arılarının filocoğrafi örüntülerine dair inceleme, Anadolu çaprazının temel rolü. *Turkish Journal of Forestry | Türkiye Ormancılık Dergisi, 17*(0), 1. doi:10.18182/tjf.65861

Mutun, S., & Dinç, S. (2015). Twelve oak gall wasp species (Hymenoptera, Cynipidae) new to the Turkish fauna. *Turkish Journal of Zoology, 39*, 962–964. doi:10.3906/zoo-1409-7

Mutun, S., & Dinç, S. (2019). The Anatolian Diagonal and Paleoclimatic Changes Shaped the Phylogeography of Cynips quercus (Hymenoptera, Cynipidae). *Annales Zoologici Fennici*, *56*(1–6), 65. doi:10.5735/086.056.0107

Myers, N., Mittermeier, R. A., Mittermeier, C. G., da Fonseca, G. A., & Kent, J. (2000). Biodiversity hotspots for conservation priorities. *Nature*, *403*(6772), 853–858. doi:10.1038/35002501 PMID:10706275

Noroozi, J., Talebi, A., Doostmohammadi, M., Rumpf, S. B., Linder, H. P., & Schneeweiss, G. M. (2018). Hotspots within a global biodiversity hotspot - areas of endemism are associated with high mountain ranges. *Scientific Reports*, *8*(1), 10345. doi:10.103841598-018-28504-9 PMID:29985437

Noroozi, J., Zare, G., Sherafati, M., Mahmoodi, M., Moser, D., Asgarpour, Z., & Schneeweiss, G. M. (2019). Patterns of endemism in Turkey, the meeting point of three global biodiversity hotspots, based on three diverse families of vascular plants. *Frontiers in Ecology and Evolution*, *7*, 159. Advance online publication. doi:10.3389/fevo.2019.00159

O'Hara, J. E., Henderson, S. J., & Wood, D. M. (2020). *Preliminary checklist of the Tachinidae (Diptera) of the world. Version 2.1.* World Checklist of Teh Tachinidae. http://www.nadsdiptera.org/ Tach/ WorldTachs/ Checklist/ Worldchecklist.html

Oğuzoğlu, Ş., Avci, M., Şenal, D., & Karaca, İ. (2017). *Anatis ocellata* (Linnaeus, 1758) (Coleoptera: Coccinellidae)'nın Türkiye'deki ilk kaydı. *Türkiye Entomoloji Bülteni*, *7*(2), 197–202. doi:10.16969/entoteb.331402

Okay, A. I. (2008). Geology of Turkey: A Synopsis. *Anschnitt*, *21*, 19–42.

Okay, A. I., Zattin, M., Özcan, E., & Sunal, G. (2020). Uplift of Anatolia. *Turkish Journal of Earth Sciences*, *29*(5), 696–713. doi:10.3906/yer-2003-10

Otte, D., Spearman, L., & Stiewe, M. B. D. (2022). *Mantodea Species File Online. Version 5.0/5.0 (Online Database).* Mantodea Home Page. http://mantodea.speciesfile.org

Özbek, H. (2016). Türkiye'de arı (Hymneoptera: Apoidea: Apiformes) avcısı olan Asilidae (Diptera) türleri. *Uludağ Arıcılık Dergisi*, *15*(1), 10–15. doi:10.31467/uluaricilik.377559

Özbek, H., & Terzo, M. (2016). Distribution data for the Tribes Ceratinini and Allodapini (Hymenoptera: Apidae) with a checklist of the Subfamily Xylocopinae of Turkey. *Acta Entomologica Serbica*, *21*, 93–112.

Özdan, A., & Gürbüz, M. F. (2019). Ichneumonidae (Hymenoptera) fauna of Kovada Lake National Park, Isparta, Turkey. *Turkiye Entomoloji Dergisi*, 301–312. doi:10.16970/entoted.537395

Özdikmen, H. (2017). A complete list of invasive alien longhorned beetles species for Turkey (Coleoptera: Cerambycidae). *Munis Entomology & Zoology*, *12*(2), 508–515.

Oztas, T., Koc, A., & Comakli, B. (2003). Changes in vegetation and soil properties along a slope on overgrazed and eroded rangelands. *Journal of Arid Environments*, *55*(1), 93–100. doi:10.1016/S0140-1963(02)00267-7

Öztemiz, S., & Doğanlar, M. (2015). Invasive plant pests (Insecta and Acarina) of Turkey. *Munis Entomology & Zoology*, *10*(101), 144–159.

Parolly, G. (2020). The Anatolian high-mountain ranges – Plant diversity between two seas. I: J. Noroozi (Ed.), Plant Biogeography and Vegetation of High Mountains of Central and South-West Asia (Vol. 17, pp. 215–286). Springer International Publishing. doi:10.1007/978-3-030-45212-4_7

Parratt, S. R., Walsh, B. S., Metelmann, S., White, N., Manser, A., Bretman, A. J., Hoffmann, A. A., Snook, R. R., & Price, T. A. R. (2021). Temperatures that sterilize males better match global species distributions than lethal temperatures. *Nature Climate Change*, *11*(6), 481–484. doi:10.103841558-021-01047-0

Pausas, J. G., & Keeley, J. E. (2019). Wildfires as an ecosystem service. *Frontiers in Ecology and the Environment*, *17*(5), 289–295. doi:10.1002/fee.2044

Pintilioaie, A.-M., Spaseni, P., Jurjescu, A., & Rădac, I. A. (2021). First record of the alien mantid *Hierodula tenuidentata* (Insecta: Mantodea) in Romania. *Travaux Du Muséum National d'Histoire Naturelle. Grigore Antipa*, *64*(1), 37–49. doi:10.3897/travaux.64.e65489

Prather, C. M., Pelini, S. L., Laws, A., Rivest, E., Woltz, M., Bloch, C. P., Del Toro, I., Ho, C.-K., Kominoski, J., Newbold, T. A. S., Parsons, S., & Joern, A. (2013). Invertebrates, ecosystem services and climate change. *Biological Reviews of the Cambridge Philosophical Society*, *88*(2), 327–348. doi:10.1111/brv.12002 PMID:23217156

Rampini, M., Russo, C. D., Taylan, M. S., Gelosa, A., & Cobolli, M. (2012). Four new species of *Dolichopoda* Bolivar, 1880 from Southern Sporades and Western Turkey (Orthoptera, Rhaphidophoridae, Dolichopodainae). *ZooKeys201*, 43–58. doi:10.3897/zookeys.201.2609

Raven, P. H., & Wagner, D. L. (2021). Agricultural intensification and climate change are rapidly decreasing insect biodiversity. *Proceedings of the National Academy of Sciences of the United States of America, 118*(2), e2002548117. doi:10.1073/pnas.2002548117 PMID:33431564

Ricklefs, R. E. (2007). History and diversity: Explorations at the intersection of ecology and evolution. *American Naturalist, 170*(S2, Suppl 2), S56–S70. doi:10.1086/519402 PMID:17874385

Riedel, M., Diller, E., & Çoruh, S. (2018). New contributions to the Ichneumoninae (Hymenoptera, Ichneumonidae) from Turkey. *Journal of Entomological Research Society, 20*(1), 59–72.

Riedel, M., Kolarov, J., Çoruh, S., & Özbek, H. (2014). A contribution to the Mesochorinae (Hymenoptera: Ichneumonidae) of Turkey. *Zoology in the Middle East, 60*(3), 217–221. doi:10.1080/09397140.2014.939815

Rögl, F., & Steininger, F. F. (1983). Vom Zerfall der Tethys zu Mediterran und Paratethys. *Annalen Des Naturhistorischen Museums Wien Serie A, 85*(April), 135–163.

Roy, H. E., Bacher, S., Essl, F., Adriaens, T., Aldridge, D. C., Bishop, J. D. D., Blackburn, T. M., Branquart, E., Brodie, J., Carboneras, C., Cottier-Cook, E. J., Copp, G. H., Dean, H. J., Eilenberg, J., Gallardo, B., Garcia, M., García-Berthou, E., Genovesi, P., Hulme, P. E., & Rabitsch, W. (2019). Developing a list of invasive alien species likely to threaten biodiversity and ecosystems in the European Union. *Global Change Biology, 25*(3), 1032–1048. doi:10.1111/gcb.14527 PMID:30548757

Saghaei, N., Hayat, R., Ostovan, H., & Fallahzadeh, M. (2010). Prey of some robber flies (Diptera: Asilidae) in fars province, Iran. *Journal of Entomological Research Society, 12*(1), 17–26.

Salur, A., Darilmaz, M. C., & Bauernfeind, E. (2016). An annotated catalogue of the mayfly fauna of Turkey (Insecta, Ephemeroptera). *ZooKeys, 620*, 67–118. doi:10.3897/zookeys.620.9405

Sands, D. P. A. (2018). Important issues facing insect conservation in Australia: Now and into the future. *Austral Entomology, 57*(2), 150–172. doi:10.1111/aen.12342

Sarikaya, O., & Ibis, H. M. (2016). Predatory species of bark beetles in the pine forests of izmir region in Turkey with new records for Turkish Fauna. *Egyptian Journal of Biological Pest Control, 26*(3), 651–656.

Şengör, A. M. C., Özeren, M. S., Keskin, M., Sakınç, M., Özbakır, A. D., & Kayan, İ. (2008). Eastern Turkish high plateau as a small Turkic-type orogen: Implications for post-collisional crust-forming processes in Turkic-type orogens. *Earth-Science Reviews*, *90*(1–2), 1–48. doi:10.1016/j.earscirev.2008.05.002

Şenkul, Ç., & Kaya, S. (2017). Türkiye endemik bitkilerinin coğrafi dağılışları. *Türkiye Coğrafya Dergisi, 69*, 109–120. doi:10.17211/tcd.322515

Şenol, Ö., Akyildirim Beğen, H., Görür, G., & Demirtaş, E. (2015). New additions and invasive aphids for Turkey's aphidofauna (Hemiptera: Aphidoidea). *Turkish Journal of Zoology*, *39*, 39–45. doi:10.3906/zoo-1308-12

Şenol, Ö., Görür, G., & Beğen, H. A. (2017). Recent findings on aphid fauna from east and south eastern parts of Turkey. *Biological Diversity and Conservation*, *10*(3), 76–78.

Seven, S. (2014). A new species of blue from Turkey, "*Neolycaena soezen*" Seven, sp. n. (Lepidoptera: Lycaenidae). *SHILAP Revista de Lepidopterologia*, *42*(166), 311–317.

Sevgili, H. (2004). *A revision of Turkish species of Isophya Brunner von Wattenwyl (Orthoptera: Tettigoniidae: Phaneropterinae)*, 397–397.

Sevgili, H. (2016). Spermatophore weight and sperm number in the southernmost species of the genus *Poecilimon* Fischer, 1853 (Orthoptera: Phaneropterinae). *Acta Biologica Turcica*, *29*(3), 83–88.

Sevgili, H., Caglar, S. S., & Saglam, I. K. (2010). Re-evaluation of the genus *Phonochorion* (Orthoptera: Tettigoniidae: Phaneropterinae). *European Journal of Entomology*, *107*(4), 631–645. doi:10.14411/eje.2010.072

Sevgili, H., & Çıplak, B. (2000). The Orthoptera of Şanliurfa province from the Mesopotamian part of Turkey. *The Italian Journal of Zoology*, *67*(2), 229–240. doi:10.1080/11250000009356316

Sevgili, H., Çiplak, B., Heller, K. G., & Demirsoy, A. (2006). Morphology, bioacoustics and phylogeography of the *Isophya major* group (Orthoptera: Tettigoniidae: Phaneropterinae): A species complex occurring in Anatolia and Cyprus. *European Journal of Entomology*, *103*(3), 657–671. doi:10.14411/eje.2006.086

Sevgili, H., Heller, K.-G., & Reinhold, K. (2008). A re-assessment of the Poecilimon syriacus group (Orthoptera Tettigonioidea, Phaneropteridae) based on bioacoustics, morphology and molecular data. *Insect Systematics & Evolution*, *39*(4), 361–379. doi:10.1163/187631208788784309

Sevgili, H., Önal, H., & Yiğit, A. (2015). Mating behavior and spermatophore characteristics in two closely related bushcricket species of the genus *Phonochorion* (Orthoptera: Phaneropterinae). *Journal of Insect Behavior*, *28*(4), 369–386. doi:10.100710905-015-9509-1

Sevgili, H., Şirin, D., Heller, K.-G., & Lemonnier-Darcemont, M. (2018). Review of the *Poecilimon (Poecilimon) zonatus* species group and description of new species from Turkey with data on bioacoustics and morphology (Orthoptera: Phaneropterinae). *Zootaxa*, *4417*(1), 1–62. doi:10.11646/zootaxa.4417.1.1 PMID:30313607

Sevgili, H., & Yılmaz, K. (2022). Contributions of citizen scientists to monitoring alien species: the case study on Giant Asian Mantes, *Hierodula tenuidentata* and *H. patellifera* (Mantodea: Mantidae). *Zoology in the Middle East*, *68*(4), 1–9. doi: 10.1080/09397140.2022.2145802

Sileshi, G., & Mafongoya, P. L. (2006). The short-term impact of forest fire on soil invertebrates in the Miombo. *Biodiversity and Conservation*, *15*(10), 3153–3160. doi:10.100710531-005-5411-z

Sipahiler, F. (1996). Studies on the Trichoptera fauna of southern Anatolia. *Part 1. Biological Sciences and Living Resources*, *16*(16), 293–312.

Sipahiler, F. (2005). In K. Tanida (Ed.), *A checklist of the caddisflies of Turkey (Trichoptera)* (pp. 393–405).

Sipahiler, F. (2008). Zoogeographical characteristics of the Trichoptera fauna of Turkey. *Ferrantia*, *55*, 93–109.

Şirin, D., Sevgili, H., & Taylan, M. S. (2018). *Distribution of endemic Ensifera species in Anatolia*. II. *European Congress on Orthoptera Conservation.*

Şirin, D., Taylan, M. S., Sevgili, H., & Mol, A. (2019). Bioacoustics review of Anatolian species of the predatory bush-cricket genus *Saga* (Orthoptera: Tettigoniidae: Saginae) with the description of a new species. *Zootaxa, 4664*(1), zootaxa.4664.1.3. doi:10.11646/zootaxa.4664.1.3

Skuhravá, M. (2016). Checklist of gall midges (Diptera: Cecidomyiidae) of Turkey. *Acta Societatis Zoologicae Bohemicae*, *80*(2–3), 197–199.

Smith, C. I., & Farrell, B. D. (2005). Phylogeography of the longhorn cactus beetle Moneilema appressum LeConte (Coleoptera: Cerambycidae): was the differentiation of the Madrean sky islands driven by Pleistocene climate changes? *Molecular Ecology*, *14*(10), 3049–3065. doi:10.1111/j.1365-294X.2005.02647.x PMID:16101773

Snyder, W. E., & Evans, E. W. (2006). Ecological Effects of Invasive Arthropod Generalist Predators. *Annual Review of Ecology, Evolution, and Systematics*, *37*(1), 95–122. doi:10.1146/annurev.ecolsys.37.091305.110107

Soyhan, T., Başer, S., & Nazari, V. (2013). First record of *Cacyreus marshalli* Butler, 1898 (Lycaenidae) from Turkey. *Nota Lepidopterologica*, *36*(2), 189–190.

Stewart, J. R., Lister, A. M., Barnes, I., & Dalén, L. (2010). Refugia revisited: individualistic responses of species in space and time. *Proceedings. Biological Sciences / the Royal Society*, *277*(1682), 661–671. doi:10.1098/rspb.2009.1272

Stork, N. E. (2018). How many species of insects and other terrestrial arthropods are there on earth? *Annual Review of Entomology*, *63*(1), 31–45. doi:10.1146/annurev-ento-020117-043348 PMID:28938083

Taşdemir, A., & Ustaoğlu, M. R. (2005). Göller Bölgesi İçsularının Chironomidae ve Chaoboridae (Diptera) Faunasının Taksonomik Yönden İncelenmesi. *Su Ürünleri Dergisi*, *22*(3), 377–384.

Tavşanoğlu, Ç., & Serter Çatav, Ş. (2012). Seed size explains within-population variability in post-fire germination of *Cistus salviifolius*. *Annales Botanici Fennici*, *49*(5–6), 331–340. doi:10.5735/085.049.0604

Taylan, Mehmet Sait, Russo, C. D., Rampini, M., & Ketmaier, V. (2013). Molecular systematics of the genus *Troglophilus* (Rhaphidophoridae, Orthoptera) in Turkey: mitochondrial 16S rDNA evidences. *ZooKeys257*, 33–46. doi:10.3897/zookeys.257.4133

Taylan, Mehmet Sait, Mol, A., & Şirin, D. (2015). Description of a new species of the genus *Troglophilus* krauss, 1879 (Orthoptera: Rhaphidophoridae) from Northern Anatolia, Turkey. *Journal of Entomological Research Society*, *17*(2), 51–57.

Taylan, M. S., di Russo, C., & Cobolli, M. (2011). The Dolichopodainae and Troglophilinae cave crickets of Turkey: An update of taxonomy and geographic distribution (Orthoptera, Rhaphidophoridae). *Zootaxa*, *2829*(1), 59. doi:10.11646/zootaxa.2829.1.3

Taylan, M. S., Russo, C. D., Cobolli, M., & Rampini, M. (2012). New species of the genus *Troglophilus* Krauss, 1879 (Orthoptera: Rhaphidophoridae) from Western and Southern Anatolian caves, Turkey. *Zootaxa*, *3597*(1), 33. doi:10.11646/zootaxa.3597.1.4

Taylan, M. S., & Şirin, D. (2016). Speciation of the genus *Dolichopoda* in Anatolia with reference to the role of ancient central lake system. *Insect Systematics & Evolution, 47*(3), 267–283. doi:10.1163/1876312X-47032143

Tekinoğlu, V., & Çalıskan, H. (2019). The Crane Flies (Tipulidae: Diptera) fauna of the Uludağ Mountain (Turkey). *Biological Diversity and Conservation, 12*(2), 57–64. doi:10.5505/biodicon.2019.32042

Tezcan, S. (2020). Analysis of the insect fauna of Turkey and suggestions for future studies. *Munis Entomology & Zoology Journal, 15*(2), 690–710.

Tilman, D., Clark, M., Williams, D. R., Kimmel, K., Polasky, S., & Packer, C. (2017). Future threats to biodiversity and pathways to their prevention. *Nature, 546*(7656), 73–81. doi:10.1038/nature22900 PMID:28569796

Tougeron, K., Brodeur, J., Le Lann, C., & Baaren, J. (2020). How climate change affects the seasonal ecology of insect parasitoids. *Ecological Entomology, 45*(2), 167–181. doi:10.1111/een.12792

Toussaint, E. F. A., Sagata, K., Surbakti, S., Hendrich, L., & Balke, M. (2013). Australasian sky islands act as a diversity pump facilitating peripheral speciation and complex reversal from narrow endemic to widespread ecological supertramp. *Ecology and Evolution, 3*(4), 1031–1049. doi:10.1002/ece3.517 PMID:23610642

Trakel Web Page. (2021). *TRAKEL kelebek fotoğrafları, kelebek türleri, kelebekler hakkında detaylı Türkçe bilgi.* Trakel.

Tschopp, A., Riedel, M., Kropf, C., Nentwig, W., & Klopfstein, S. (2013). The evolution of host associations in the parasitic wasp genus *Ichneumon* (Hymenoptera: Ichneumonidae): convergent adaptations to host pupation sites. *BMC Evolutionary Biology, 13*(1), 74. doi:10.1186/1471-2148-13-74 PMID:23537515

Tunç, İ., & Hastenpflug-Vesmanis, A. (2016). Records and checklist of Thysanoptera in Turkey. *Turkish Journal of Zoology, 40*, 769–778. doi:10.3906/zoo-1512-37

Tuncer, C., Özdemir, İ. O., & Kushiyev, R. (2020). Türkiye fındık bahçelerinde yeni zararlı türler: *Xylosandrus germanus* Blandford (Col.: Curculionidae: Scolytınae), *Metcalfa pruinosa* Say (Hem.: Flatidae), *Croesus septentrionalis* Linnaeus (Hym.: Tenthredinidae) ve *Anoplophora chinensis* Forster (Col.: Cerambycidae). *Black Sea Journal of Agriculture, 3*(1), 74–81.

Tunçyürek-Soydanbay, M. (1978). Türkiye'de bitki zararlısı bazı böceklerin doğal düşman listesi, Kısım II. *Turkiye Entomoloji Dergisi, 2*(2), 61–92.

Ülgentürk, S. (2016). A new introduced wax scale to Turkey: *Ceroplastes ceriferus* (Fabricius) (Hemiptera: Coccidae). *Mun. Ent. Zool., 11*(2), 359–362.

Uma, R., & Sevgili, H. (2015). Spermatophore allocation strategy over successive matings in the bushcricket *Isophya sikorai* (Orthoptera, Phaneropterinae). *Ethology Ecology and Evolution, 27*(2), 129–147. doi:10.1080/03949370.2014.896830

Ünal, M. (2016). Pamphagidae (Orthoptera: Acridoidea) from the Palaearctic Region: taxonomy, classification, keys to genera and a review of the tribe Nocarodeini I.Bolívar. *Zootaxa, 4206*(1), zootaxa.4206.1.1. doi:10.11646/zootaxa.4206.1.1

Ünal, M. (2021). *Turkish Orthoptera Site*.

Underwood, E. C., & Quinn, J. F. (2010). Response of ants and spiders to prescribed fire in oak woodlands of California. *Journal of Insect Conservation, 14*(4), 359–366. doi:10.100710841-010-9265-7

Uslu, E., & Bakış, Y. (2012). Geographic distribution of Turkish Oaks. *Dendrobiology (Poznan), 67*, 41–48.

Uzun, I., & Bayır, A. (2009). Horticultural Biodiversity in Turkey. *Bulletin UASVM Horticulture, 66*, 536–543.

Verves, Y., Barták, M., & Kubík, Š. (2018). Checklist of flesh flies of Turkey (Diptera, Sarcophagidae). *ZooKeys, 743*, 95–136. doi:10.3897/zookeys.743.22022

Verves, Y., Barták, M., Kubík, Š., & Civelek, H. S. (2017). New records of Sarcophagidae from Turkey (Diptera). *ZooKeys, 703*, 129–158. doi:10.3897/zookeys.703.12377

Wagener, S. (2006). Butterfly diversity and protection in Turkey. *Bonner Zoologische Beitrage, 54*(1), 3–23.

Wainright, C. A., Muhlfeld, C. C., Elser, J. J., Bourret, S. L., & Devlin, S. P. (2021). Species invasion progressively disrupts the trophic structure of native food webs. *Proceedings of the National Academy of Sciences of the United States of America, 118*(45), e2102179118. doi:10.1073/pnas.2102179118 PMID:34725150

Westphal, M. I., Browne, M., MacKinnon, K., & Noble, I. (2008). The link between international trade and the global distribution of invasive alien species. *Biological Invasions, 10*(4), 391–398. doi:10.100710530-007-9138-5

Wiemers, M., Balletto, E., Dincă, V., Faltynek Fric, Z., Gerardo Lamas, Lukhtanov, V., Munguira, M. L., Chris A M van Swaay, Vila, R., Vliegenthart, A., Wahlberg, N., & Verovnik, R. (2018). An updated checklist of the European Butterflies (Lepidoptera, Papilionoidea). *ZooKeys811*, 9–45. doi:10.3897/zookeys.811.28712

Wietzke, A., Westphal, C., Gras, P., Kraft, M., Pfohl, K., Karlovsky, P., Pawelzik, E., Tscharntke, T., & Smit, I. (2018). Insect pollination as a key factor for strawberry physiology and marketable fruit quality. *Agriculture, Ecosystems & Environment*, *258*, 197–204. doi:10.1016/j.agee.2018.01.036

Wilson, R. J., & Maclean, I. M. D. (2011). Recent evidence for the climate change threat to Lepidoptera and other insects. *Journal of Insect Conservation*, *15*(1–2), 259–268. doi:10.100710841-010-9342-y

Winn, M. I., & Pogutz, S. (2013). Business, ecosystems, and biodiversity. *Organization & Environment*, *26*(2), 203–229. doi:10.1177/1086026613490173

Yalçın, M., Akçay, Ç., Taşçıoğlu, C., & Yüksel, B. (2019). Insect species damaging industrial wood in western black sea region of turkey. *Drvna Industrija*, *70*(2), 183–191. doi:10.5552/drvind.2019.1818

Yamaç, A., Gilli, E., Tok, E., & Törk, K. (2021). *History, Archaeology and Caves* (Vol. 1). Natural sciences in the caves of turkey. I: Caves and Karst of Turkey. Springer International Publishing., doi:10.1007/978-3-030-65501-3_3

Yilmaz, K., & Sevgili, H. (2023). The genus *Holaptilon* Beier, 1964: Discussion on the poorly known boxer mantis genus, with a new species (Mantodea, Gonypetidae). *Zootaxa*, *5231*(4), 427–444. doi:10.11646/zootaxa.5231.4.5 PMID:37045135

Yu, D. S., van Achterberg, C., & Horstmann, K. (2016). *Taxapad 2016, Ichneumonoidea 2015.*

Zachos, J., Pagani, M., Sloan, L., Thomas, E., & Billups, K. (2001). Trends, rhythms, and aberrations in global climate 65 Ma to present. *Science*, *292*(5517), 686–693. doi:10.1126cience.1059412 PMID:11326091

Zaitsev, A. S., Gongalsky, K. B., Malmström, A., Persson, T., & Bengtsson, J. (2016). Why are forest fires generally neglected in soil fauna research? A mini-review. *Applied Soil Ecology*, *98*, 261–271. doi:10.1016/j.apsoil.2015.10.012

APPENDIX

Figure 1. Examples for habitat diversity from Anatolia. There is a positive relationship between habitat-plant diversity/biomass production and the richness of species diversity (Cardinale et al., 2007). All photos were captured by H. Sevgili.

Figure 2. Various types of habitats of Anatolia. Karacadağ is a mountain in the Southeast Anatolian Region with a shield type volcano structure. This volcano was active in the late Pleistocene-Quaternary (Canpolat, 2005). This mountain, located between Şanlıurfa and Diyarbakır-Mardin, has the hottest and the most arid climate of the region. Biodiversity of Karacadağ Mt. is under threat due to extreme drought and overgrazing in summer and autumn while the peaks of this mountain are covered with snow in winter. In addition, settlements, and the fact that habitats with unique vegetation and biodiversity are rapidly being destroyed for agricultural purposes, pose a great threat. This mountain is the only refuge for some cold-tolerant insect species in the region. There are important differences between the insect fauna of the northern and southern parts of the Taurus Mountains, extending from the East-West direction along the Mediterranean. Many alpine and cold-like endemic species are localized in alpine and subalpine regions in the northern parts of the mountains (Anti-Taurus). All photos were captured by H. Sevgili

Figure 3. All of the land parts constituting Anatolia did not uplifted above sea waters as a whole. Four important plateaus and elevations became land parts in different geologic timescales. By Okay et al. (2020).

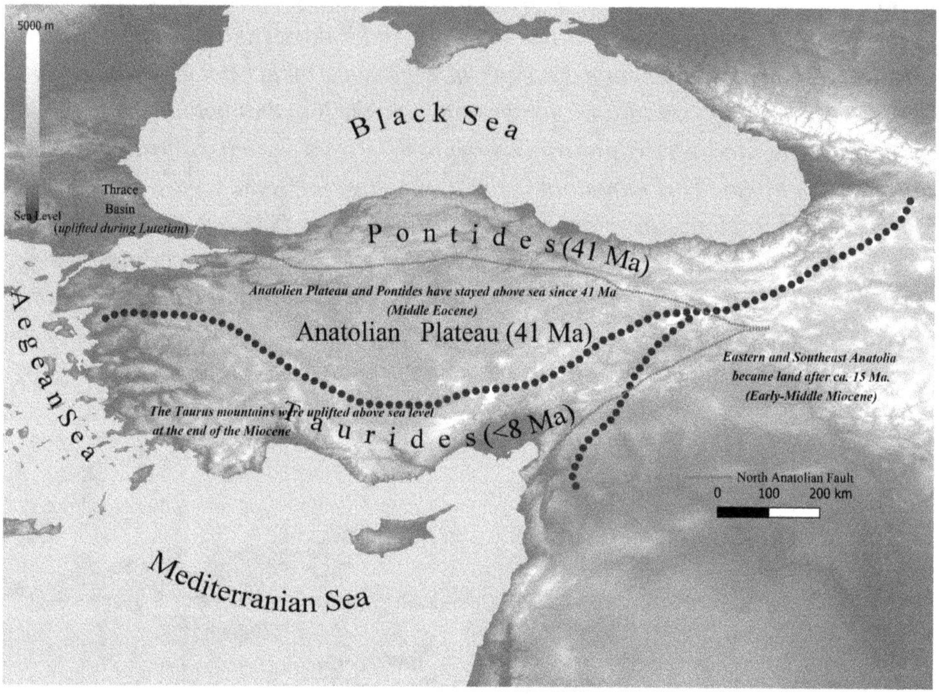

Figure 4. National Parks of Turkey. National parks are managed by the General Directorate of Nature Conservation and National Parks.

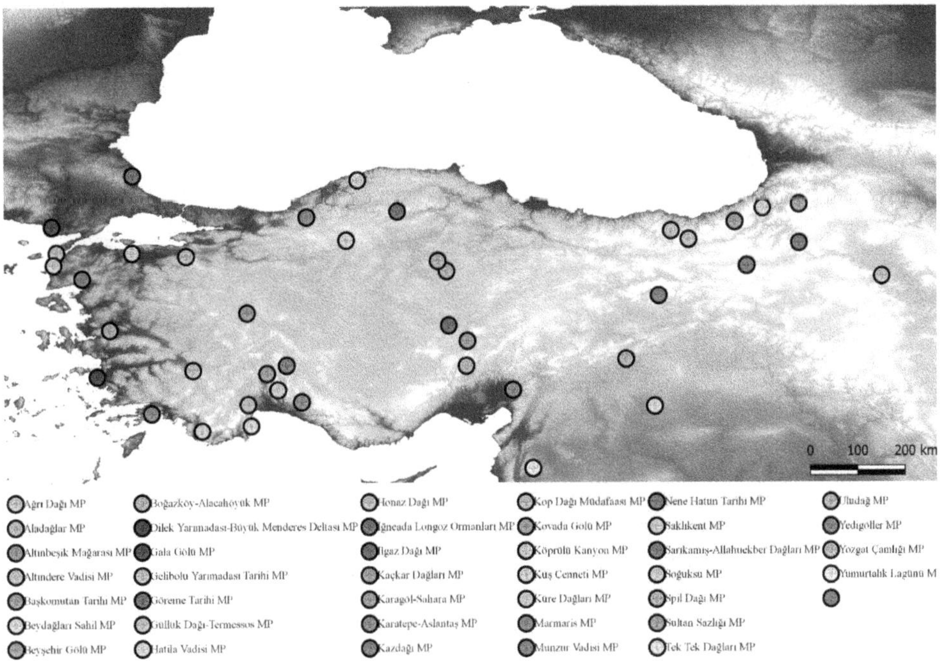

Figure 5. The micropterous Rove beetle genus Sunius (Coleoptera: Staphylinidae), which is endemic to the Mediterranean region of Turkey. The majority of its species has spread along the Taurus line across tall mountain ranges (The map is modified from (Anlaş, 2020)). This group of insects, which has very limited mobility, evolved locally as endemic species across narrow areas. This particular range is a unique location that has the highest rate of endemism. This is a fact for many other plant and insect species. The colored regions on the map represent mountainous areas and the lines represent streams and rivers.

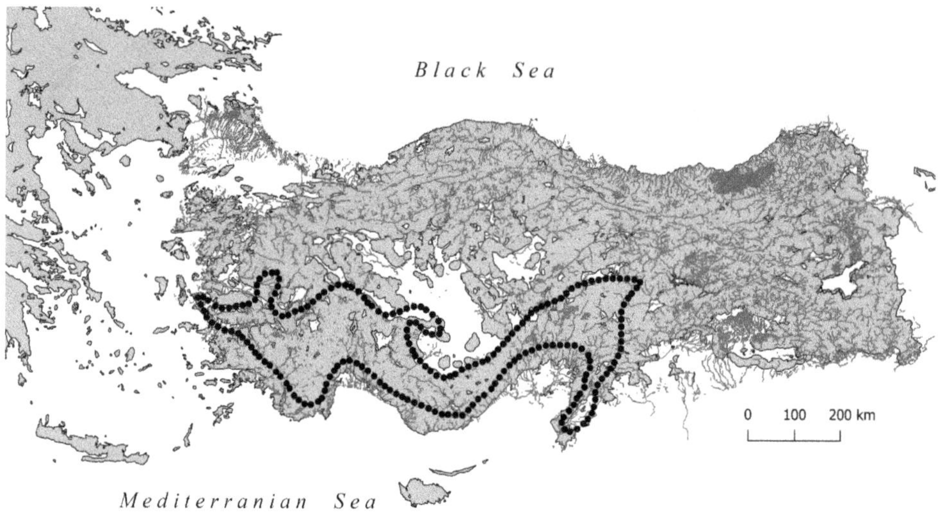

Figure 6. The areas where the biological diversity of Anatolia is the richest. The Anatolian Diagonal was the distribution route for the high mountain populations of the Caucasian and Mediterranean populations as well as acting as a barrier between the Eastern and Western flora -fauna.

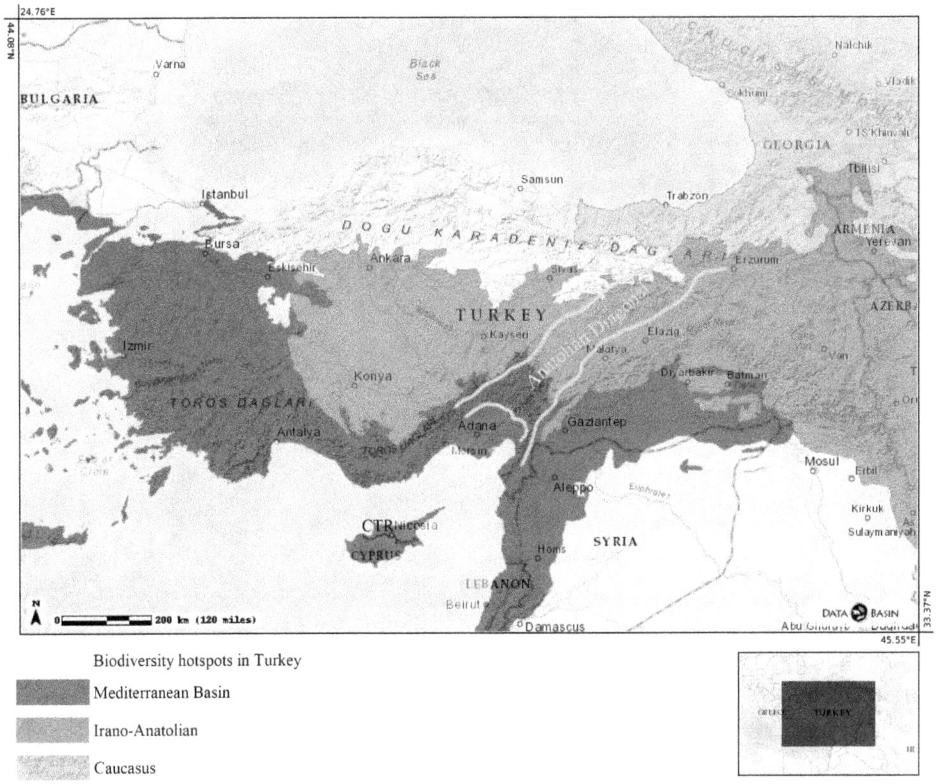

Figure 7. The diversity and distribution of endemic cave cricket species (Orthoptera: Raphidophoridae: the genera Dolichopoda and Troglophilus) in Turkey (the locality data from (Di Russo et al., 2007; Rampini et al., 2012; M S Taylan & Şirin, 2016; Mehmet Sait Taylan et al., 2012, 2015; M S Taylan et al., 2011).

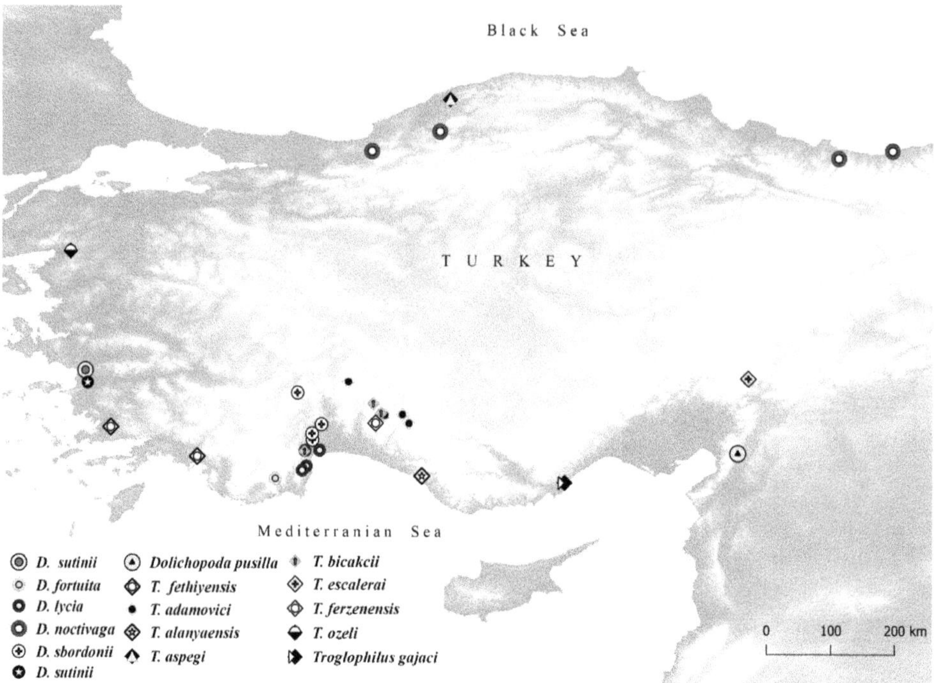

Figure 8. Dolichopoda (Dolichopoda) noctivaga Di Russo and Rampini, 2007 (Orthoptera: Raphidophoridae), which spreads in the Black Sea region and uses the old houses in the forest as a habitat (Tunca, Ardeşen, Rize, Photo: H. Sevgili).

Figure 9. Neolycaena soezen Seven, 2014. Recently discovered butterfly from Eskişehir and Ankara in Centre Anatolia. I captured this beautiful couple in the steppes of Çankırı.

Figure 10. Mayflies (Ephemeroptera) are an aquatic insect order. They are represented widely in rivers of Turkey. Almost every region of Turkey is represented by a large number of endemic species. 24 endemic species in total have been listed (Kazancı and Türkmen 2012; Salur et al. 2016). The map was prepared based on provincial and type locality of the species using the checklists of the authors (Kazancı and Türkmen 2012; Salur et al. 2016).

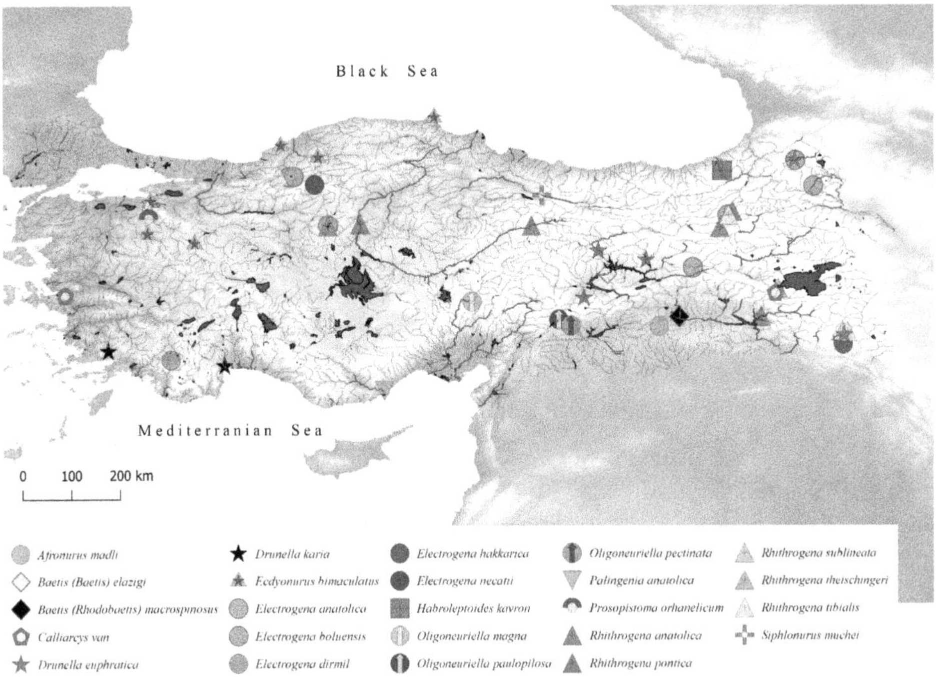

Figure 11. Saga ephippigera (female) and S: longicaudata (male). S. longicaudata known from Southeastern Mediterranian of Anatolia and endemic to Turkey. Photos by H. Sevgili from Elazığ and Tarsus, Mersin, Turkey.

Figure 12. Distribution of the genus Saga in Turkey. According to the literature (Şirin et al., 2019), there is no distribution of the species in the Black Sea region of Turkey. Several species spread in the Balkans in Europe. It is understood that the evolutionary origins of the species are located in Anatolia.

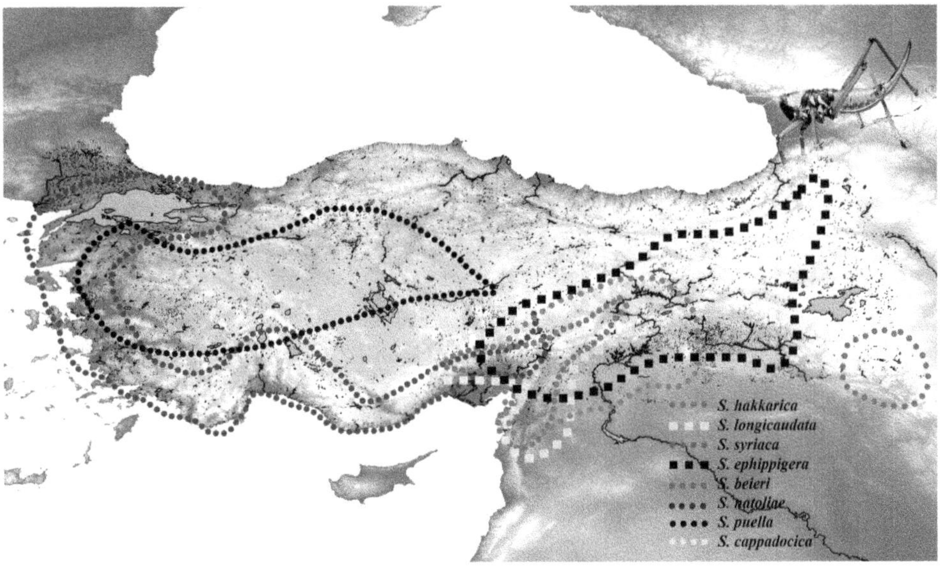

Figure 13. A common species (Machimus sp.) in Anatolia from the Robber flies, the successful predator group of insects (photo capture by H. Sevgili)

Figure 14. A simple model of the speciation pattern in insects in distinctive topography, climate, and vegetation, differentiation of ancestral populations from the effects of tectonic evolution, and glacial and interglacial periods during the Pleistocen. The spread strategies of cold-loving and temperate species have been different, especially during the inter-glacial periods. Within a heterogeneous geography covered by tall mountains like Anatolia, populations with limited mobility have become isolated on sky islands as a result of the warm climate. Herbivorous populations in temperate zones, on the other hand, have shown allopatric/sympatric speciation by specializing in a way to become suitable with different host plants in the Anatolian steppes (S: Species).

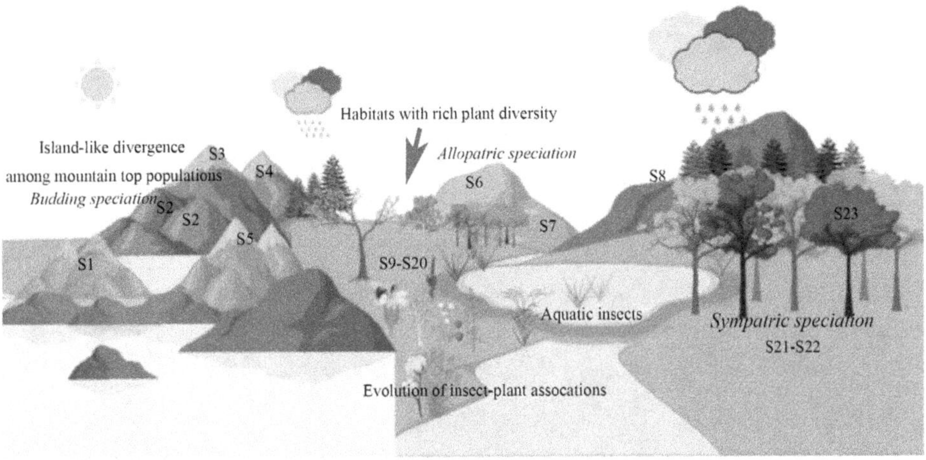

Figure 15. Two endemic bush-cricket species are known from the same habitat in subalpine zones of Anti Taurus (Ermenek) Photos by H. Sevgili.

Figure 16. Distribution of the Poecilimon zonatus species group on tall mountains across the Anatolian Diagonal towards the Taurus Way (locality data by Sevgili et al. (2018)). P. varicornis is isolated on the mountains of Lebanon as a relict population and shows a disjunct distribution from the species group.

Figure 17. A large number of endemic species such as the represented Pamphagidae (toad grasshopper) species that live in subalpine and alpine zones on the mountains of Turkey, have been isolated from each other along a mountain range, and this has diversified these species. The loss of wings in both sexes has severely limited the movement range of species, and gene exchange between local populations has been interrupted by geographic barriers. Because these highland species have adapted to the cold climate unlike other species living at the same altitude, they overwinter in the last stage of nymphs. All photos were captured by H. Sevgili.

Figure 18. The distribution of endemic species of the genus Carabus in Turkey. It is noteworthy that endemic species are mostly distributed in the southern and high mountain regions of the Black Sea region and Northeast Anatolia. These species, which are extremely restricted in their mobility and who like cold, have uplift from the sea much earlier than other regions of the country and evolved into terrestrial habitats and retreated to the isolated high parts of the mountains during the interglacial periods, causing the interruption of gene flow causing the interruption of gene flow which promotes the budding and allopatric speciation.

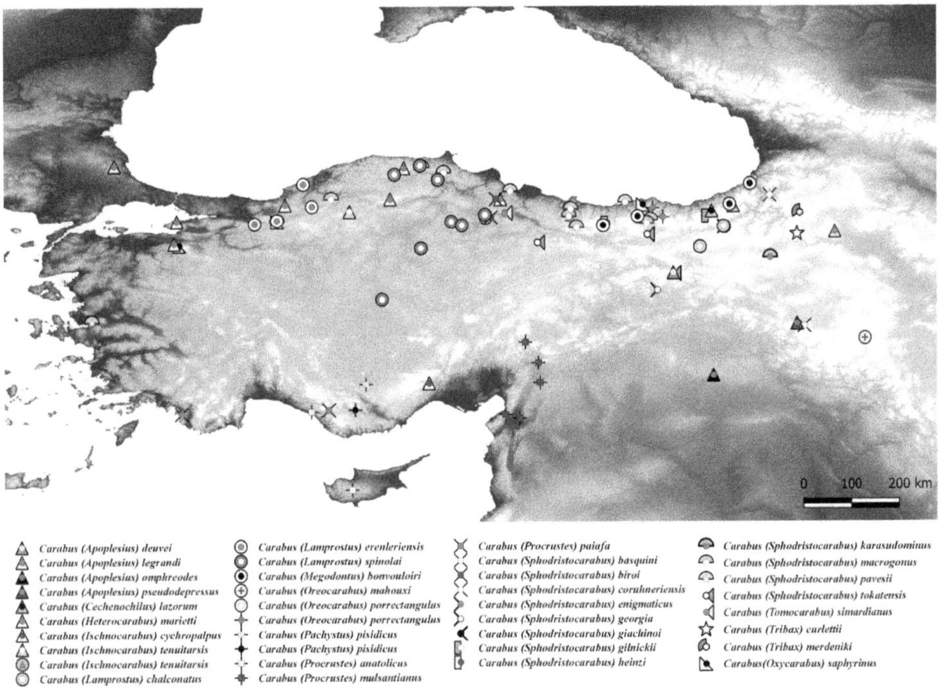

Figure 19. Orosanga japonica in the year 2007 is estimated to have entered Turkey through Georgia (Akıner et al., 2019). They were also sighted at the coast of Southeast Bulgaria in 2011. It has also been documented that this invasive species, which can cause serious damage on the fresh branches of almost all kinds of plants within a 500 m altitude range from the sea, is the natural prey of some bird species (Booted Warbler) and dragonflies (Suthern Skimmer) (Karataş et al., 2020).

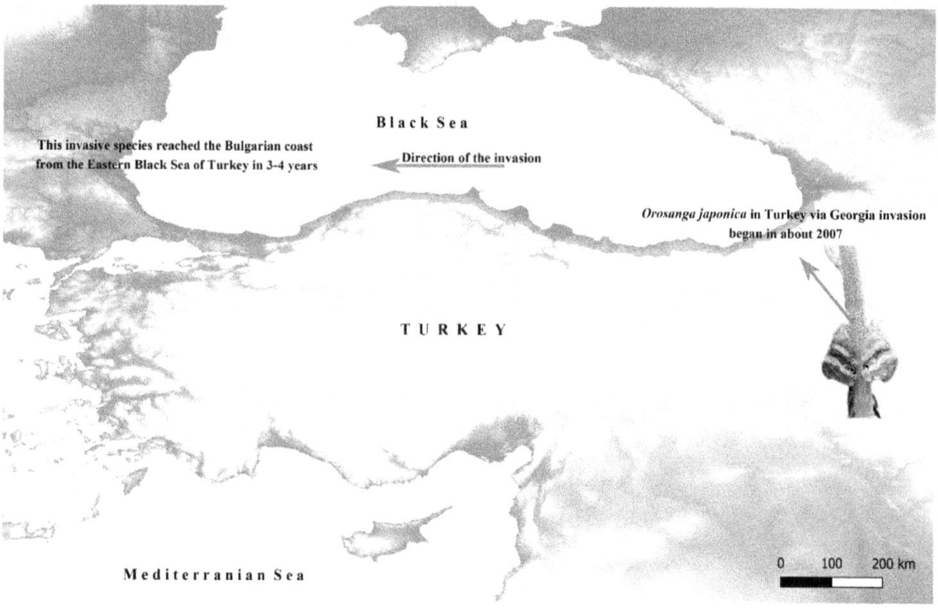

Figure 20. The natural steppes of the country are now fragmented by activities such as agriculture, urbanization, while other small and isolated parts are also under the influence of overgrazing

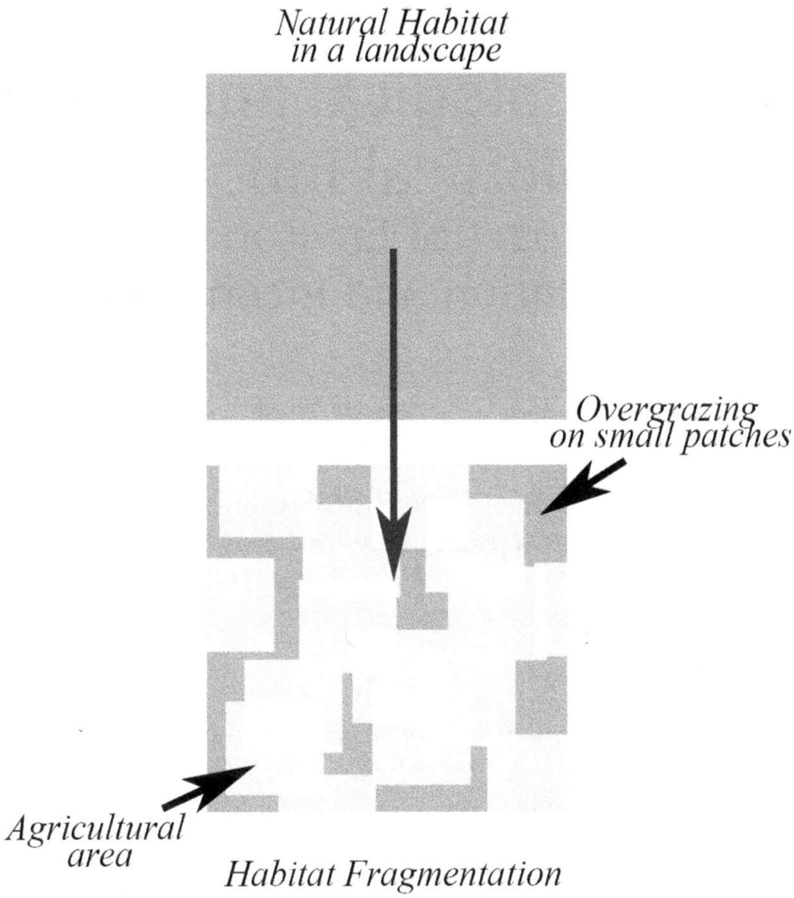

Chapter 5
Socio-Ecological and Green Technological Innovation:
The Impact of Environmental Regulation and Incentives

José G. Vargas-Hernandez
Budapest Centre for Long-term Sustainability, Hungary

Omar C. Vargas-González
https://orcid.org/0000-0002-6089-956X
Tecnológico Nacional de México, Ciudad Guzmán, Mexico

ABSTRACT

This study intends to analyze the impact of environmental regulation and incentives in socio-ecological and green technological innovation. It departs from the assumption that environmental regulation affects the performance of socio-ecological and green technological innovation in organizations. The method employed is critical analytical and reflective based on the theoretical, conceptual, and methodological literature. It is concluded that the analysis confirms that there is a direct relationship between the incentives of environmental regulation and the socio-ecological and green technological innovation in organizations.

INTRODUCTION

The rapid spread at global scope of innovative technologies has modernized the economic sectors that are contributors of environmental pollution and other health risk hazards with limiting findings in technological innovations and ecological

DOI: 10.4018/978-1-6684-9034-1.ch005

solutions. The development of organizational socio-ecology is based on theoretical methodological and empirical research. The abundance of research in population socio-ecology is accompanied by criticisms of ecological research and organizational socio-ecology regarding the abandonment of organizations (Amburgey & Rao, 1996).

Research can be advanced at the intersection between sustainability and population socio-ecology (Salimath, & Jones, 2011). The theoretical proposal builds on the theory of organizational choice and the theory of resource dependency (Pfeffer and Salancik, 1978) and the ecological theory of organizations that allow concentrating on the specificities of the sector as exercises of elections.

The theoretical framework of organizational socio-ecology and sociological theory applies ecological principles to organizational analysis. The theoretical framework of organizational socio-ecology and sociological theory is a tool to formulate and implement socioecological principles to organizational strategies for turning around low-performance organizations, inducing the emergence of an economic and social environment and with an impact on non-linear development.

The organisational socio-ecology theory assumes that there is convergence between the paradigm of organizational socio-ecology and sociological perspectives leading to organizational research. The fundamental assumption of organizational socio-ecology is the differentiation of organizational populations. organizational socio-ecology emphasizes issues that are still under discussion, and they conclude with areas of progress and fertilization with other disciplines of organizational thought. Organizational socio-ecology is linked to other areas with the incorporation of methodological innovations such as strategic simulation models.

The forces of the organizational socio-ecology context give rise to criticism of the assumption that organizational destiny is a controllable variable. Organizational socio-ecology is dynamic and is based on the methodology of event history analysis in the analysis of temporal processes.

The theory of population socio-ecology is a tool for the theoretical and empirical analysis of organizational phenomena. The population of socio-ecology has several theoretical currents. Population socio-ecology is a perspective of convergent intersection with sustainability that integrates dimensions, levels of analysis and results. Routine socio-ecology deals at the organizational and inter-organizational analysis levels. The socio-ecology of routines at the organizational level, and the notion organizational character (Birnholtz, Cohen & Hoch, 2007) is defined as the ability to regenerate a coherent socio-ecology of patterns of action.

Research in organizational socio-ecology looks at the interactions of organizational populations. Organizational socio-ecology research intersects sustainability. The application of principles and assumptions of organizational socio-ecology to analyze the sustainability of organizations. The theory of organizational socio-ecology is based on a social Darwinism of organizational populations The organizational socio-ecology theory analyzes the alterations, difficulties, and restrictions of the organizational populations in application conditions (Dimaggio y Powell, 1983).

Research in the ecological theory of population is consolidated in the theory of population socio-ecology. The research focuses on the socio-ecology of populations in coordination and convergence with organizational sustainability considering that there are several areas related to the sustainability of populations that potentially contribute to population sustainability (Salimath, & Jones, 2011). The variables gas emissions, energy consumption, green technology innovation, institutional quality, and economic growth and urbanization are all interdependent and cointegrated.

There is a relationship between environmental regulation and green innovation. Values are assigned to environmental regulations leading to green innovation performance. When the environmental regulation system is weak, firms tend to emphasize maximization of profits, environmental taxes and engage in treatments to expand production scale and balance the regulation cost.

Green innovation research focuses on green product innovation and environmental management systems, the development of technological systems adopted by organizations, and the adoption of innovative technologies dealing with the integration of environmentally friendly practices. Environmental science and technology and strong normative and regulatory institutions support the transformation towards green innovation of firms.

This analysis on the impact of environmental regulation and incentives in socio-ecological and green technological innovation assumes that environmental regulation affects the performance of socio-ecological and green technological innovation in organizations. The study begins with some implications of the organizational socio-ecology to develop a framework for the analysis followed by the organizational green technological innovation leading to determine the impact of the environmental regulation policies and incentives. Finally, some conclusions are presented based on the analysis.

ORGANIZATIONAL SOCIO-ECOLOGY

Although the term ecology is not used Kremser, Pentland & Brunswicker (2017) they theorize the interdependence traced from the dynamics of social phenomena while sets of practices (Nicolini & Monteiro, 2017). The dynamic interplay of ecological and social systems leads to a socio-ecological systems (Berkes et al., 2003; Folke et al., 2002; Gunderson & Holling, 2002).

Organizational socio-ecology has implications for organizational theories. The emergence of models that value resources such as the resource-based theory, the stakeholder theory and theory of organizational socio-ecology that value external influences. Firms are restricted by internal factors of firms such as capital and technology and external factors such as inclusiveness, participants, and bureaucratic factors in the context of environmental regulation policies (Hu et al., 2020; Chen et al., 2021). Strict environmental policies lead small firms to be inclined to reduce expenditures to enhance pollution control rather than green technologies development. Organizational socio-ecology theory studies organizational size, sustaining that small organizations have more problems to capitalize, so they are prone to dissolve.

Organizational socio-ecology breaks the assumption of organization- environment by focusing the analysis not on the individual organization but on the aggregate propositions at the population level. In the classical approach to environment, as in socio-ecology theory and contingency theory, it is assessed with objective data to plan and adjust the organization to context. The contingency theory advocates organizational change based on the ability to adapt and adjust an alignment between the organization and the environment. The specific environment refers to the specific conditions of each organization, related to its sector, geography, stakeholders, etc. (Cadena-Roa, 2010). The intermediate environment in organizational socio-ecology theory refers to the conditions of organizations grouped by geographic area, theme, etc. (Hannan & Freeman, 1977, 1989).

Organizations compete for resources and resource positioning in organizational socio-ecology. The niche width theory is a category of organisational socio-ecology based on the survival opportunities of environmental changes of organizations (Hannan and Freeman, 1977). Ecological contingencies imposed by the environment require the fulfilment of components of organizational socio-ecology.

Organizational socio-ecology analyzes the characteristics of the vital manifestations of organizational populations, such as disappearance considering individual situations such as their size and age that are associated with survival (Aldrich & Auster, 1986). Population socio-ecology analyzes the sample as the entire population over time periods.

Organizational socio-ecology as a social demand on its organizations that exceed internal limits and affect the influence of human beings on the climate and natural environment. The internal social responsibility of organizations is introduced in the fields of organizational socio-ecology with the application of socio-ecological and environmental measures (Oller Alonso, 2021).

Building an organizational socio-ecology to learn as a management model of professional development must be based more on guiding than on control (Le Boterf, 1999). The organizational socio-ecology framework uses its knowledge to formulate and implement turn around organizational strategies. Organizational socio-ecology requires consensus for the cumulative generation of knowledge.

The socio-ecology of routines at the firm level, as well as the notion organizational character (Birnholtz, Cohen & Hoch, 2007) defined as the ability of participants to regenerate a coherent socio-ecology of patterns of action. Routine socio-ecology is analogous to the idea of a cluster of routines. Organizational routines are understood like ecologies (Pentland, 2004). Organizational socio-ecology of routines, which are often coupled and linked with actions, artifacts, and people. Routines in socio-ecology has the macro and micro levels of analysis of actors and activities.

The interdependence between routines challenges the scholars to meet the interactions in a socio-ecology between routines considering that routines never perform in a vacuum (Howard-Grenville et al. 2016) but are ubiquitous in organizations (Becker, 2008), configured in an interdependent network (Feldman et al., 2016; Pentland, 2011; Sele & Grand, 2016; Spee, Jarzabkowski, & Smets, 2016). Routines are intertwined by actions, artifacts and people linked to other routines. Ecological routines are co-located at a particular level of macro and micro analysis of actors and activities (Howard-Grenville et al., 2016).

An ecosystems adaptive governance and social-ecological systems (Dietz et al., 2003; Eckerberg & Joas, 2004; Folke et al., 2005; Ostrom, 2005). Space changes the stability properties of ecological interactions. The implications of chaos-induced dispersal in continuous-time models for socio-ecology are important (Sole & Bascompte, 1995). Elimination of species affected by habitual loss and fragmentation from spatial ecological patterns provides direct criteria (Hannan, 2005).

The practice of sustainability is related in convergence with the socio-ecology of the population. Population socio-ecology is a tool for the analysis of macro organizational phenomena that traces the use in theoretical and empirical research (Salimath, & Jones, 2011). Population socio-ecology supports organizational sustainability and explores the convergence that links the theory and practice of the organizational socio-ecology of populations. Population socio-ecology has areas of convergence with organizational sustainability for the analysis of population sustainability (Salimath, & Jones, 2011).

The industrial structure has effects on regional economic growth and ecological welfare (Zhao et al., 2020; Huang et al., 2021). Technological innovation infrastructure system improves the innovation efficiency by protecting the property rights. For rapid economic growth, most developing countries are dependent on non-renewable energy consumption, leading to significant CO 2 releases with adverse effects on environmental quality.

Improving the well-being via development increases output accompanied by the substantial consumption of fossil fuel with harmful influence on the environment. Energy consumption releasing more carbon is the major contributor to environmental footprint. The firms prefer to scale up production rather than use high-tech equipment exacerbating environmental pollution and weakening motivation (Bian et al., 2021). Many studies are used to evaluate environmental performance (Zhou et al., 2008;

Sueyoshi & Goto 2012; Wu et al., 2019). The pollution emissions characteristics leading to the environmental performance affect the green development efficiency.

ORGANIZATIONAL GREEN TECHNOLOGICAL INNOVATION

The concept of green ergonomics is in the context of green technological innovation including the development of the products and services aimed at achieving eco-benefits and reducing the organizational environmental impacts (Bernroider, 2002; Damanpour, 1992). Green innovative ergonomics focuses on contributions to enhancing the human-nature connection and diminishing the environmental impact, by linking design of innovative products and services aimed to reduce emissions and pollution with other pro-nature benefits (Mandavilli, Rys, & Russell, 2008; Sanquist, 2008; Hilliard & Jamieson, 2008).

Organizations focus on technological innovation advances to implement green innovation characteristics that may affect the adoption and use (Sneideriene & Rugine, 2019; Kousar, Sabri, Zafar, & Akhtar, 2017; Weng & Lin, 2011; Lin & Ho, 2011). The government encourages organizations to adopt and use green technology innovation with technical assistance and provides financial support and training for adoption.

Organizations adopt technological innovation before use with elevated levels of environmental uncertainty that affects green technology innovation (Zailani, Iranmanesh, Nikbin, & Jumadi, 2014). The adoption of green technology innovation reduces negative organizational impacts of technological systems processing and of the internal administrative processing. Organizations may be able to try to adopt the green technology innovation several times before the final adoption (Baines, Brown, Benedettini, & Ball, 2012).

The green innovation of firms is related to the life cycles. Green organizational innovation adoption leads to offer ecological and reliable product and services environmentally friendly at competitive prices delivered on time (Thatte, 2007). Differences in the firm's ownership structure, life cycle, size, etc., affect the environmental regulations affecting green innovation. Environmental regulations have a positive direct relationship with green technological innovation. Some inconsistencies may lie on the measures of green innovation performance, such as using granted-patents and pollutant reduction (Amore & Bennedsen, 2016; Fujii & Managi 2016). Triability is the organizational innovation characteristic to adopt any trial of innovation before the use and implementation because the degree of uncertainty (Rogers, 2003).

Research finds that relative advantage is related to green innovation adoption The relative advantage is the perception of the characteristic of organizational technology innovation in terms of price, quality, satisfaction, use and life span over its costs. Perception of organizational relative advantage is positively related to green innovation. If the organizational perception of relative advantage of innovation is greater than the cost and better that the current technology, the organization may adopt the innovation. (Lin & Ho, 2011; Rogers, 2003; Kousar et al., 2017; Grandon & Pearson, 2004).

The innovation management is related to internal and external practices. Innovation externalities are being solved by means of government R&D subsidies (Bi et al., 2016; Li et al., 2018; Liu et al., 2020). Investments on research and development, human capital, and gross domestic product per capita have effects on the development of green innovation. Investments on research and development, has effects on human capital and gross domestic product and on the development of green innovation. High investments in research and development, human capital, and gross domestic product per capita, the green innovation performance improves.

Investments in research and development enhances the firm technological innovation capabilities. Green investment is the ratio between fiscal expenditure on energy conservation and environmental protection and total fiscal expenditure. Environmental protection is a social responsibility of governments (Li et al., 2015; Jiang et al., 2020; Wang & Jiang, 2021).

Regional economic development is a macroeconomic operation related to the investment level of innovation (Grossman & Krueger, 1995). Local governments may invest and encourage firms to develop innovative activities through technology spillovers from foreign investments and competition incentives.

Complexity of innovation is the degree of organizational perception of technology innovation as being difficult to use. If the complexity of innovation increases, the adoption decrease. The level of complexity increases with the difficulties of innovation diffusion and knowledge sharing. The perception of organizational complexity is negatively related to green innovation. Organization tends to adopt technology innovation when innovation is diffused, and knowledge shared. Innovation requires skills and efforts increasing the level of complexity while the level of adoption decreases (Kousar et al., 2017; Etzion, 2007; Rogers, 2003; Tornatzky & Klein, 1982; Deng & Ji, 2015; Bradford & Florin, 2003).

There is a lack of green technological innovations in developing countries compared to developed nations (Li, Pan, Kim, Linn, & Chiang, 2015). A systematic review by Taklo, Tooranloo, & Shahabaldini (2020) of multiple industrial sectors and regions on the emergence of green technologically innovative advances for enhancing technical systems and enrichment of the customer experience.

Technological innovation is adopted after observing the impacts (Shahrul & Normah, 2015; Tan & Eze, 2008). Use of innovation that is visible to potential adoption reflects the result of the characteristic observability (Rogers, 2003). The perception of observability is positively related to green innovation. Observability is a significant characteristic of technological innovation because enables organizations to identify the benefits of innovation with high possibilities to be adopted by organizations (Hatimtai & Hassan, 2018; Rogers, 2003).

Green innovation is categorized in green management, green process, and green product (Chen, Shih, Shyur, & Wu, 2012; Chen & Hsu, 2009). Green technology innovation capabilities of firms, green product innovation capabilities followed by green process innovation capabilities and governance technology capabilities are being promoted by firms. Green innovation capabilities, green product innovation, green process innovation, and governance technology capabilities must have to be implemented based on the green technology innovation capabilities of firms (Wang et al., 2019). Organizational technological innovations compatible with capabilities of the organization have positive effects on the environment (Etzion, 2007). The perception of organizational compatibility is positively related to green technological innovation.

The innovation of green process is related to updates in systems, methods and processes and process innovation is aimed to produce eco-friendly goods and services to meet the ecological concerns and environmental targets (Antonioli, Mancinelli, & Mazzanti, 2013). The indicators of a green process innovation capabilities are the strengthening transformation and enhancing protection design processes, pollution elimination and green process review. Actions of government environmental regulation has effects on the investments of pollution elimination. Pollution control benefits are greater than those of reduction of production promoting firms to invest in green innovation activities

The green technology innovation capabilities of firms are related to the elements of green product innovation, process innovation and technology governance. Organizations must intensify green technology innovation to meet the technical specifications and standards of green emissions set by governments, improve the technology governance capabilities through wastes and the innovation capabilities of green products, processes, and develop green technology innovation (Li, 2019; Wang et al., 2018; Saud et al., 2019).

Organizational learning and innovation thrive in informal settings (Gunderson, 1999). The internal fit of a learning organization has a positive relationship with organizational innovation, and this with organizational performance. The organizational learning intention has effects on innovation capability and this on organizational performance (Teece, 2016).

Organizational learning guides organizational innovation in knowledge-intensive organizations. Knowledge based view extends the resource-based view (Rahim, Eladwiah, & Rahman, 2013) and it is being used in various research of stakeholder theory, open innovation, big data, data management, outsourcing and offshoring

(Hoskisson, Gambeta, Green, & Li, 2018; Santoro, Vrontis, Thrassou, & Dezi, 2018; Xu, Frankwick, & Ramirez, 2016 and Pereira, Munjal, & Ishizaka, 2019). Individual and organizational learning lead to innovation enhancing organizational performance. Organizational learning guides organizational innovation (Visser & Togt, 2016).

The ad-hocratic organization structure is one that is flexible, self-renewing, organic, innovative, and collaborative which can fuse different disciplines into a project team avoiding the traps of bureaucratic structures such as the labor division and coordination is the result of interactions (Toffler, 1970; Mintzberg, 1979).

From the perspective of motivation, green innovation should be separated from technological innovation ability. Organizational behaviors are classified based on their motivations as symbolic strategic environmental actions for cooperation and substantive environmental actions, or substantive strategies and cooperation (Delmas & Montes Sancho, 2010; Neumann, 2021; Truong et al., 2021).

ENVIRONMENTAL REGULATION POLICIES AND INCENTIVES

The mechanism of environmental regulation policy has influence on the motivation of green innovation firms to differentiate between the green innovation behaviors in regulation and regulated objects of government enterprise. Government environmental support increases the adoption of green innovation (Lin & Ho, 2010). Government environmental support refers to the systems that take the form of environmental regulations, policies, standards, and interventions that tend to reduce the environmental effect and encourage firms to engage in environmental technology innovation (Eiadat, Kelly, Roche, & Eyadat, 2008). Government environmental support motivates firms to adopt environmental methods of production (Gadenne, Kennedy, & McKeiver, 2009).

Mechanisms of environmental regulations policies affect organizational green innovation in different green innovations behaviors. State-owned enterprises assume responsibilities than non-state-owned do not assume in implementing the green innovation development under constraints of environmental regulations (Zhou et al., 2017; Bai et al., 2019). State-owned enterprises improve after the implementing of environmental regulations since constrained environmental regulations policy, firms respond faster to policies on green innovation due to the responsibility and use of scarce resources.

Environmental regulation policies cause small firms to enhance pollution control mora than implementing green technological innovation. Growing firms have access to abundant funds for R&D investments in green innovation to improve environment-friendly production and green technology (Cao & Cao, 2019). Environmental policy formulation must consider the specific measures for pollution control capacities of firms according to its firm´s ownership structure, size, life cycle, etc.

Policy on environmental regulations considering the value of carbon trading, it has a strong incentive effect in improving firm green innovation. Openness has no significant impact on green innovation because there is an indirect effect between openness and green innovation openness. That technological innovation in organizations reduces carbon emissions and enhances environmental quality (Dauda et al., 2021; Ko et al., 2021). Green innovation significantly lessens energy consumption and CO 2 gas emissions.

Carbon emission trading market must be liberalized positioning the emissions of firms in price mechanisms to stimulate green technological innovation activities. Carbon finance is the ratio between carbon dioxide emissions and gross domestic product. A higher level of CO2 emissions and environmental pollution as the causal links requires the establishment of effective institutional quality to green innovation. The causality link is established between the energy consumption and the institutional quality. Thus, there is a relationship between green innovation and institutional quality.

Research has found causal relationships among the variables are determined from CO2 emissions to economic growth, CO 2 emissions to green innovation and institutional

quality, urbanization and energy consumption to economic growth and institutional quality. From institutional quality to green innovation. Research also has found a bidirectional link between economic growth and gas emissions. Institutional quality and urbanization in developing countries may increase CO2 emissions.

The influence of environmental regulations on green technological innovation of growing and mature firms achieves the highest impact while the green technological innovation of startup firms has severe financial constraints, struggle to meet capital needs, face elevated level of risk-taking leading to achieve the lowest impact on investments in green innovation technologies (Shahzad et al., 2019).

The mechanism between capital misallocation, green development, and technological innovation leads to more efficient socio-ecology environment. Incorporating capital allocation into technological innovation in an analytical framework promotes regional green development efficiency. Regarding heterogeneity, the effect of capital allocation is significant. Capital allocation has a positive relationship with technological innovation enhancing the input of the firms with an impact on green development efficiency. Capital allocation enhances green development efficiency leading to reducing the level of technological innovation. The impact of capital allocation on technological innovation is positive (Schmidt et al., 2020; Gao et al., 2019).

Government intervention on capital misallocated may lead to deterioration of the environmental. Prices distortion of capital market, capital allocation cannot be optimized (Zhang & Lu., 2017). Firms have difficulties in obtaining the capital required for technological innovation of the firm leading to weaken the motivation to conduct research and development (Gao et al., 2019). If the price of capital controlled by government, firms tend to rent-seeking to obtain low-cost factors crowing research and development funds and reducing innovation efficiency (Murphy et al., 1993). The intervention of external factors may destroy the price of capital mechanism lowering costs of production factors for polluting industries. Technological innovation has a mediating role with capital allocation on green development efficiency.

Technological innovation increases the use of clean processes reducing the environmental pollution (Hua et al. 2018). Technological innovation stimulates firms to use efficient resources, adopt equipment and promote the upgrading of industries (Sun et al. 2020b). Technological innovation tends to facilitate green economy development having a mediating role on capital allocation on green economic efficiency while capital allocation improves the green development efficiency. Technological innovation has a mediating role with capital allocation and enhances green development efficiency.

Green finance addressing the financial gap is a relevant financial instrument that contributes to green innovation development of firms (Luo et al., 2021). Green finance motivates firms to invest in green technological innovation and is the driving force for green transformation of polluted firms through the engagement in green innovation activities. Environmental regulation improves the effects of green finance in green innovation. Environmental regulations strengthen the impact of green innovation projects and the effects of green finance.

Firms tend to adopt conservative strategies to invest in green technological projects with low-risk and intense environmental regulations (Boubakri et al. 2013; Chiu and Lee 2020). Green finance and strict environmental regulations have positive effects on green innovation. The moderation effects of environmental regulation advance the effects of green finance on green innovation. Environmental regulations at low levels leads to increasing of innovation performance as green finance performance increases.

An emphasis on green innovation outcomes by the firms requires more efficient green finance in producing green innovation. Green innovation output can be achieved by urging firms to use green finance and investments and avoiding time-consuming projects. Environmental regulation improves the effects of green finance in green innovation. Green finances are contingent to the implementation of environmental regulations and can improve with more intense regulations.

Large firms with more available resources tend to improve the quality of green technological innovation. Large firms have more mature development of organizational systems achieving high innovation management (Yu et al., 2022b). Large firms count on access to abundant innovation resources such as capital, R&D, patents, technological talents (Gupeng & Xiangdong, 2012; Hu et al., 2017). The green patent application is a measure that has effects of green finance in driving green technological innovation. The use of green patent application to measure innovative behaviors is not comprehensive for the different motivations on green innovation choices.

Large firms promote green technological innovation even if they are constrained by environmental regulations and small firms that are more restricted by capital, technology, talent, etc., have more difficulties in promoting green innovation. Firms' behaviors must be constrained by environmental regulations to develop green technological innovation. The implementation of an intense environmental regulations improves green innovation considering constraints of firms may not lead to transformation of technologically innovative production. Stringent environmental regulations tend to strength the motivation of firms to pursue technological innovation.

Green innovation as dependent variable can be measured by the number of patent applications associated to energy intensity (Wurlod & Noailly, 2018). Green patent applications indicate the advance of green innovation (Tolliver et al. 2021) and they are leading to innovation activities without institutional preferences and time-lag, used as indicators to evaluate green innovation (Fujii & Managi 2016; Cho & Sohn, 2018; Fujii 2016).

Environmental regulation has a positive impact on green innovation patents and green utility model patents, although from the perspective of influence, the impact of environmental regulation on green innovation may decrease if firms have not eliminated strategic green innovation. Green technology innovation strategies must be formulated and implemented with sensitivity considering cultural differences that have an impact on innovation processes (Westwood & Low, 2003).

Environmental regulations influence government subsidies to eliminate compliance costs improving green innovation with minimal impact on green innovation behaviors. Government subsidies on green innovation must be classified in the diverse types to determine the effect on the choices of innovative behaviors with different motivations. Government innovation subsidies policies must be more targeted and flexible according to the qualifications of firms in need and the appropriateness of subsidies, matching the development stage at the life cycle to enhance core competitiveness and realize positive effects on green technological innovation.

Regulatory capture and government subsidies are moderating variables. Government subsidies supplying funds provide firms to conduct green innovation aligned to the government goals of technological innovation (Boeing, 2016; Bai et al., 2019). Government subsidies promote green innovation; however, regulatory capture has a negative role in the regulatory process of environmental regulation promoting green innovation of firms. Increasing green subsidies lead to firms to have more resources and funds to invest in green technology development enhancing green product, process innovation and technology governance capabilities (Saud et al., 2019; Lin et al., 2020). The regulatory capture behavior requires more accurate indicators to provide verification of results on the impact of green innovation effects.

The regulatory capture leads to different green innovation behaviors. The regulatory capture has a distorting effect on the high quality substantive green innovation incentives leading to economic consequences of the environmental regulations. Regulatory capture has a negative effect on environmental regulation in promoting firms as the explanation for the different green innovation behaviors. Regulatory captures may lapse for regulatory stringency as the leading cause for environmental accidents intentionally balancing profits (Graham et al., 2011; Steinzor, 2012). Firms taking a profit maximization orientation may have more potential benefits from regulatory capture behavior than those having intense motivational behavior for regulatory capture (Li et al., 2021; Zhang et al., 2021).

Environmental regulations tend to influence green innovation of growing and mature firms at the highest level, while the startup stage has the lowest impact due to the available resources in capital, technology, and talent.

CONCLUSION

In general terms, this analysis concludes that there is a direct relationship between the incentives of environmental regulation and the socio-ecological and green technological innovation in organizations. The socio-ecology of the organizational population has implications for sustainability. Institutional reforms are essential to reduce ecological emissions and achieve climate change goals.

Firms realize high quality green innovation behavior to promote green technology to obtain a competitive advantage. Green innovation firms are higher during the growth and maturity periods when they improve green technological innovation and develop a keen sense of social responsibility for more active innovation activities. The role of technological innovation needs to be strengthened to improve green development efficiency. The organization must implement and scrutinize green product innovation capabilities implementation, green process innovation capabilities, and technology governance capabilities.

Environmental regulation affects the firm´s green innovation through the mechanisms of innovation compensation and compliance costs. The scientific design of environmental regulation policy tools must be strengthened to monitor the emission reduction and conduct green innovation. Environmental regulation intensity must be enhanced according to the specific situations of firms. Firms that bear excessive environmental costs must be inclined to regulatory capture rather than improve green technological environmental protection which may destroy the incentive mechanism with extended time and higher costs.

Governments in policymaking and stakeholders must facilitate organizational green strategies to invest in technological innovation projects aimed to reduce environmental pollution. Policymaking need to strengthen institutional quality and enact more effective policies to address the environmental challenges. The different impacts of institutional environments must take the research model. Institutional environment is relevant in factor allocation and environmental performance

Firm differences in ownership, life cycle, size, etc., have a significant impact on the environmental regulation green innovation incentives. Environmental regulations and green finance have effects on green technological innovation. The green finance levels on green innovation have heterogeneous impacts with a moderating effect on environmental regulations. Green finance, green innovation and environmental regulations tend to be stable with fixed effects.

Green finance and environmental regulations have effects on green innovation. Environmental regulation is a moderator is used to examine the effects. Regional green innovation can be measured by the level of contribution of green finance and the green patent application. Governments must support with fiscal incentives and financial subsidies the development of innovative firms encouraging more flows of capital into the green technology innovations and clean processes.

Research findings suggest that subsidies do not explain the gap between the green technological innovation behaviors.

Future research in applying population socio-ecology principles and assumptions to examine organizational sustainability in organizations.

Future research for measurement methods must separate the individual effects of the mechanisms of government subsidies and regulatory capture on green innovation behavior with different motivations.

REFERENCES

Aldrich, H., & Auster, E. R. (1986). Even dwarfs started small: Liabilities of age and size and their strategic implications. In B. M. Staw & L. L. Cummings (Eds.), *Research in Organizational Behavior* (Vol. 8, pp. 165–198). JAI Press.

Amburgey, T., & Rao, H. (1996). Organizational ecology: Past, present, and future directions. *Academy of Management Journal*, *39*(5), 1265–1286. doi:10.2307/256999

Amore, M. D., & Bennedsen, M. (2016). Corporate governance and green innovation. *Journal of Environmental Economics and Management*, *75*, 54–72. doi:10.1016/j.jeem.2015.11.003

Antonioli, D., Mancinelli, S., & Mazzanti, M. (2013). Is environmental innovation embedded within high- performance organizational changes? The role of human resource management and complementarity in green business strategies. *Research Policy*, *42*(4), 975–988. doi:10.1016/j.respol.2012.12.005

Bai, Y., Song, S., Jiao, J., & Yang, R. (2019). The impacts of government R&D subsidies on green innovation: Evidence from Chinese energy-intensive firms. *Journal of Cleaner Production*, *233*(9), 819–829. doi:10.1016/j.jclepro.2019.06.107

Baines, T., Brown, S., Benedettini, O., & Ball, P. (2012). Examining green production and its role within the competitive strategy of manufacturers. *Journal of Industrial Engineering and Management*, *5*(1), 53–87. doi:10.3926/jiem.405

Becker, M. C. (Ed.). (2008). Handbook of organizational routines. Cheltenham, UK: Edward Elgar Publishing Becker, M. C. (Ed.). (2008). Handbook of organizational routines. Cheltenham, UK: Edward Elgar Publishing. doi:10.4337/9781848442702

Berkes, F., Folke, C., & Colding, J. (Eds.). (2003). *Navigating Social-Ecological Systems: Building Resilience for Complexity and Change*. Cambridge University Press.

Bernroider, E. (2002). Factors in SWOT analysis applied to micro, small-to-medium, and large software enterprises: An Austrian study. *European Management Journal*, *20*(5), 562–573. doi:10.1016/S0263-2373(02)00095-6

Bi, K., Huang, P., & Wang, X. (2016). Innovation performance and influencing factors of low-carbon technological innovation under the global value chain: A case of Chinese manufacturing industry. *Technological Forecasting and Social Change*, *111*, 275–284. doi:10.1016/j.techfore.2016.07.024

Bian, Y., Wu, L., Bai, J., & Yang, Y. (2021). Does factor market distortions inhibit green economic growth? *World Econ Papers*, *2*, 105–119.

Birnholtz, J. P., Cohen, M. D., & Hoch, S. V. (2007). Organizational character: On the regeneration of Camp Poplar Grove. *Organization Science*, *18*(2), 315–332. doi:10.1287/orsc.1070.0248

Boeing, P. (2016). The allocation and effectiveness of China's R&D subsidies—Evidence from listed firms. *Research Policy*, *45*(9), 1774–1789. doi:10.1016/j.respol.2016.05.007

Boubakri, N., Cosset, J. C., & Saffar, W. (2013). The role of state and foreign owners in corporate risk-taking: Evidence from privatization. *Journal of Financial Economics, 108*(3), 641–658. doi:10.1016/j.jfineco.2012.12.007

Bradford, M., & Florin, J. (2003). Examining the role of innovation diffusion factors on the implementation success of enterprise resource planning systems. *International Journal of Accounting Information Systems, 4*(3), 205–225. doi:10.1016/S1467-0895(03)00026-5

Cadena-roa, J. (Coordinador) (2022). *Contribuciones para el estudio del desempeño asociativo resultados de la puesta a prueba del protocolo para la evaluación de asociaciones.* Universidad Nacional Autónoma de México Centro de Investigaciones Interdisciplinarias en Ciencias y Humanidades México.

Cao, X., Deng, M., Song, F., Zhong, S., & Zhu, J. (2019). Direct and moderating effects of environmental regulation intensity on enterprise technological innovation: The case of China. *PLoS One, 14*(10), e0223175. doi:10.1371/journal.pone.0223175 PMID:31589643

Chen, C. C., Shih, H. S., Shyur, H. J., & Wu, K. S. (2012). A business strategy selection of green supply chain management via an analytic network process. *Computers & Mathematics with Applications (Oxford, England), 64*(8), 2544–2557. doi:10.1016/j.camwa.2012.06.013

Chen, H. L., & Hsu, W. T. (2009). Family ownership, board independence, and R&D investment. *Family Business Review, 22*(4), 347–362. doi:10.1177/0894486509341062

Chen, Z., Zhang, X., & Chen, F. (2021). Do carbon emission trading schemes stimulate green innovation in enterprises? Evidence from China. *Technological Forecasting and Social Change, 168*, 120744. doi:10.1016/j.techfore.2021.120744

Chiu, Y. B., & Lee, C. C. (2020). Effects of financial development on energy consumption: The role of country risks. *Energy Economics, 90*, 104833. doi:10.1016/j.eneco.2020.104833

Cho, J. H., & Sohn, S. Y. (2018). A novel decomposition analysis of green patent applications for the evaluation of R&D efforts to reduce CO2 emissions from fossil fuel energy consumption. *Journal of Cleaner Production, 193*, 290–299. doi:10.1016/j.jclepro.2018.05.060

Damanpour, F. (1992). Organizational size and innovation. *Organization Studies, 13*(3), 375–402. doi:10.1177/017084069201300304

Dauda, L., Long, X., Mensah, C. N., Salman, M., Boamah, K. B., Ampon-Wireko, S., & Dogbe, C. S. K. (2021). Innovation, trade openness and CO2 emissions in selected countries in Africa. *Journal of Cleaner Production, 281*, 125143. doi:10.1016/j.jclepro.2020.125143

Delmas, M. A., & Montes-Sancho, M. J. (2010). Voluntary agreements to improve environmental quality: Symbolic and substantive cooperation. *Strategic Management Journal, 31*(6), 575–601.

Deng, Q., & Ji, S. (2015). Organizational green IT adoption: *Concept and evidence. Sustainability (Basel), 7*(12), 16737–16755. doi:10.3390u71215843

Dietz, T., Ostrom, E., & Stern, P. L. (2003). The Struggle to Govern the Commons. *Science, 302*(5652), 1907–1912. doi:10.1126cience.1091015 PMID:14671286

DiMaggio, P., & Powell, W. W. (1983). The iron cage revisited: Collective rationality and institutional isomorphism in organizational fields. *American Sociological Review, 48*(2), 147–160. doi:10.2307/2095101

Eckerberg, K., & Joas, M. (2004). Multi-level Environmental Governance: A Concept under Stress? *Local Environment, 9*(5), 405–412. doi:10.1080/1354983042000255315

Eiadat, Y., Kelly, A., Roche, F., & Eyadat, H. (2008). Green and competitive? An empirical test of the mediating role of environmental innovation strategy. *Journal of World Business, 43*(2), 131–145. doi:. jwb.2007.11.012 doi:10.1016/j

Etzion, D. (2007). Research on organizations and the natural environment, 1992-present: A review. *Journal of Management, 33*(4), 637–664. doi:10.1177/0149206307302553

Feldman, M. S. (2016). Past, Present, and future. In J. Howard-Grenville, C., Rerup, A., Langley, & H., Tsoukas (Eds.). Organizational Routines: How They are Created, Maintained, and Changed (pp. 23-46) Oxford: Oxford University Press

Folke, C., Carpenter, S., Elmqvist, T., Gunderson, L., Holling, C. S., & Walker, B. (2002). Resilience and Sustainable Development: Building Adaptive Capacity in a World of Transformations. *Ambio, 31*(5), 437–440. doi:10.1579/0044-7447-31.5.437 PMID:12374053

Folke, C., Hahn, T., Olsson, P., and Norberg, J. (2005). Adaptive Governance of Social–ecological Systems. *Annual Review of Environment and Resources*, in press (will be published in October 2005).

Fujii, H. (2016). Decomposition analysis of green chemical technology inventions from 1971 to 2010 in Japan. *Journal of Cleaner Production, 112*, 4835–4843. doi:10.1016/j.jclepro.2015.07.123

Fujii, H., & Managi, S. (2016). Research and development strategy for environmental technology in Japan: A comparative study of the private and public sectors. *Technological Forecasting and Social Change*, *112*, 293–302. doi:10.1016/j.techfore.2016.02.012

Gadenne, D. L., Kennedy, J., & McKeiver, C. (2009). An empirical study of environmental awareness and practices in SMEs. *Journal of Business Ethics*, *84*(1), 45–63. doi:10.100710551-008-9672-9

Gao, X., Lyu, Y., Shi, F., Zeng, J., & Liu, C. (2019). The impact of financial factor market distortion on green innovation efficiency of high-tech industry. *Ekoloji*, *28*(107), 3449–3461.

Graham, B., Reilly, W., Beinecke, F., Boesch, D., Garcia, T., Murray, C., & Ulmer, F. (2011). *Deep water: The Gulf oil disaster and the future of offshore drilling. Report to the President*. National Commission on the BP Deepwater Horizon Oil Spill and Offshore Drilling.

Grossman, G. M., & Krueger, A. B. (1995). Economic growth and the environment. *The Quarterly Journal of Economics*, *110*(2), 353–377. doi:10.2307/2118443

Gunderson, L. (1999). Resilience, Flexibility and Adaptive Management: Antidotes for Spurious Certitude? *Conservation Ecology*, *3*(1), 7. http://www.consecol.org/vol3/iss1/art7. doi:10.5751/ES-00089-030107

Gunderson, L. H., & Holling, C. S. (Eds.). (2002). Panarchy; Understanding Transformations in Human and Natural Systems, Island, Washington, DC.

Gupeng, Z., & Xiangdong, C. (2012). The value of invention patents in China: Country origin and technology field differences. *China Economic Review*, *23*(2), 357–370. doi:10.1016/j.chieco.2012.02.002

Hannan, M. (2005). Ecologies of Organizations: Diversity and Identity. *The Journal of Economic Perspectives*, *19*(1), 51–70. doi:10.1257/0895330053147985

Hannan, M. T., & Freeman, J. (1989). *Organizational Ecology*. Harvard University Press. doi:10.4159/9780674038288

Hannan, M. T., & Freeman, J. H. (1977). The Population Ecology of Organizations. *American Journal of Sociology*, *82*(5), 929–963. doi:10.1086/226424

Hatimtai, M. H., & Hassan, H. (2018). The relationship between the characteristics of innovation towards the effectiveness of ICT in Malaysia productivity corporation. *Malaysian Journal of Communication*, *34*(1), 253–269. doi:10.17576/JKMJC-2018-3401-15

Hilliard, A., & Jamieson, G. A. (2008). Winning solar races with interface design. *Ergonomics in Design*, *16*(2), 6–11. doi:10.1518/106480407X312374

Hoskisson, R. E., Gambeta, E., Green, C. D., & Li, T. X. (2018). Is my firm-specific investment protected? Overcoming the stakeholder investment dilemma in the resource-based view. *Academy of Management Review*, *43*(2), 284–306. doi:10.5465/amr.2015.0411

Howard-Grenville, J., Rerup, C., Langley, A., & Tsoukas, H. (2016). Introduction: Advancing a process perspective on routines by zooming out and zooming in. In J. Howard-Grenville, C., Rerup, A., Langley, & H., Tsoukas (Eds.). Organizational routines: how they are created, maintained, and changed (pp.1-22.) Oxford: Oxford University Press.

Hu, A. G. Z., Zhang, P., & Zhao, L. (2017). China as number one? Evidence from China's most recent patenting surge. *Journal of Development Economics*, *124*, 107–119. doi:10.1016/j.jdeveco.2016.09.004

Hua, Y., Xie, R., & Su, Y. (2018). Fiscal spending and air pollution in Chinese cities: Identifying composition and technique effects. *China Economic Review*, *47*, 156–169. doi:10.1016/j.chieco.2017.09.007

Huang, M., Ding, R., & Xin, C. (2021). Impact of technological innovation and industrial-structure upgrades on ecological efficiency in China in terms of spatial spillover and the threshold effect. *Integrated Environmental Assessment and Management*, *17*(4), 852–865. doi:10.1002/ieam.4381 PMID:33325155

Jiang, Z., Wang, Z., and Lan, X. (2021). How Environmental Regulations Affect Corporate Innovation? the Coupling Mechanism of Mandatory Rules and Voluntary Management. *Technol. Soc., 65*, 101575. doi:. 2021.101575 doi:10.1016/j.techsoc

Ko, Y.-C., Zigan, K., & Liu, Y.-L. (2021). Carbon capture and storage in South Africa: A technological innovation system with a political economy focus. *Technological Forecasting and Social Change*, *166*, 120633. . techfore. 2021. 120633 doi:10.1016/j

Kousar, S., Sabri, P. S. U., Zafar, M., & Akhtar, A. (2017). Technological factors and adoption of green innovation- moderating role of government intervention: A case of SMEs in Pakistan. *Pakistan Journal of Commerce and Social Sciences*, *11*(3), 833–861.

Kremser, W., Pentland, B. T., & Brunswicker, S. (2017). *The Continuous Transformation of Interdependence in Networks of Routines*. Presented at European Group for Organizational.

Li, J., Pan, S. Y., Kim, H., Linn, J. H., & Chiang, P. C. (2015). Building green supply chains in eco-industrial parks towards a green economy: Barriers and strategies. *Journal of Environmental Management*, *162*, 158–170. doi:10.1016/j.jenvman.2015.07.030 PMID:26241931

Li, N. B. (2019). Environmental Regulation and Corporate Green Technology Innovation - a Conditional Process Analysis. *Inn. Mong. Soc. Sci. Chin. Ed*, *40*(6), 109–115. doi:10.14137/j.cnki.issn1003-5281.2019.06.016

Li, S., Song, X., & Wu, H. (2015). Political connection, ownership structure, and corporate philanthropy in China: A strategic-political perspective. *Journal ofBusiness Ethics*, *129*(2), 399–411. doi:10.100710551-014-2167-y

Li, Z., Liao, G., Wang, Z., & Huang, Z. (2018). Green loan and subsidy for promoting clean production innovation. *Journal of Cleaner Production*, *187*, 421–431. doi:10.1016/j.jclepro.2018.03.066

Lin, C. Y., & Ho, Y. H. (2010). The influences of environmental uncertainty on corporate green behavior: An empirical study with small and medium-size enterprises. *Social Behavior and Personality*, *38*(5), 691–696. doi:10.2224bp.2010.38.5.691

Lin, Y. X., Sha, K. C., & Wang, J. (2020). Practical Experience and Inspiration of Foreign Emission Permit System. *Environ. Impact. Assess. 42* (1), 14–16.

Liu, J., Zhao, M., & Wang, Y. (2020). Impacts of government subsidies and environmental regulations on green process innovation: A nonlinear approach. *Technology in Society*, *63*, 101417. doi:10.1016/j.techsoc.2020.101417

Liu, Y., Wang, A., & Wu, Y. (2021). Environmental regulation and green innovation: Evi-dence from China's new environmental protection law. *Journal of Cleaner Production*, *297*, 126698. doi:10.1016/j.jclepro.2021.126698

Luo, Y., Salman, M., & Lu, Z. (2021). Heterogeneous impacts of environmental regulations and foreign direct investment on green innovation across different regions in China. *The Science of the Total Environment*, *759*, 143744. doi:10.1016/j.scitotenv.2020.143744 PMID:33341514

Mandavilli, S., Rys, M. J., & Russell, E. R. (2008). Environmental impact of modern roundabouts. *International Journal of Industrial Ergonomics*, *38*(2), 135–142. doi:10.1016/j.ergon.2006.11.003

Mintzberg, H. (1979). *The structuring of organizations: A synthesis of the research*. Prentice Hall.

Murphy, K. M., Shleifer, A., & Vishny, R. W. (1993). Why is rent-seeking so costly to growth? *The American Economic Review*, *83*(2), 409–414. https://www.jstor.org/stable/2117699

Neumann, T. (2021). Does it pay for new firms to be green? An empirical analysis of when and how different greening strategies affect the performance of new firms. *Journal of Cleaner Production*, *317*, 128403. doi:10.1016/j.jclepro.2021.128403

Nicolini, D., & Monteiro, P. (2017). The practice approach: For a praxeology of organizational and management studies. In A. Langley & H. Tsoukas (Eds.), *The Sage Handbook of Process Organization Studies*. Sage.

Oller Alonso, M. (2021). *La Responsabilidad Social Corporativa de las empresas del mármol en la comarca del Almanzora, Almería (2019-2021): análisis de sus estrategias de comunicación integral.* Doctorado en Gestión de la Información y de la Comunicación en las Organizaciones Universidad de Murcia Escuela Internacional de Doctorado.

Ostrom, E. (2005). Understanding Institutional Diversity, Princeton University Press, Princeton New Jersey, USA.

Pentland, A. (2004, October). *Social Dynamics: Signals and Behavior. International Conference on Developmental Learning*, Salk Institute, San Diego, CA

Pentland, B. T., Hærem, T., & Hillison, D. (2011). The (N)Ever-Changing World: Stability and Change in Organizational Routines. *Organization Science*, *22*(6), 1369–1383. doi:10.1287/orsc.1110.0624

Pereira, V., Munjal, S., & Ishizaka, A. (2019, October). Outsourcing and offshoring decision making and its implications for businesses- A synthesis of research pursuing five pertinent questions. *Journal of Business Research*, *103*, 348–355. doi:10.1016/j.jbusres.2019.07.009

Pfeffer, J., & Salancik, G. (1978). *The External Control of Organizations: A Resource Dependence Perspective*. Harper & Row.

Rahim, A., Eladwiah, R., & Abdul Rahman, A. (2013). Resource-based framework of green IT capability toward firms 'competitive advantage. *Resource (Saint Joseph, Mich.)*, *6*, 18–2013.

Rogers, E. M. (2003). Elements of diffusion. *Diffusion of Innovations*, *5*, 1–38.

Salimath, M., & Jones, R. III. (2011). Population ecology theory: Implications for sustainability. *Management Decision*, *49*(6), 874–910. doi:10.1108/00251741111143595

Sanquist, T. F. (2008). Human factors and energy use. *HFES Bulletin*, *51*(11), 1–3.

Santoro, G., Vrontis, D., Thrassou, A., & Dezi, L. (2018). The Internet of Things: Building a knowledge management system for open innovation and knowledge management capacity. *Technological Forecasting and Social Change*, *136*, 347–354. doi:10.1016/j.techfore.2017.02.034

Saud, S., Chen, S., Haseeb, A., Khan, K., & Imran, M. (2019). The Nexus between Financial Development, Income Level, and Environment in Central and Eastern European Countries: A Perspective on Belt and Road Initiative. *Environmental Science and Pollution Research International*, *26*(16), 16053–16075. doi:10.100711356-019-05004-5 PMID:30968296

SchmidtC.SchneiderY.SteffenS.StreitzD. (2020). Capital misal-location and innovation. *Available at* SSRN 3489801. doi:10.2139/ssrn.3489801

Sele, K., & Grand, S. (2016). Unpacking the dynamics of ecologies of routines: Mediators and their generative effects in routine interactions. *Organization Science*, *27*(3), 722–738. doi:10.1287/orsc.2015.1031

Shahrul, N. S., & Normah, M. (2015). Digital version newspaper: Implication towards printed newspaper circulation in Malaysia. *Malaysian Journal of Communication*, *31*(2), 687–701.

Shahzad, F., Lu, J., & Fareed, Z. (2019). Does firm life cycle impact corporate risk taking and performance? *Journal of Multinational Financial Management*, *51*, 23–44. doi:10.1016/j.mulfin.2019.05.001

Sneideriene, A., & Rugine, H. (2019). Theoretical approach on the green technologies development. *Regional Formation and Development Studies*, *2*(28), 124–134.

Sole, R. V., & Bascompte, J. (1995). Spatial Self Organization Self-Organization. Complex Ecosystems. Princeton University Press.

Spee, P., Jarzabkowski, P., & Smets, M. (2016). The influence of routine interdependence and skillful accomplishment on the coordination of standardizing and customizing. *Organization Science*, *27*(3), 759–781. doi:10.1287/orsc.2016.1050

Steinzor, R. (2012). The case for abolishing centralized white house regulatory review. *Michigan Journal of Environmental & Administrative Law*, *1*(1.1), 209. doi:10.36640/mjeal.1.1.case

Studies: EGOS Le Boterf, G. (1999). *L'ingénierie des compétences*. París: Éditions d'Organisation.

Sueyoshi, T., & Goto, M. (2012). Efficiency-based rank assessment for electric power industry: A combined use of data envelopment analysis (DEA) and DEA-discriminant analysis (DA). *Energy Economics*, *34*(3), 634–644. doi:10.1016/j.eneco.2011.04.001

Sun, X., Zhou, X., Chen, Z., & Yang, Y. (2020b). Environmental efficiency of electric power industry, market segmentation and technologi-cal innovation: Empirical evidence from China. *The Science of the Total Environment*, *706*, 135749. doi:10.1016/j.scitotenv.2019.135749 PMID:31940733

Taklo, S. K., & Tooranloo, H. S. & Shahabaldini parizi, Z. (. (2020). Green Innovation: A Systematic Literature Review. *Journal of Cleaner Production*, *2020*(7). Advance online publication. doi:10.1016/j.jclepro.2020.122474

Tan, K. S., & Eze, U. C. (2008). An empirical study of internet-based ICT adoption among Malaysian SMEs. *Communications of the IBIMA*, *1*, 1–12.

Teece, D. J. (2016). Dynamic capabilities and entrepreneurial management in large organizations: Toward a theory of the (entrepreneurial) firm. *European Economic Review*, *86*, 202–216. doi:10.1016/j.euroecorev.2015.11.006

Thatte, A. A. (2007). *Competitive advantage of a firm through supply chain responsiveness and SCM practices* [Doctoral dissertation, The University of Toledo].

Toffler, A. (1970). *Future shock*. Random House.

Tolliver, C., Fujii, H., Keeley, A. R., & Managi, S. (2021). Green innovation and finance in Asia. *Asian Economic Policy Review*, *16*(1), 67–87. doi:10.1111/aepr.12320

Tornatzky, L. G., & Klein, K. J. (1982). Innovation characteristics and innovation adoption-implementation: A meta-analysis of findings. *IEEE Transactions on Engineering Management*, *29*(1), 28–45. doi:10.1109/TEM.1982.6447463

Truong, Y., Mazloomi, H., & Berrone, P. (2021). Understanding the impact of symbolic and substantive environmental actions on organizational reputation. *Industrial Marketing Management*, *92*, 307–320. doi:10.1016/j.indmarman.2020.05.006

Visser, M., & Togt, K. (2016). Learning in Public Sector Organizations: A Theory of Action, Approach. *Public Organization Review*, *16*(2), 235–249. doi:10.100711115-015-0303-5

Wang, F. Z., Jiang, T., & Guo, X. C. (2018). Government Quality, Environmental Regulation and Corporate Green Technology Innovation. *Sci. Res. Manage. 39* (1), 26–33. CNKI:SUN:KYGL.0.2018-01-004.

Wang, K., & Jiang, W. (2021). State ownership and green innovation in China: The con-tingent roles of environmental and organizational factors. *Journal of Cleaner Production, 314*, 128029. doi:10.1016/j.jclepro.2021.128029

Wang, Y., Sun, X., & Guo, X. (2019). Environmental regulation and green productivity growth: Empirical evidence on the Porter Hypothesis from OECD industrial sectors. *Energy Policy, 132*, 611–619. doi:10.1016/j.enpol.2019.06.016

Weng, M. H., & Lin, C. Y. (2011). Determinants of green innovation adoption for small and medium-size enterprises (SMES). *African Journal of Business Management, 5*(22), 9154–9163.

Westwood, R., & Low, D. R. (2003). The multicultural muse: Culture, creativity, and innovation. *International Journal of Cross Cultural Management, 3*(2), 235–259. doi:10.1177/14705958030032006

Wu, H. Y., Tsai, A., & Wu, H. S. (2019). A HYBRID MULTI-CRITERIA DECISION ANALYSIS APPROACH FOR ENVIRONMENTAL PERFORMANCE EVALUATION: AN EXAMPLE OF THE TFT-LCD MANUFACTURERS IN TAIWAN. [EEMJ]. *Environmental Engineering and Management Journal, 18*(3), 597–616. doi:10.30638/eemj.2019.056

Wurlod, J. D., & Noailly, J. (2018). The impact of green innovation on energy intensity: An empirical analysis for 14 industrial sectors in OECD countries. *Energy Economics, 71*, 47–61. doi:10.1016/j.eneco.2017.12.012

Xu, Z., Frankwick, G. L., & Ramirez, E. (2016). Effects of big data analytics and traditional marketing analytics on new product success: A knowledge fusion perspective. *Journal of Business Research, 69*(5), 1562–1566. doi:10.1016/j.jbusres.2015.10.017

Yu, Z., Khan, S. A. R., Ponce, P., & Jabbour, A. B. L. (2022). Factors affecting carbon emissions in emerging economies in the context of a green recovery: Implications for sustainable development goals. *Technological Forecasting and Social Change, 176*, 121417. doi:10.1016/j.techfore.2021.121417

Zailani, S., Iranmanesh, M., Nikbin, D., & Jumadi, H. B. (2014). Determinants and environmental outcome of green technology innovation adoption in the transportation industry in Malaysia. *Asian Journal of Technology Innovation, 22*(2), 286–301. doi:10.1080/19761597.2014.973167

Zhang, D. G., & Lu, Y. Q. (2017). Impact of market segmentation on energy efficiency. *China Popul Resour Environ, 27*(1), 65–72.

Zhang, W., He, L., & Yuan, H. (2021). Enterprises' decisions on adopting low-carbon technology by considering consumer perception disparity. *Technovation, 102238.* doi:10.1016/j.technovation.2021.102238

Zhao, X., Shang, Y., & Song, M. (2020). Industrial structure distortion and urban ecological efficiency from the perspective of green entre-preneurial ecosystems. *Socio-Economic Planning Sciences*, *72*, 100757. doi:10.1016/j.seps.2019.100757

Zhou, K. Z., Gao, G. Y., & Zhao, H. (2017). State Ownership and firm innovation in China: An integrated view of institutional and efficiency logics. *Administrative Science Quarterly*, *62*(2), 375–404. doi:10.1177/0001839216674457

Zhou, P., Ang, B. W., & Poh, K. L. (2008). Measuring environmental performance under different environmental DEA technologies. *Energy Economics*, *30*(1), 1–14. doi:10.1016/j.eneco.2006.05.001

Chapter 6
Studies of Some Techniques for Conservation of Forest and Wildlife

Ram Manoher Ahirwar
Government PG College, Tikamgarh, India

ABSTRACT

Forests are large areas of land with trees. Overall, forests are the natural habitat of large-scale wildlife, growth of trees, shrubs, and different varieties of plants which, unfortunately, are dwindling every year. Hence, conservation of forest is an important responsibility that all of us must undertake. Almost 31% of the earth's land surface is covered by forests. According to a 2015 report, 23% of India's land cover consists of the forest region. Trees and plants in the forest are an integral part of the ecosystem. It sustains life on the planet, provides clean air and shelter. Also, forests help conserve biodiversity. Apart from keeping a check on the global temperature, forests also contribute towards preventing soil from getting eroded and sheltering more than 80% of the animal species and terrestrial biodiversity. They also aid in improving the socioeconomic aspects of a country. Wildlife refers to the animal species that are not domesticated. So, any living organism that lives in the forest region is associated with wildlife.

DOI: 10.4018/978-1-6684-9034-1.ch006

INTRODUCTION

Just like a collection of buildings devoid of human inhabitants does not make a city, a framework of trees alone does not make a forest ecosystem. Whereas the plant world is mostly sedentary, silent, and full of utilitarian hues of green for photosynthesis, wildlife enlivens forests with movement, sound, and color. Wildlife heavily influences forests via behaviors, including pollination, seed dispersal, and herbivory. Importantly, proper management of forest wildlife is contingent on a sound understanding of forest wildlife ecology.

Aldo Leopold, a forester, and wildlife conservationist, postulated that people have a right to use and manage forests wisely and the responsibility of doing so in a way that recognizes the intrinsic value people share with other species and the ecosystems that contain them. By this definition, Leopold captured the true essence of forest wildlife management; he identified that the conservation of forest products and wildlife could be simultaneously achieved. Indeed, forest management, when appropriately implemented, often provides the most practical means by which to manage forest wildlife habitat.

Forest management, especially timber harvest, alters succession and changes plant structure and composition, which in turn affects wildlife habitat. The ecological changes that accompany forest management in some part simulate the effects of natural disturbance on forest ecosystems, and therefore may affect wildlife similarly. Importantly, long-standing natural disturbance regimes have been altered by humans through means such as wildfire suppression and urbanization. As such, the natural disturbance regimes to which many wildlife species adapted to over millennia are far less common in today's modern world. As described earlier, different wildlife species associate with different successional stages.

Hence, successional changes brought upon by timber harvest create or enhance habitat for some species while simultaneously diminishing habitat suitability for others. In other words, there are always "winners" and "losers." For example, large-scale removal of trees in a mature stand via clearcutting shifts the vegetation community from late successional to early successional, in turn benefitting wildlife adapted to the conditions therein. Concurrently, late-successional wildlife, especially canopy dwellers, largely will be eliminated in the stand until the canopy redevelops. Because habitat requirements vary among wildlife species, providing habitat for all wildlife species at small spatial and temporal scales is implausible (Grodsky, Moorman, & Russell, 2015).

Therefore, forest wildlife managers should identify focal wildlife species or communities before initiating forest management practices and account for local variations in wildlife–habitat relationships or habitat planning for wildlife must be conducted at large spatial scales (i.e., thousands of hectares) (Leopold 1933). In this research article, we focus on the forest and wildlife management.

Conservation of Forest

Conservation of forest is the practice of planting more trees and maintaining the forested areas for the sustainability for future generations. Forests are an important natural resource and are beneficial to humans in several ways. But due to increasing deforestation activities, it has become essential to conserve forests throughout the world. Deforestation is the permanent destruction or loss of forests for the expansion of lands for agriculture, livestock, etc. The process of destructing forests for the expansion of agricultural land is referred to as shifting cultivation.

Following are some of the major steps for conservation of forest:

1. With the advent of industrialization, several trees have been cut at an alarming rate for raw materials and various other purposes. This felling of trees can be regulated by selective cutting, clear-cutting and shelterwood cutting.
2. Forest fires are one of the common causes of loss of forests. Sometimes the forest land is set on fire to make the land available for commercial purposes. Once cleared, there can be no vegetation. Natural forest fires are also responsible for the destruction of huge forest covers. Latest firefighting techniques should be adopted to conserve the forest. However, forest fires are an important part of the ecosystem and it helps replenish nutrients in the soil from dead and decaying matter.
3. More trees should be planted to increase the forest cover. Trees should be selected according to the geographical conditions of a particular region and proper care should be taken during the growth of trees.
4. Prevention of exploitation of forestry and forest products is necessary for the conservation of forest.
5. The existing forests should be protected from diseases by spraying chemicals, antibiotics, or development of pest-resistant strains of trees.

Conservation of Wildlife

Wildlife is a very important and vital aspect of our ecosystem. Listed below are a few of the reasons stating the importance of wildlife conservation:

- **Important for their medicinal values**: More than one-third of our pharmaceutical needs are catered by wild plants. Forests provide great scope for breakthroughs in the field of medical science and technology along with the requirements for the large-scale manufacture of antibiotics and other medicines for therapeutic uses.

- **Helps keep our environment healthy:** They are responsible for maintaining temperatures globally, thereby fighting against the greenhouse effect and in turn preventing the sea levels to rise sharply.
- **Important to maintain ecological balance:** the interdependence of plants and animals is very essential
- **Economical importance:** Fossil fuels obtained from forests help in the economic growth of the country, which in turn contributes towards a better standard of living.
- **Helps preserve biodiversity:** There are thousands of species that take shelter in these huge forests.
- Microorganisms in wildlife take part in nitrogen fixation, thus, bringing about an increase in the levels of soil fertility.

Reasons Why Forests and Wildlife Matter

There is an absolute need to save our flora and fauna in the forests for the following reasons:

- It is all about breathing: Without oxygen, no living organisms can survive. Roughly 20.9% of earth's gases consist of oxygen and dependence on oxygen is higher, hence plants are essential.
- It plays a specific and significant role in the carbon cycle by absorbing carbon dioxide during photosynthesis.
- It is home for different species: Modernization has caused deforestation. This has led many animals to become homeless, leading to the death of these animals in most cases because of unfavorable conditions.
- Provides employment: Forests provide a source of livelihood to millions around the world. Raw materials that can later be sold come from these forests.
- Forests provide wood: Wood serves as an important raw material for manufacturing industries as well as commercial purposes not only animals and plants, millions of people still live in the forests and in its vicinity. They survive from the abundant resources that forests have to offer.
- Rescue during natural disasters: Vegetation in forests helps during floods as they seep in and absorb all the water, thereby preventing damage to infrastructure.
- Helps regulate climatic conditions: Forests play a key role in bringing about a balance in the atmospheric temperature.

- Biodiversity at its best: To get a closer view of life, visit a forest. Forests, as we know are home to thousands of species.
- Forests also provide great scope for exploration for science enthusiasts.
- Entertainment: Forest areas these days have been turned into adventure spots to help people get a first-hand experience of wildlife.

Key Points on Conservation of Forest and Wildlife

- Deforestation also referred to as degradation or loss of forests needs to be controlled or hopefully stopped.
- Fuelwood and timber should be used wisely and not wasted.
- Forest fire needs to be monitored and damage control measures need to be taken at the right time. Every year, lakhs of trees are destroyed because of forest fires. 2018 saw one of the deadliest wildfires ever in the history of California causing severe damage to property and life.
- Measures and interest need to be channeled towards regrowing trees and increasing the land cover for forests. Reforestation and afforestation need to be promoted throughout the world.
- Proper forest management bodies need to be enforced to monitor wildlife actions and update on the latest.
- Care and habitation need to be provided to extinct species and efforts need to be taken to save them from becoming extinct.

Some Queries!

What steps should be taken to conserve forests?

The forests can be conserved in the following manner:

- Introducing afforestation programs.
- Controlling forest fires.
- Proper utilization of forest resources.
- Proper care should be taken to protect the forests from pests and diseases.

Why is it beneficial to conserve forests?

Forests provide us with a variety of resources such as wood, timber, fibre and other raw materials for domestic as well as commercial purposes. They absorb the harmful carbon dioxide gas and maintain the earth's temperature. They also hold the soil firmly and prevent erosion of soil. This is why it is beneficial to conserve forests.

How can we conserve wildlife?

Wildlife can be conserved by:

- Developing protective areas such as national parks, wildlife sanctuaries to protect the animals in their natural habitat.
- The endangered and vulnerable species can be kept in captivity in places such as zoos and bred to increase their population.
- The cutting of forests should be strictly prohibited. It is an abode of a variety of wild animals.
- Hunting of wild animals should be banned.

Why is it beneficial to protect wildlife?

Wild animals form a part of the food chain. They are responsible for maintaining a balance in the ecosystem. In order to maintain an ecological balance, the wildlife needs to protect.

What are the reasons that have led to wildlife destruction?

Wildlife destruction is caused due to:

- Increasing demand for hides and meat has led to the hunting of animals.
- Cutting down of forests has led to the scarcity of food and space that has killed many animals.
- Frequent floods and earthquakes are the natural causes of wildlife destruction.

What will happen if all the forests are destroyed?

Forests are the major part of an ecosystem. They are an abode to various plants and animals. They maintain environmental balance by inhaling carbon dioxide and regulating rainfall. They provide food and space to animals that help to maintain the ecological balance. If all the forests are destroyed, it will have a severe impact on the environmental balance. There will be frequent earthquakes and floods. The temperature of the earth will rise enormously. Life on earth will be impossible.

CONCLUSION

Proper forest management is contingent on a sound understanding of forest wildlife ecology. Forests provide habitat—food, cover, and water—for many wildlife species and consequentially are among the most biologically diverse terrestrial ecosystems. Wildlife–habitat relationships form the foundation for understanding ecological interactions between forest wildlife and their environment. Specifically, forest wildlife management involves five interrelated ecological principles with implications on wildlife habitat: (1) disturbance, (2) succession, (3) structure (vegetation and deadwood), (4) composition, and (5) landscape ecology.

The true essence of forest wildlife management is captured by the idea that conservation of forest products for human consumption and forest wildlife can be simultaneously achieved. Forest management practices, including final harvest, regeneration, and site preparation, thinning, prescribed burning, and herbicide application, have direct effects on the five ecological principles outlined earlier and consequentially influence forest wildlife habitat. Successful forest wildlife management is often achieved by implementing forest management practices that facilitate habitat conditions for target wildlife species. The fundamental concepts presented in this article generally apply to all forests.

REFERENCES

Grodsky, S. M., Moorman, C. E., & Russel, K. R. (2015). Forest wildlife management. In G. LaRocque (Ed.), *Ecological forest management handbook* (pp. 47–85). CRC Press. doi:10.1201/b19150-5

Leopold, A. (1933). *Game management.* Scribner.

Chapter 7
Unraveling the Decline:
Exploring Global Patterns in Shorebird Populations

Vishal M. Makwana
https://orcid.org/0000-0001-6343-9449
Zoology Department, Sir P.P. Institute of Science, M.K. Bhavnagar University, India

Dipak Kumar Mahida
https://orcid.org/0000-0003-1299-4623
Department of Biochemistry and Forensic Science, Gujarat University, India

Pravinsang P. Dodia
Zoology Department, Sir P.P. Institute of Science, M.K. Bhavnagar University, India

ABSTRACT

Shorebirds, or Waders, are called sentinels of global environmental change. Many of these species breed in the arctic and subarctic and migrate back to tropical and temperate regions in their non-breeding seasons. Presently, more than half of shorebird populations with documented trends are in decline and migratory species are more inclined to be impacted. The extent and taxonomic range of the shorebird populations' decreases, along with facts showing that the Arctic is experiencing a higher rate of temperature rise than the rest of the world, indicate global warming as the probable cause of these declines. Primary factors responsible for their decline are degraded habitat quality, which has been influenced by anthropogenic activities and climate change. This highlights the necessity to pinpoint the drives of challenges in order to implement successful conservation measures. In this chapter, the authors aimed to understand the perspective on how and where the habitats of shorebirds are changing and factors responsible for their declines.

DOI: 10.4018/978-1-6684-9034-1.ch007

1. INTRODUCTION

1.1 Background

"The restlessness of shorebirds, their kinship with distance and swift seasons, the wistful signal of their voices down the longest coastlines of the world make them, for me, the most affecting of wild creatures. I think of them as birds of the wind, as wind birds."

~ Peter Matthiessen (Matthiessen, 1973)

The Shorebirds, also known as Waders, a diverse group of birds, belong to the order Charadriiformes, which encompasses 386 species worldwide divided into 19 families. Among these families, Laridae boasts the highest number of species with 101, followed closely by Scolopacidae with 97 species and Charadriidae with 71 species. Unfortunately, many shorebird species face significant threats to their survival. According to the IUCN Red List of threatened species (IUCN, 2023), a comprehensive assessment of the conservation status of various organisms, the shorebird group has been greatly affected. Among the shorebirds, seven species have been declared extinct, meaning that they no longer exist in the wild. Additionally, eleven species are classified as Critically Endangered, which indicates they face an extremely high risk of extinction in the wild. Furthermore, fifteen species are considered Endangered, signifying a very high risk of extinction (HBW and BirdLife International, 2022). The detailed list of the order Charadriiformes along with their IUCN status is given in Table 1.

Table 1. Total number of species comprising 19 families of the order Charadriiformes (Shorebirds)

Family	English Name	No. of Species	IUCN						
			EX	CR	EN	VU	NT	LC	DD
Alcidae	Auks	25	1	0	2	4	3	15	0
Burhinidae	Thick-knees	10	0	0	0	0	2	8	0
Charadriidae	Plovers	71	0	3	1	4	8	54	1
Chionidae	Sheathbills	2	0	0	0	0	0	2	0
Dromadidae	Crab-plover	1	0	0	0	0	0	1	0
Glareolidae	Coursers, Pratincoles	17	0	1	0	0	2	14	0
Haematopodidae	Oystercatchers	11	1	0	1	0	1	8	0
Ibidorhynchidae	Ibisbill	1	0	0	0	0	0	1	0
Jacanidae	Jacanas	8	0	0	1	0	0	7	0
Laridae	Gulls, Terns, Skimmers	101	0	1	4	9	8	79	0
Pedionomidae	Plains-wanderer	1	0	0	1	0	0	0	0
Pluvianellidae	Magellanic Plover	1	0	0	0	0	1	0	0
Pluvianidae	Egyptian Plover	1	0	0	0	0	0	1	0
Recurvirostridae	Avocets, Stilts	7	0	1	0	0	0	6	0
Rostratulidae	Painted-snipes	3	0	0	1	0	0	2	0
Scolopacidae	Sandpipers, Snipes, Phalaropes	97	5	3	4	6	20	59	0
Stercorariidae	Skuas	7	0	0	0	0	0	7	0
Thinocoridae	Seedsnipes	4	0	0	0	0	0	4	0
Turnicidae	Buttonquails	18	0	2	0	2	0	13	1

(EX-Extinct, CR-Critically Endangered, EN-Endangered, VU-Vulnerable, NT-Near Threatened, LC-Least Concern, DD-Data Deficient)

Shorebirds are found in diverse habitats ranging from intertidal mudflats, sandy beaches, and rocky coastlines to freshwater wetlands, grasslands, ploughed fields, and flooded agricultural lands. These adaptable birds have successfully colonized a wide array of ecosystems worldwide. One of the defining characteristics of shorebirds is their specialized feeding behaviour. They primarily rely on a diet composed of molluscs, small crustaceans, marine worms, and insects. This diet variation reflects the vast array of feeding strategies and adaptations observed among different species of shorebirds.

Figure 1. Some typical attributes of a shorebird (In picture: Calidris minuta)
photo credit: Mr. Ashok Mashru

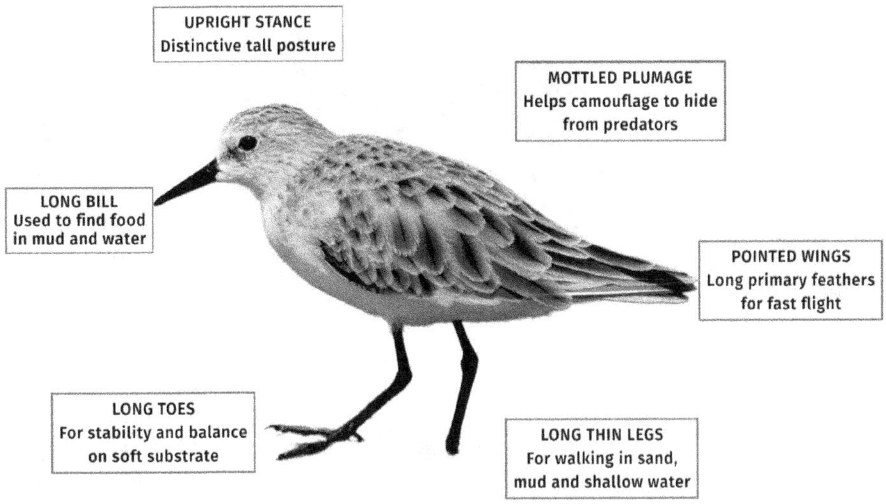

In terms of physical characteristics, shorebirds exhibit remarkable diversity (Figure 1). They come in various sizes, ranging from small sandpipers to larger species like godwits and curlews. This size diversity corresponds to differences in bill shapes and leg lengths, allowing each species to exploit their preferred food sources in their specific habitats. For instance, species with long, probing bills can effectively reach deep into mud or sand to extract hidden invertebrates, while those with shorter bills may specialize in catching insects or surface-dwelling prey (Figure 2). Similarly, leg lengths play a crucial role in navigating different types of terrain, such as wading through shallow waters or traversing open fields.

Figure 2. Different species of shorebirds exhibiting the diversity of shapes and sizes. A - Numenius arquata, B - Tringa totanus, C - Calidris ferruginea, D - Calidris alpina, E - Pluvialis fulva, F - Arenaria interpres.
(photo credits: Mr. Ashok Mashru)

Shorebird migrations are awe-inspiring feats that showcase the remarkable endurance and navigational abilities of these avian travellers. Among the shorebird species that engage in extraordinary migrations are Red Knots (*Calidris canutus*), Buff-breasted Sandpipers (*Tryngites subruficollis*), and Hudsonian Godwits (*Limosa haemastica*) (Burger et al., 2012; Senner, 2012; Faria et al., 2023). These species breed in the Arctic regions and undertake astonishing journeys to reach their wintering grounds at the southern tip of South America. These shorebirds migrate over great distances every year, covering around 20,000 miles total. This amazing voyage often entails crossing several continents, travelling through many ecosystems, and overcoming a variety of environmental obstacles. The necessity to escape adverse weather conditions during breeding and non-breeding seasons as well as the availability of food supplies are the driving forces behind these migrations. These shorebirds use a mix of their intrinsic navigational abilities and acquired actions to get where they are going throughout their protracted migrations. They are able to make such arduous journeys because to incredible physiological adaptations. This entails having efficient flight mechanisms, specialised fuel storage options, and the capacity to control metabolism for extended durations of flight. These shorebirds halts at several stopover sites throughout their migratory routes in order to rest and refuel to replenish their energy levels. These resting sites are vital to the success of their migrations because they provide them essential food resources to sustain their travel. The survival and reproductive success of these migratory shorebird populations may be significantly impacted by the loss or deterioration of these vital stopover sites.

1.2 Importance of Shorebird Populations

Shorebirds have earned the title of "sentinels of global environmental change" due to their remarkable attributes that make them highly sensitive indicators of ecological shifts worldwide (Piersma & Lindstrom 2004). These birds have a widespread distribution, inhabiting various regions across the globe, and undertake extensive migration journeys during their annual cycles. Their intricate utilization of different habitats further contributes to their significance as environmental indicators. The unique adaptations and ecological niches occupied by shorebirds contribute to their global distribution and success. Their ability to exploit a wide range of habitats and food resources underscores their importance within ecosystems. Shorebirds contribute to ecosystem functioning by regulating prey populations, nutrient cycling, and promoting habitat diversity. Moreover, they are indicators of the overall health and integrity of wetlands, coastal areas, and other vital ecosystems they inhabit. By understanding the diverse ecological roles of shorebirds and the specific adaptations they possess, we gain insights into their vulnerability to environmental changes and the factors contributing to their population declines. A substantial proportion of shorebird species engage in breeding activities within the arctic and subarctic regions during their breeding seasons. These remote and often harsh environments serve as critical nesting grounds for these birds. However, as a result of their migratory behaviour, they leave these northern breeding areas and migrate back to tropical and temperate regions for their non-breeding seasons (Colwell, 2010). This cyclic movement is a remarkable feat, as it allows them to exploit the resources available in different habitats at different times of the year. The existence of shorebirds in both arctic-subarctic and temperate-tropical habitats is very crucial. Some populations reproduce exclusively in the arctic and subarctic regions, while many other populations are found in temperate and tropical regions as well. This flexible distribution shows how adaptable and versatile shorebirds are as they move across various environments. Surprisingly, the group of shorebirds boasts some of the planet's most northerly breeding populations of terrestrial vertebrates (Delany et al. 2009). Their adaptability and resilience are shown by their capacity to effectively reproduce in the harsh arctic and subarctic environments. These shorebirds have adapted to take benefit of the brief but productive summer in these high-latitude regions.

The complexity of these long-distance migrations may be better understood by considering the ecological connectivity between breeding and non-breeding habitats. Researchers and environmentalists can gain valuable insights about the effects of climate change by studying shorebirds. These birds act as sensitive indicators, capturing changes in environmental factors such habitat availability, habitat quality, and climate. Monitoring the migration patterns, breeding success rates and population trends of shorebirds allows researchers to identify and analyse the impacts of environmental changes on these avian species. Therefore, shorebirds are essential for improving our understanding of larger ecological processes and developing conservation plans to save the fragile ecosystems on our planet. The main goal of this chapter is to develop a thorough knowledge of the changes affecting shorebirds' breeding and non-breeding habitats from a global viewpoint. Researchers and scientists intended to learn more about the processes and elements behind the loss of shorebird populations by looking at these changes.

2. DISTRIBUTION AND MIGRATION OF SHOREBIRD POPULATIONS

A simple but persuasive explanation for why shorebirds undertake migrations requiring such astonishing endurance can be answered by a simple yet compelling motivation: food. This migratory behaviour is best shown by the context of Arctic breeding grounds (Anderson et al., 2023). These breeding regions have practically constant sunshine throughout the month of June, providing an amazing 24 hours of daylight. Shorebirds benefit greatly from this prolonged daylight since it gives them twice as much time to carry out vital activities like feeding and energy storage. Because there are more daylight hours available, shorebirds may spend more time foraging, which is essential for refuelling their energy stores. This abundance of food is essential for getting them ready for the subsequent stage of their migration journey. These hemispheric travellers continue to be propelled by the want for food, which controls their behaviour and determines when they leave for the next leg of their trip. These migrating shorebirds are always conscious of the need to continue their journeys in search of adequate habitats and ample food supplies, therefore the time for their next migration is constantly drawing near. They may tailor their foraging and energy-storage techniques to the changing light cycle and other environmental signals, ensuring that they are prepared for the impending migration. Shorebirds' extraordinary endurance and survival tactics are highlighted by the astounding adaptations and intrinsic instincts they exhibit in response to shifting climates. By capitalizing on favourable environmental conditions and prioritizing their nutritional needs, these birds are able to navigate vast distances and endure the challenges posed by their ever-changing habitats.

The hemispheric travels of these remarkable avian species align with the occurrence of a reliably abundant food supply at each stopover along their migration routes. Take, for example, the case of Red Knots (*Calidris canutus rufa*). Their nesting season in the Arctic coincides with the brief period of insect abundance during the summer months. As this vital food source begins to dwindle, they embark on their southward journey to capitalize on the bountiful offerings of the tidal flats along the Atlantic coast (Thibault and Levisen, 2013). The timing of their northbound migration is meticulously synchronized with the spawning of horseshoe crabs, particularly in the Delaware Bay. Red Knots depend on horseshoe crab eggs, which are crucial source of protein and fat for them. The Red Knots gain twice as much weight after eating these nutrient-rich eggs, giving them the energy, they need for their final push before heading to the Arctic to reproduce. The availability of these very nourishing horseshoe crab eggs at certain stopover locations is a major factor in the migratory patterns of Red Knots (Thibault and Levisen, 2013). They strategically plan their migration so they may arrive at the ideal location and moment to feast on this plentiful and vital food source. This complex synchronisation between food availability and migratory timings is evidence of these birds' evolutionary adaption and their capacity to exploit seasonal resources for survival and successful reproduction.

Bar-tailed Godwits (*Limosa lapponica*) (Figure 3) have captured global attention due to their remarkable annual migration, spanning over 30,000 km, which involves three trans-oceanic flights ranging from 5,000 to 13,000 km each. This extraordinary feat has intrigued researchers worldwide, as demonstrated by studies conducted by researchers (Gill et al., 2009, Battley et al., 2012; Conklin et al., 2017). Particularly noteworthy is their non-stop flight from post-breeding staging sites in western Alaska to non-breeding sites in New Zealand and eastern Australia, which holds the record for the longest continuous flight ever documented by a land bird (Conklin et al., 2017). The Bar-tailed Godwits' endurance and capabilities challenge our understanding of avian endurance.

Figure 3. Bar-tailed Godwit (Limosa lapponica)
(Photo credit: Mr. Ashok Mashru)

Just as Nearctic shorebirds escape the cold winters of the northern hemisphere, several Austral shorebird species also undertake migratory journeys to flee the cold winters of the southern hemisphere. These characteristic austral species, such as the Rufous-chested Dotterel (*Charadrius modestus*) and Least Seedsnipe (*Thinocorus rumicivorus*), are just a couple of examples among the Patagonian breeders that migrate northward after the breeding season (Isacch and Martínez, 2003). These migratory shorebirds exhibit fascinating movement patterns as they navigate across vast distances. The Rufous-chested Dotterel and Least Seedsnipe, among others, depart their Patagonian breeding grounds and embark on a northward migration, seeking milder climates and more favourable conditions for their wintering period. They may escape the chilly, resource-scarce settings that characterise the southern hemisphere winter by relocating to more temperate areas.

The Magellanic Plover (*Pluvianellus socialis*), one of these austral shorebirds, makes a fascinating argument (Lishman and Nol, 2012). This mysterious species migrates altitudinally, leaving the marshes and lagoons of the Andean plateaus where they breed behind to spend the winter at the estuaries of the southern Atlantic coast. These movements allow them to take advantage of the milder coastal habitats, which provide ideal climate and food supplies during the winter. It is interesting that many of these austral shorebirds rely on the same stopover locations as their Nearctic relatives. Regardless of their hemispheric origin, they must pause at these important stopover locations to rest and refuel throughout their travels. These resting places are essential to meeting the energy needs of the shorebirds' trips, whether they are austral shorebirds flying north or Nearctic shorebirds flying south.

3. TRENDS AND PATTERNS IN SHOREBIRD POPULATION CHANGES

3.1 Historical Perspective

The negative effects of market hunting have been a major factor in the reduction of shorebird populations since the 1800s. Shorebirds were severely threatened at this time because they were the object of economic exploitation. The continuous hunting that led to the apparent extinction of the Eskimo Curlew (*Numenius borealis*) in the Americas is one dramatic example (Roberts et. al, 2010). The demands for feathers, meat, and other items at this time caused market hunting pressures that had a disastrous effect on shorebird populations. However, after the end of the market hunting period, shorebirds continued to confront new and continuing difficulties as a result of continuous habitat alterations caused by human activity (Pfister et. al, 1992). Shorebird populations have been negatively impacted by the vast habitat modification, which includes the conversion of wetlands, loss of coastal and intertidal regions, and disturbance of natural ecosystems (Stroud et. al, 2006). Critical habitats have been fragmented and deteriorated as a result of these changes, making it harder to find suitable breeding grounds, stopover sites, and wintering grounds.

3.2 Current Status: Recent Population Changes

The population decline of shorebirds has escalated alarmingly in recent decades. It can be attributed to several reasons, including habitat loss and degradation, climate change, pollution, disturbance, and the synergistic effects of several stressors. The combination of these threats has put tremendous pressure on shorebird populations, which were already at risk, escalating the challenges they confront to survive and reproduce. Since more than half of the shorebird populations with documented trends are experiencing a decline is a matter of concern (BirdLife International, 2021). This alarming statistic emphasises how urgent the situation is and shows how much more research and conservation efforts are required. The downtrend in shorebird populations is not limited to migratory routes or specific regions; it affects all migratory flyways. This indicates that the difficulties these birds confront are widespread and not limited to a certain geographic location. Additionally, it has been noted that migratory species are especially susceptible to the effects of environmental changes since they rely on a variety of habitats during their annual cycles (BirdLife International, 2021).

It is thought that widespread and rapid changes in land use, which have an impact on both their intertidal feeding habitat and coastal roosting places, are primarily to blame for the considerable drop in shorebird populations travelling along the East Asian-Australasian Flyway (Gallo-Cajiao et. al, 2020). The natural coastal ecosystems in the Inner Gulf of Thailand, which serve as critical stopover and wintering sites for these birds, have been greatly altered. Instead, artificial wetlands such as salt pans and aquaculture ponds have largely replaced the original coastal ecology in this region (Green et. al, 2015). During their non-breeding periods, a majority of migratory shorebird species, as well as many non-migratory species, rely on coastal wetlands (Colwell, 2010). These coastal wetlands serve as crucial habitats where shorebirds roost on open supratidal regions during high tide and feed primarily on benthic macroinvertebrates in intertidal areas during low tide (Colwell, 2010). These specific ecological requirements make shorebirds highly dependent on healthy coastal habitats with intact intertidal zones. Any changes or disruptions in these coastal regions can have significant consequences for shorebird populations. Coastal regions, which are vital for shorebird survival during their non-breeding periods, are also experiencing global changes that have the potential to negatively impact shorebird populations as a whole (Pearce-Higgins et. al., 2017). These global changes include factors such as sea-level rise, habitat degradation, pollution, and climate change. As coastal regions undergo transformations, the availability of suitable roosting sites and food resources for shorebirds may be compromised, leading to generalized population declines across multiple species. As per the IUCN Red List of Threatened Species (IUCN, 2023), the list of shorebird species falling under the categories- Extinct, Critically Endangered and Endangered is given in Table 2.

*Table 2.The list of extinct, critically endangered, and endangered shorebird species as per IUCN (*Probably Extinct)*

IUCN	Family	English Name	Scientific Name
Extinct (EX)	Haematopodidae	Canarian Oystercatcher	*Haematopus meadewaldoi*
	Scolopacidae	Tahiti Sandpiper	*Prosobonia leucoptera*
	Scolopacidae	Moorea Sandpiper	*Prosobonia ellisi*
	Scolopacidae	Christmas Sandpiper	*Prosobonia cancellata*
	Scolopacidae	South Island Snipe	*Coenocorypha iredalei*
	Scolopacidae	North Island Snipe	*Coenocorypha barrierensis*
	Alcidae	Great Auk	*Pinguinus impennis*
Critically Endangered (CR)	Recurvirostridae	Black Stilt	*Himantopus novaezelandiae*
	Charadriidae	Southern Red-breasted Plover	*Charadrius obscurus*
	Charadriidae	Javan Lapwing	*Vanellus macropterus**
	Charadriidae	Sociable Lapwing	*Vanellus gregarius*
	Scolopacidae	Eskimo Curlew	*Numenius borealis**
	Scolopacidae	Slender-billed Curlew	*Numenius tenuirostris*
	Scolopacidae	Spoon-billed Sandpiper	*Calidris pygmaea*
	Turnicidae	New Caledonian Buttonquail	*Turnix novaecaledoniae**
	Turnicidae	Buff-breasted Buttonquail	*Turnix olivii*
	Glareolidae	Jerdon's Courser	*Rhinoptilus bitorquatus*
	Laridae	Chinese Crested Tern	*Thalasseus bernsteini*
Endangered (EN)	Haematopodidae	Chatham Islands Oystercatcher	*Haematopus chathamensis*
	Charadriidae	Shore Plover	*Thinornis novaeseelandiae*
	Pedionomidae	Plains-wanderer	*Pedionomus torquatus*
	Rostratulidae	Australian Painted-snipe	*Rostratula australis*
	Jacanidae	Madagascar Jacana	*Actophilornis albinucha*
	Scolopacidae	Far Eastern Curlew	*Numenius madagascariensis*
	Scolopacidae	Great Knot	*Calidris tenuirostris*
	Scolopacidae	Tuamotu Sandpiper	*Prosobonia parvirostris*
	Scolopacidae	Spotted Greenshank	*Tringa guttifer*
	Laridae	Indian Skimmer	*Rynchops albicollis*
	Laridae	Peruvian Tern	*Sternula lorata*
	Laridae	Black-fronted Tern	*Chlidonias albostriatus*
	Laridae	Black-bellied Tern	*Sterna acuticauda*
	Alcidae	Marbled Murrelet	*Brachyramphus marmoratus*
	Alcidae	Guadalupe Murrelet	*Synthliboramphus hypoleucus*

3.3 Notable Species-Specific Population Shifts

The Chukchi Sea coast and the Asian side of the Bering Sea are the exclusive breeding grounds for the Spoon-billed Sandpiper (SbS) (*Calidris pygmaea*), a small bird species that migrates to Southeast Asia during the winter (Lappo et. al., 2012; Zöckler et. al., 2016). The spoon-billed sandpiper population has significantly declined in recent years, leading to its classification as critically endangered on the IUCN Red List in 2008 (Tomkovich et. al., 2002; Syroechkovski et. al., 2010; Zöckler et. al., 2010). Currently, it is estimated that there are only 120 (Zöckler et. al., 2016) to 210-228 breeding pairs (Clark et. al., 2018; Green et. al., 2021) of this species worldwide. Efforts to protect the spoon-billed sandpipers and their habitat have gained international attention in the East Asian-Australasian Flyway (EAAF) in order to reverse the population decline (Zöckler et. al., 2016; Peng et. al., 2017; Menxiu et. al., 2012; Clark et. al., 2014; Chang et. al., 2019). Since these efforts began, the annual decline rate has decreased from 26% in the 2000s (Zöckler et. al., 2010) to 8-10% per year (Green et. al., 2021; Zöckler et. al., 2020; Chowdhury et. al., 2022). In 2001, the largest known breeding population of spoon-billed sandpipers was discovered in the area of Meinypil'gyno Village in Southern Chukotka, Russia. However, it has shrunk from a minimum of 80 pairs in 2003 to a maximum of 12 pairs in 2022, although remains the largest breeding population (Zöckler et. al., 2010; Tomkovich et. al., 2016).

The North American population of American Oystercatchers (*Haematopus palliatus*) is one of the species with the highest risk of extinction if considerable conservation efforts are not made, according to the 2014 State of the Birds Watch List. The National Audubon Society has designated the American oystercatcher as a "yellow" species, signifying a serious conservation issue on a national scale. In addition, it is regarded as a "species of special concern" in several states along the East Coast of US, including New Jersey (Clay et. al., 2014).

Despite the fact that the population of Ruddy Turnstones (*Arenaria interpres*) appears to be decreasing (Simmons et. al., 2015), the decline is not considered rapid enough to meet the criteria for being classified as "vulnerable" based on the population trend criterion (>30% decline over ten years or three generations).

4. MAJOR FACTORS INFLUENCING SHOREBIRD DECLINES

The decrease in shorebird populations is evident across various taxa and has a significant taxonomic range. This shows that a broad range of species are being affected by the decreases rather than just certain groups of shorebirds. Shorebird breeding grounds are going through alarming ecological changes, such as altered breeding habitats and in prey-predator interactions (Aharon-Rotman et al., 2015; Wauchope et al., 2017). These changes are undoubtedly substantial and give rise to concerns about how potentially they can affect shorebird populations (Aharon-Rotman et al., 2015; Weiser et al., 2018a; Weiser et al., 2018b). There are additional factors causing the population decline of shorebirds in addition to the indirect evidence connecting higher temperatures to it. One important factor is degraded habitat quality, which has been impacted by both anthropogenic activities and climate change (Joulami et al., 2019; Koleček et al., 2021). The ecosystems that shorebirds depend on for breeding, foraging, and resting have been negatively impacted by human activities like as habitat destruction, pollution, and disturbance. The wider effects of climate change, including as rising sea levels and altered weather patterns, may also worsen the quality of habitat and interfere with shorebirds' access to critical resources.

4.1 Climate Change

Climate change is an urgent and present reality, and its effects are profoundly impacting shorebirds and their habitats. Shorebirds, with their dependence on coastal wetlands, the Arctic tundra, saline lakes in arid regions, and highly specialized food resources during migration, are particularly susceptible to the changes unfolding across the globe. One of the major consequences of climate change for shorebirds is the loss of habitat due to rising sea levels and increased coastal erosion (Sutherland et. al., 2012). As sea levels rise, crucial coastal wetlands and intertidal habitats, which serve as vital foraging and roosting grounds for shorebirds, are in danger of being submerged or degraded. This loss of habitat can have severe consequences, populations may become fragmented and displaced, and limiting their access to vital resources. Additionally, climate change brings hotter and drier circumstances to certain regions, posing difficulties for shorebirds that rely on these habitats for breeding and migration. The availability and quality of freshwater ecosystems, such as saline lakes and inland wetlands, are at risk as water sources diminish or become more saline due to increased evaporation and shifting precipitation patterns (Sutherland et. al., 2012). This reduction in suitable breeding grounds and stopover locations can disrupt shorebirds' lifecycle and hinder their competence to successfully reproduce and refuel their energy stores during migration. The warming Arctic brought on its own set of challenges, such as the encroachment of shrubs into traditional nesting sites. As the Arctic area warms at an accelerated rate, shrubs expand their range, changing the open tundra that shorebirds rely on for nesting. This shrub encroachment can result in to habitat degradation and decreased nesting success rates, further threatening the survival of shorebird species that breed in the Arctic.

Another crucial consequence of climate change is the disruption of the timing of migration and the availability of food supplies (Sutherland et. al., 2012). As seasons change and become less predictable, the timing of crucial ecological events, such as the emergence of insect prey base or the accessibility of certain plants and invertebrates, may go out of sync with the arrival of migrating shorebirds. This mismatch can have negative effects on shorebirds' capability to find adequate food during critical stages, leading to reduced breeding success and compromised energy stores for long-distance migrations.

4.1.1 Global Warming

One key factor contributing to the reduction of shorebird populations is global warming, which is proposed as the probable decisive cause (BirdLife International, 2021). Scientific data shows that the Arctic region is experiencing a faster rate of temperature rise compared to the rest of the planet (IPCC, 2013). This is crucial because several shorebird species reproduce in the Arctic and subarctic regions. The elevated warming in these important breeding grounds poses a risk to the birds' nesting success, as it alters the timing and availability of key resources needed for successful reproduction. Global warming affects shorebirds in manifold ways. Increasing temperatures can cause alterations in habitat suitability, disrupting the availability of food supplies and nesting sites. Shifts in climatic patterns may also influence the timing of migration and breeding, leading to a mismatch between the birds' life cycle events and resource availability. Such ecological mismatches can negatively impact reproductive success and survival rates (Sutherland et. al., 2012). The connection between the declines in shorebird populations, the extensive nature of the declines across all flyways, and the accelerated warming in the Arctic strongly suggests that global warming is a key driver of the observed declines. These findings emphasize the urgent need for global action to mitigate climate change and implement effective conservation measures to protect shorebird populations and their habitats. By addressing the root causes of global warming and implementing sustainable practices, we can strive to reverse the current downward trend and ensure the long-term survival of these remarkable and ecologically important birds.

4.1.2 Sea Level Rise

The rise in sea levels, a global driver of changes in coastal environments, has led to the erosion and submergence of shorebird feeding habitats (Galbraith et al., 2002; Nicholls and Cazenave, 2010). As sea levels increase, the encroachment of water onto previously exposed areas diminishes the availability of intertidal zones where shorebirds find their prey. The loss of these vital foraging grounds poses a substantial challenge for shorebirds, as it lessens their access to food supplies during critical phases of their annual cycles. Furthermore, the growth and expansion of salt-tolerant plants in shorebird roosting sites further intensifies the challenges faced by these birds (Jackson et al., 2021). The spread of such vegetation reduces the accessibility of open grounds where these birds traditionally roost, rest, and engage in behavioural activities such as preening and social interactions. The growth of salt-tolerant vegetation changes the structure and composition of these sites, making them less inhabitable for shorebird use. Sea levels are rising as a result of climate change, posing a threat to coastal regions with low elevation and the wildlife inhabiting there, such as Piping Plovers (*Charadrius melodus*). Scientists (Seavey et. al., 2011) studied the habitat of plovers on the barrier islands of Suffolk County, New York, and concluded that increasing sea levels could lead to a decrease in breeding parts if the plover habitat cannot relocate further inland due to human activities. As deteriorated beaches expand, the likelihood of conflicts emerging between plovers and recreational activities will also rise. In the incident of a major hurricane, up to 95% of the plover habitat could be flooded, indicating the potential impact of powerful storms on their habitat.

The consequences of climate change for shorebirds are disastrous. Predictions (Galbraith et. al., 2014) state that 90% of the 52 species that regularly breed in North America are anticipated to face a higher risk of extinction. This includes 10 imperilled species facing even greater threats in addition to 28 species already identified as being at high risk of extinction. These alarming projections indicate the dire need for conservation efforts to mitigate the effects of climate change on them and preserve their populations. Addressing climate change needs global efforts to cut greenhouse gas emissions, adapt to shifting weather conditions, and protect and restore critical ecosystems. By taking decisive steps to mitigate climate change and implementing actions to conserve shorebird habitats, we can alleviate the stresses these extraordinary birds face and protect their survival in the face of this global crisis.

4.2 Anthropogenic Activities and Habitat loss

4.2.1 Hunting Pressures

Illegal hunting poses a grave threat to the survival of shorebirds. These birds, both juvenile and adult, are targeted and killed illegally for variety of purposes, including consumption, sport, or trade. The scale of this hunting activity has had a negative impact on the survival rates of these species. When individuals are unlawfully hunted, it disrupts the natural reproductive cycle, impeding the production of new progenies and reducing the overall population size (Sutherland et. al., 2012). Furthermore, habitat degradation has also contributed greatly to the decline of these long-distance migratory shorebirds. The destruction of their habitats, including crucial coastal regions and wetlands, diminishes the availability of suitable breeding grounds and stopover sites. Therefore, the survival rate of both juvenile and adult birds is compromised, as they struggle to locate suitable habitats for feeding, nesting, and resting along their migratory routes.

4.2.2 Human Activities and Habitat Destruction

Anthropogenic activities have also contributed to the challenges encountered by shorebirds in coastal areas worldwide. Over the past few decades, usage of coastal areas and their surroundings has intensified, leading to several adverse impacts on shorebirds (Pearce-Higgins et al., 2017; Sutherland et al., 2012). Disturbance caused by humans, such as recreational activities, can interrupt foraging and resting behaviours, leading to increased energy expenditure, and reduced foraging efficacy for shorebirds. Pollution, including coastal runoff and marine debris, poses further threats to the health and survival of shorebirds. These pollutants can contaminate their habitats and food sources, potentially causing adverse health effects. Furthermore, illegal hunting and competition for food sources can further strain shorebird populations. Hunting activities that target shorebirds directly impact their survival and breeding success, especially when taking place in crucial stop-over or wintering sites. Additionally, competition for scarce food resources due to human exploitation can restrict foraging opportunities and increased competition amongst individuals. This puts extra pressure on intertidal and supratidal habitats, affecting the availability and quality of food supplies for shorebirds. The cumulative impacts of these human-induced disturbances and pressures, coupled with the effects of rising sea levels and altered coastal environments, poses substantial challenges to shorebird populations. To mitigate these challenges and ensure the conservation of these species, it is crucial to implement measures that lowers the rate of habitat degradation, minimize human disturbance, and promote sustainable management practices in coastal areas. By protecting these significant habitats and minimizing anthropogenic impacts, we can provide shorebirds with the necessary environments to thrive and sustain healthy populations.

The danger of these amazing birds is also made clear by the need of conserving a network of locations as a key element of shorebird conservation. Shorebirds often demonstrate a behaviour of gathering in huge numbers at certain staging places along their trips to rest and recharge, prompted by their migratory tendencies. However, since they depend on certain locations for rest and recharging, they are very vulnerable to the destruction or loss of these vital ecosystems. These staging locations' importance cannot be emphasised. They act as crucial resting places, supplying shorebirds with the materials they need to refuel for the arduous demands of migration. Shorebirds may rest and get ready for the next difficult leg of their hard trip in these habitats since they are abundant in food sources, shelter, and good roosting locations. The loss of even one of these vital staging places, however, might have disastrous repercussions, even causing the decrease or even extinction of an entire population within a flyway (Pfister et. al., 1992). Given the interconnectedness of these flyways, which cross continents and travel through many nations, the loss of even one essential habitat has the potential to upset the delicate balance and proper operation of the whole network. A staging region may be destroyed or degraded for a number of reasons, such as habitat loss brought on by urbanisation, land conversion for agriculture, pollution, coastal development, or changes to hydrological patterns (Sutherland et al., 2012). Shorebirds may suffer long-term effects from human activities that damage or destroy these habitats, putting both their survival and the integrity of their flyways at danger.

Any species' population growth depends on a fine balance between two essential factors: productivity and survival. A worrying trend of population loss has been seen in long-distance migratory shorebird species, especially in the East Asian-Australasian Flyway (EAAF). Illegal hunting and habitat damage are two important causes that are mostly to blame for this reduction. The population dynamics of long-distance migrating shorebirds, notably in the EAAF, have been severely impacted by the combined impacts of illegal hunting and habitat loss (Studds et. al., 2017). These elements not only reduce both juvenile and adult bird survival rates, but they also prevent the breeding of future generations. To address these challenges and restore the delicate balance between production and survival, urgent and effective conservation actions are required to protect these extraordinary species over the long term.

It's critical to understand that the loss of shorebird populations is a complex problem caused by a number of ecological reasons. The severity of the issue is highlighted by the extent of the reductions, which calls for extensive conservation initiatives to address the root causes. We can work to reverse the trend of decline and ensure the long-term viability of these remarkable bird species by mitigating the effects of climate change and anthropogenic activities that harm their habitats and putting conservation measures in place to protect and restore essential shorebird habitats.

5. CONSERVATION EFFORTS AND FUTURE DIRECTIONS

5.1 Knowledge Gaps in Understanding Shorebird Population Changes

Like many other migratory species, shorebirds exhibit distinctive behavioural patterns that require frequent movement between designated places at various stages of their life cycles. The need to migrate between "roost locations" and "foraging grounds," which serve as safe havens during high tide when the intertidal regions where they normally feed become flooded, is a vital part of their travels. The fact that shorebirds make seasonal long-distance migrations further emphasises how important these roosting and feeding locations are.

The various dietary preferences and foraging behaviours shown by various shorebird species are one of the elements impacting the effects of habitat degradation (Touhami et al., 2020). Because each species has distinct nutritional needs and unique foraging habits, changes in their environments and the availability of food may have diverse effects on various species. Understanding these little differences is crucial to understanding how habitat degradation affects various shorebird populations. In order to identify changes in species distribution and offer insight on the quick changes taking place in crucial ecosystems, long-term population surveys have been crucial (Anderson et al., 2023). The information on population trends, species abundance, and changes in species composition over time is significant information that is provided by these surveys. Researchers may learn more about how changing environmental circumstances impact shorebird populations and the ecosystems they rely on by analysing these long-term records.

When trying to pinpoint and treat the precise environmental changes causing the declines in migratory shorebird populations, this provides a substantial barrier. This challenge derives from the fact that migratory animals go through multiple life phases that take place in numerous places scattered over enormous geographic regions. It is difficult to identify the precise causes of the population drop due to the complexity of their life cycles. It is essential to understand the socioeconomic and ecological reasons driving the risks these species face in order to properly protect them. With this knowledge, researchers and conservationists may create focused plans to lessen the effects and encourage population rebound. The most urgent problems facing migratory animals may be addressed by conservation efforts by understanding the environmental changes and their causes. However, owing to the participation of several jurisdictions, conservation action implementation might be difficult. Even though the risks to a certain migratory species are well known, unfriendly laws in one country may thwart conservation efforts that have been started in another. Even if the threats to a certain migratory species are evidently established, unfavourable legislation in one jurisdiction may hamper conservation actions initiated in another. This legal complexity creates barriers and can slow down or impede conservation actions, making it more challenging to address the threats facing migratory animals effectively. Overcoming these obstacles requires collaborative efforts between different jurisdictions, conservation organizations, researchers, and policymakers. By working together, it is possible to overcome legal barriers and develop comprehensive conservation strategies that protect migratory species throughout their entire range. Coordinated actions based on scientific evidence and effective legislation can help ensure the long-term survival and well-being of these animals, even in the face of complex environmental changes.

While much attention has been given to the broader long-distance migrations of shorebirds, the preservation of these smaller-scale roosting and foraging migrations has received relatively less focus. However, it is essential for conservationists to pay attention to both roost and foraging locations because the condition of one can influence the utilization of the other. For example, if there is disturbance or disruption at a roost location, shorebirds may underutilize nearby foraging areas. On the other hand, habitat degradation or a decrease in food availability in foraging grounds can lead birds to explore additional feeding opportunities in their roost locations. Therefore, it becomes clear that maintaining a network of stopover sites, which are the transitional locations between breeding and wintering grounds, is just as critical as ensuring the spatial connection between roosting and feeding grounds. By protecting and maintaining these sites, conservation efforts can enable the successful completion of their migratory journeys. These stopover sites provide the necessary resources and resting areas for the shorebirds to refuel their energy, get ready for the next leg of their journey, and ultimately increase their chances overall of survival. Protecting the network of stopover locations needs a comprehensive understanding of the ecological requirements of shorebirds throughout their migratory routes. It involves identifying and conserving suitable roosting and foraging sites, considering the potential connections and dependencies between these sites, and reducing any threats that may impair the natural functioning of these regions. By recognizing the importance of both roost and foraging sites and implementing conservation actions to protect them, we can support the long-term sustainability of shorebird populations and the ecosystems they rely on.

Salt pans are valuable resources for migrating shorebirds. However, as development demands rise, maintaining salt pans throughout a vast region can be difficult. In order to incentivize operators to protect these salt pans, financial incentives may be required. While shorebirds do not much rely on semi-traditional aquaculture practices, drained ponds can still provide them with resting and foraging areas (Green et. al., 2015). It is important for operators to regularly empty their ponds to offer additional habitat for shorebirds. Both traditional aquaculture and salt pan management systems should avoid using nets and pond liners, as these can be detrimental to the well-being of shorebirds. By optimizing the management of aquaculture ponds and salt pans, a more practical, cost-effective, and geographically expansive approach to shorebird conservation can be achieved. This approach goes beyond solely relying on protected areas and focuses on maximizing the potential of existing aquaculture and salt pan systems to provide suitable habitats for shorebirds. In the presence of potentially more profitable land-use options, environmentalists will face challenges in promoting the preservation of salt pans. To address conflicts arising from land use and to formulate conservation policies that balance the needs of shorebirds with the opportunities and limitations of sustaining human livelihoods, it is crucial to integrate ecological analyses with a comprehensive comprehension of the economic factors that drive change. This integration will enable the development of timely and fair conservation strategies that effectively protect shorebirds while considering the economic realities and incentives associated with alternative land-use options.

5.2 Emerging Research Areas and Technologies

5.2.1 Headstarting (HS)

By implementing a strategy known as Headstarting (HS), which involves incubating eggs and rearing chicks in captivity while protecting them from predation during critical stages, it is possible to enhance the productivity of certain species. While headstarting has been often used in reptile conservation efforts, its use in avian species has been far less prevalent. Many wader species, including the American Oystercatcher (*Haematopus palliates*) (Collins et. al., 2016), Black Stilt (*Himantopus novaezelandiae*) (van Heezik et. al., 2009), Black-tailed Godwit (*Limosa limosa*) (Hiscock et. al., 2018), Piping Plover (*Charadrius melodus*) (Powell et. al., 1997), and Snowy Plover (*Charadrius nivosus*) (Neuman et. al., 2013), have seen success with headstarting. These conservation efforts have featured a variety of tactics, from moving eggs to a safe area for natural incubation by animals without parents (such maleo birds) to a year-long process of captive upbringing after artificial incubation. Headstarting has proven effective in enhancing egg production by stimulating the development of replacement clutches, as well as boosting the survival rates of both eggs and chicks. Additionally, it assists as a protective measure during years with unfavourable climatic conditions, preventing the potential collapse of populations due to severe environmental obstacles on breeding and non-breeding grounds (Boyd and Piersma, 2001). Moreover, Headstarting can help address the sex bias in chick survival triggered by environmental factors, thus balancing the sex ratio of first-breeders and enabling faster population growth.

5.2.2 Geolocator Studies

Researchers (Stutchbury et. al., 2009), in 2009 successfully deployed a geolocators lighter than ever before on Purple Martins gave emergence to the hope for an instrument small enough for waders at last. Geolocators can reveal more information than just locations of the birds, including flight speeds, flight paths, day/night flight patterns, weather/flight interactions, incubation patterns, movements between wintering grounds, new stopover locations, layover time at various stopovers, time spent wandering, and, also about whether birds prefer open or brushy habitat! There are many geolocator studies in progress on different shorebird species (Weiser et. al., 2016). However, it is crucial to ensure the consistent recapture of the same bird to avoid expending efforts in vain and causing unnecessary harm to the birds. Initially, conducting a flagging study is essential to assess the birds' faithfulness to the trapping place and enhance the trapping techniques.

5.3 Conservation Initiatives for Shorebirds

5.3.1 Convention on the Conservation of Migratory Species of Wild Animals (CMS)

CMS, an environmental treaty of the United Nations, offers a worldwide forum for the protection and responsible utilization of migratory animals and their habitats. It brings together the Range States, the nations through which migratory animals' journey, and establishes a legal framework for internationally coordinated conservation efforts. As the sole global convention devoted to the conservation of migratory species, their habitats, and migration routes, CMS collaborates with international organizations, NGOs, media partners, and corporate entities to enhance its mission.

5.3.2 Arctic Migratory Birds Initiative (AMBI)

The Arctic Migratory Birds Initiative (AMBI) is a project within the Conservation of Arctic Flora and Fauna (CAFF), aimed at enhancing the status and ensuring the long-term viability of diminishing populations of migratory birds that breed in the Arctic. AMBI collaborates with many partner countries and organizations to facilitate and initiate efforts that focus on priority species and conservation concerns in specific geographic areas. The goal is to prioritize and address the conservation requirements of migratory birds in these targeted regions.

5.3.3 Pacific Shorebird Conservation Initiative

It is collaborative efforts among partners across the Pacific Americas Flyway aim to ensure the survival of shorebird populations for present and future generations. The objective is to recognize key threats, implement effective conservation strategies, and adopt coordinated approaches to maintain and restore shorebird populations and their habitats within the Flyway. The main goal is to establish and maintain self-sustaining populations of shorebird species throughout the Flyway. Furthermore, there is a focus on enhancing resilience to the impacts of climate change and sustaining the ecosystems that offer provision for both humans and shorebirds.

5.3.4 The East Asian-Australasian Flyway Partnership (EAAFP)

The Partnership establishes a comprehensive framework across the flyway to foster dialogue, cooperation, and collaboration between a range of stakeholders dedicated to conserving migratory waterbirds and their habitats. These stakeholders include governmental bodies, site managers, technical institutions, UN agencies, development agencies, private sector entities, academic institutions, NGOs, community groups, and local communities. Through collaboration, the EAAFP Partners strive to create a Flyway Site Network that ensures sustainable management of internationally important wetlands. They contribute to research, monitoring, information exchange, and communication, education, participation, and awareness (CEPA) efforts concerning the conservation of migratory waterbirds and their habitats. In addition, capacity building initiatives and the development of flyway-wide conservation actions are prioritized.

In addition to the aforementioned initiatives, many national and international organizations are engaged in the conservation of shorebirds worldwide. Prominent among these organizations are WHSRN (Western Hemisphere Shorebird Reserve Network), the Atlantic Flyway Shorebird Initiative, Wetlands International, the International Wader Study Group, and the Global Flyway Network. These organizations play an important role in coordinating and implementing conservation efforts on a regional and global scale. WHSRN focuses on protecting critical habitats and promoting collaboration across the Western Hemisphere. The Atlantic Flyway Shorebird Initiative aims to conserve and restore shorebird populations along the Atlantic coast. Wetlands International works in the direction of the sustainable management of wetlands for the conservation of shorebird. The International Wader Study Group conducts research and promotes information of shorebird species. The Global Flyway Network facilitates international cooperation and conservation strategies for migratory birds, including shorebirds. Through their collective endeavours, these organizations contribute significantly to the conservation of shorebirds worldwide.

6. CONCLUSION

It is crucial to have a comprehensive knowledge of the causes of shorebird declines in order to apply effective conservation interventions (Ellis et al., 2021; Rashiba et al., 2022). For this, a thorough knowledge of the biological, environmental, and human-caused variables influencing shorebird habitats and populations is necessary. Taking into account elements like habitat loss, degradation, pollution, climate change, and human disruptions is part of this. Initiatives for conservation may be created to specifically address the problems impacting shorebird populations by recognising the main difficulties and their root causes. A cooperative strategy including scientists, policymakers, conservation organisations, and local people is required to develop successful conservation programmes and strategies for shorebirds. By collaborating, it is feasible to solve the many issues that shorebirds confront and put policies into place that guarantee both their long-term survival and the preservation of their ecosystems. There is an urgent need for a global perspective on important habitat modifications in order to define effective conservation priorities and design vital conservation programmes and policies for migratory shorebird species (Amano et al., 2018). This entails locating and protecting crucial habitats, putting conservation measures into place throughout the whole migratory paths, and spreading knowledge about the significance of shorebird conservation. To identify areas of concern and allocate resources for conservation efforts wisely, it is crucial to comprehend the changes taking place in shorebird habitats across the globe. Researchers may develop a thorough knowledge of the problems facing shorebird populations by looking at population-level trends, researching species-specific foraging habits, and taking the interaction between habitat degradation and changing environmental circumstances into consideration. In order to maintain and restore their habitats and appropriate conservation policies must be developed to stop the population reductions of shorebirds and ensure the survival of these remarkable migratory species by prioritising their conservation and putting focused measures in place.

ACKNOWLEDGEMENT

The authors are thankful to the Head, Zoology Department, Sir P. P. Institute of Science, M. K. Bhavnagar University for their continuous guidance and motivation. Authors would also like to acknowledge Mr. Ashok Mashru for contributing his photographs for this chapter.

REFERENCES

Aharon-Rotman, Y., Soloviev, M., Minton, C., Tomkovich, P., Hassell, C., & Klaassen, M. (2015). Loss of periodicity in breeding success of waders links to changes in lemming cycles in Arctic ecosystems. *Oikos*, *124*(7), 861–870. doi:10.1111/oik.01730

Amano, T., Székely, T., Sandel, B., Nagy, S., Mundkur, T., Langendoen, T., Blanco, D., Soykan, C. U., & Sutherland, W. J. (2018). Successful conservation of global waterbird populations depends on effective governance. *Nature*, *553*(7687), 199–202. doi:10.1038/nature25139 PMID:29258291

Anderson, C. M., Fahrig, L., Rausch, J., Martin, J. L., Daufresne, T., & Smith, P. A. (2023). Climate-related range shifts in Arctic-breeding shorebirds. *Ecology and Evolution*, *13*(2), e9797. doi:10.1002/ece3.9797 PMID:36778838

Anderson, C. M., Fahrig, L., Rausch, J., Martin, J. L., Daufresne, T., & Smith, P. A. (2023). Climate-related range shifts in Arctic-breeding shorebirds. *Ecology and Evolution*, *13*(2), e9797. doi:10.1002/ece3.9797 PMID:36778838

Battley, P. F., Warnock, N., Tibbitts, T. L., Gill, R. E. Jr, Piersma, T., Hassell, C. J., Douglas, D. C., Mulcahy, D. M., Gartrell, B. D., Schuckard, R., Melville, D. S., & Riegen, A. C. (2012). Contrasting extreme long-distance migration patterns in bar-tailed godwits Limosa lapponica. *Journal of Avian Biology*, *43*(1), 21–32. doi:10.1111/j.1600-048X.2011.05473.x

BirdLife International. (2021). *Species factsheets*. BirdLife International. http://www.birdlife.org.

Boyd, H., & Piersma, T. (2001). Changing balance between survival and recruitment explains population trends in Red Knots Calidris canutus islandica wintering in Britain, 1969-1995. *Ardea*, *89*(2), 301–317.

Burger, J., Niles, L. J., Porter, R. R., Dey, A. D., Koch, S., & Gordon, C. (2012). Migration and over-wintering of red knots (Calidris canutus rufa) along the Atlantic coast of the United States. *The Condor*, *114*(2), 302–313. doi:10.1525/cond.2012.110077

Chang, Q., Anderson, G. Q., Brides, K., Clark, J. A., Clark, N. A., Hearn, R., & Green, R. (2019). *A high proportion of the world population of the spoon-billed sandpiper occurs at Tiaozini, China, during the post-breeding Moult*. University of Cambridge.

Chowdhury, S. U., Foysal, M., & Green, R. E. (2022). Accelerating decline of an important wintering population of the critically endangered Spoon-billed Sandpiper Calidris pygmaea at Sonadia Island, Bangladesh. *Journal of Ornithology*, *163*(4), 891–901. doi:10.100710336-022-01995-0

Clark, N., Pain, D., & Green, R. (2014). Saving the spoon-billed sandpiper: An update on the conservation programme. *British Birds*, *107*(4), 7–475.

Clark, N. A., Anderson, G. Q., Li, J., Syroechkovskiy, E. E., Tomkovich, P. S., Zöckler, C., Lee, R., & Green, R. E. (2018). First formal estimate of the world population of the Critically Endangered spoon-billed sandpiper Calidris pygmaea. *Oryx*, *52*(1), 137–146. doi:10.1017/S0030605316000806

Clay, R. P., Lesterhuis, A. J., Schulte, S., Brown, S., Reynolds, D., & Simons, T. R. (2014). A global assessment of the conservation status of the American Oystercatcher Haematopus palliatus. *International Wader Studies*, *20*, 62–82.

Collins, S. A., Sanders, F. J., & Jodice, P. G. (2016). Assessing conservation tools for an at-risk shorebird: Feasibility of headstarting for American Oystercatchers Haematopus palliatus. *Bird Conservation International*, *26*(4), 451–465. doi:10.1017/S0959270916000095

Colwell, M. A. (2010). *Shorebird ecology, conservation, and management*. University of California Press. doi:10.1525/9780520947962

Conklin, J. R., Senner, N. R., Battley, P. F., & Piersma, T. (2017). Extreme migration and the individual quality spectrum. *Journal of Avian Biology*, *48*(1), 19–36. doi:10.1111/jav.01316

Delany, S., Scott, D., Dodman, T., & Stroud, D. (2009). An atlas of wader populations in Africa and Western Eurasia. *British Birds*, *102*, 639–642.

Ellis, K. S., Anteau, M. J., Cuthbert, F. J., Gratto-Trevor, C. L., Jorgensen, J. G., Newstead, D. J., Powell, L. A., Ring, M. M., Sherfy, M. H., Swift, R. J., Toy, D. L., & Koons, D. N. (2021). Impacts of extreme environmental disturbances on piping plover survival are partially moderated by migratory connectivity. *Biological Conservation*, *264*, 109371. doi:10.1016/j.biocon.2021.109371

Faria, F. A., Dias, R. A., Bencke, G. A., Bugoni, L., Senner, N. R., Almeida, J. B., Nunes, G. T., Gonçalves, M. S., & Lyons, J. E. (2023). Trends and population estimate of the threatened Buff-breasted Sandpiper *Calidris subruficollis* wintering in coastal grasslands of southern Brazil. *Bird Conservation International*, *33*, e61. doi:10.1017/S0959270923000138

Galbraith, H., DesRochers, D. W., Brown, S., & Reed, J. M. (2014). Predicting vulnerabilities of North American shorebirds to climate change. *PLoS One*, *9*(9), e108899. doi:10.1371/journal.pone.0108899 PMID:25268907

Galbraith, H., Jones, R., Park, R., Clough, J., Herrod-Julius, S., Harrington, B., & Page, G. (2002). Global climate change and sea level rise: Potential losses of intertidal habitat for shorebirds. *Waterbirds*, *25*(2), 173–183. doi:10.1675/1524-4695(2002)025[0173:GCCASL]2.0.CO;2

Gallo-Cajiao, E., Morrison, T. H., Woodworth, B. K., Lees, A. C., Naves, L. C., Yong, D. L., Choi, C. Y., Mundkur, T., Bird, J., Jain, A., Klokov, K., Syroechkovskiy, E., Chowdhury, S. U., Fu, V. W. K., Watson, J. E. M., & Fuller, R. A. (2020). Extent and potential impact of hunting on migratory shorebirds in the Asia-Pacific. *Biological Conservation, 246*, 108582. doi:10.1016/j.biocon.2020.108582

Gill, R. E. Jr, Tibbitts, T. L., Douglas, D. C., Handel, C. M., Mulcahy, D. M., Gottschalck, J. C., Warnock, N., McCaffery, B. J., Battley, P. F., & Piersma, T. (2009). Extreme endurance flights by landbirds crossing the Pacific Ocean: Ecological corridor rather than barrier? *Proceedings. Biological Sciences, 276*(1656), 447–457. doi:10.1098/rspb.2008.1142 PMID:18974033

Green, J. M., Sripanomyom, S., Giam, X., & Wilcove, D. S. (2015). The ecology and economics of shorebird conservation in a tropical human-modified landscape. *Journal of Applied Ecology, 52*(6), 1483–1491. doi:10.1111/1365-2664.12508

Green, R., Syroechkovskiy, E. E., Anderson, G. Q., Chang, Q., Chowdhury, S., Clark, J. A., & Clark, N. A. (2021). *New estimates of the size and trend of the world population of the Spoon-billed Sandpiper using three independent statistical models.*

HBW and BirdLife International. (2022). *Handbook of the Birds of the World and BirdLife International digital checklist of the birds of the world. Version 7.* BirdLife International. http://datazone.birdlife.org/userfiles/file/Species/Taxonomy/HBW-BirdLife_Checklist_v7_Dec22.zip

Hedenström, A. (2010). Extreme endurance migration: What is the limit to non-stop flight? *PLoS Biology, 8*(5), e1000362. doi:10.1371/journal.pbio.1000362 PMID:20454558

Hiscock, N., Brand, E., Calvo-Carrasco, D., & Donaldson, L. (2018). *Headstarting the Black-Tailed Godwit Limosa Limosa Limosa at Key Breeding Sites for the UK Population: 2018 Annual Report.* Wildfowl & Wetlands Trust.

IPCC. (2013). *Climate change 2013: The physical science basis. Contribution of working group I to the fifth assessment report of the intergovernmental panel on climate change.* Cambridge University Press.

Isacch, J. P., & Martínez, M. M. (2003). Habitat use by non-breeding shorebirds in flooding pampas grasslands of Argentina. *Waterbirds, 26*(4), 494–500. doi:10.1675/1524-4695(2003)026[0494:HUBNSI]2.0.CO;2

IUCN. (2023). *The IUCN Red List of Threatened Species. Version 2022-2.* IUCN. https://www.iucnredlist.org.

Jackson, M. V., Fuller, R. A., Gan, X., Li, J., Mao, D., Melville, D. S., Murray, N. J., Wang, Z., & Choi, C. Y. (2021). Dual threat of tidal flat loss and invasive Spartina alterniflora endanger important shorebird habitat in coastal mainland China. *Journal of Environmental Management*, *278*, 111549. doi:10.1016/j.jenvman.2020.111549 PMID:33260073

Joulami, L., El Hamoumi, R., Daief, Z., Bazairi, H., & Lopes, R. J. (2019). Impact of shorebird predation on intertidal macroinvertebrates in a key North African Atlantic wintering site: An experimental approach. *African Journal of Marine Science*, *41*(1), 1–9. doi:10.2989/1814232X.2018.1552193

Koleček, J., Reif, J., Šálek, M., Hanzelka, J., Sottas, C., & Kubelka, V. (2021). Global population trends in shorebirds: Migratory behaviour makes species at risk. *Naturwissenschaften*, *108*(2), 9. doi:10.100700114-021-01717-1 PMID:33580336

Lappo, E. G., Tomkovich, P. S., & Syroechkovskiy, E. E. (2012). Atlas of Breeding Waders in the Russian Arctic. Publishing House "UF Ofsetnaya Pechat": Moscow, Russia.

Lishman, C., & Nol, E. (2012). Ecology and habitat selection of the Magellanic Plover (Pluvianellus socialis): A little-known Patagonian shorebird. *The Wilson Journal of Ornithology*, *124*(3), 487–496. doi:10.1676/11-035.1

Matthiessen, P. (1973). *The Wind Birds: Shorebirds of North America*. Avery. doi:10.1604/9780670770960

Menxiu, T., Lin, Z., Li, J., Zöckler, C., & Clark, N. A. (2012). The critical importance of the Rudong mudflats, Jiangsu Province, China in the annual cycle of the Spoon-billed Sandpiper Calidris pygmeus. *Bulletin - Wader Study Group*, *119*(3), 74.

Neuman, K. K., Stenzel, L. E., Warriner, J. C., Page, G. W., Erbes, J. L., Eyster, C. R., Miller, E., & Henkel, L. A. (2013). Success of captive-rearing for a threatened shorebird. *Endangered Species Research*, *22*(1), 85–94. doi:10.3354/esr00531

Nicholls, R. J., & Cazenave, A. (2010). Sea-level rise and its impact on coastal zones. *Science, 328*(5985), 1517-1520.

Pearce-Higgins, J. W., Brown, D. J., Douglas, D. J., Alves, J. A., Bellio, M., Bocher, P., Buchanan, G. M., Clay, R. P., Conklin, J., Crockford, N., Dann, P., Elts, J., Friis, C., Fuller, R. A., Gill, J. A., Gosbell, K., Johnson, J. A., Marquez-Ferrando, R., Masero, J. A., & Verkuil, Y. I. (2017). A global threats overview for Numeniini populations: Synthesising expert knowledge for a group of declining migratory birds. *Bird Conservation International*, *27*(1), 6–34. doi:10.1017/S0959270916000678

Peng, H. B., Anderson, G. Q., Chang, Q., Choi, C. Y., Chowdhury, S. U., Clark, N. A., Gan, X., Hearn, R. D., Li, J., Lappo, E. G., Liu, W., Ma, Z., Melville, D. S., Phillips, J. F., Syroechkovskiy, E., Tong, M., Wang, S., Zhang, L., & Zöckler, C. (2017). The intertidal wetlands of southern Jiangsu Province, China–globally important for Spoon-billed Sandpipers and other threatened waterbirds, but facing multiple serious threats. *Bird Conservation International*, *27*(3), 305–322. doi:10.1017/S0959270917000223

Pfister, C., Harrington, B. A., & Lavine, M. (1992). The impact of human disturbance on shorebirds at a migration staging area. *Biological Conservation*, *60*(2), 115–126. doi:10.1016/0006-3207(92)91162-L

Piersma, T., & Lindström, Å. (2004). Migrating shorebirds as integrative sentinels of global environmental change. *The Ibis*, *146*(s1), 61–69. doi:10.1111/j.1474-919X.2004.00329.x

Powell, A. N., Cuthbert, F. J., Wemmer, L. C., Doolittle, A. W., & Feirer, S. T. (1997). Captive-rearing piping plovers: developing techniques to augment wild populations. *Zoo Biology: Published in affiliation with the American Zoo and Aquarium Association, 16*(6), 461-477.

Rashiba, A. P., Jishnu, K., Byju, H., Shifa, C. T., Anand, J., Vichithra, K., Xu, Y., Nefla, A., Muzaffar, S. B., Aarif, K. M., & Rubeena, K. A. (2022). The Paradox of Shorebird Diversity and Abundance in the West Coast and East Coast of India: A Comparative Analysis. *Diversity (Basel)*, *14*(10), 885. doi:10.3390/d14100885

Roberts, D. L., Elphick, C. S., & Reed, J. M. (2010). Identifying anomalous reports of putatively extinct species and why it matters. *Conservation Biology*, *24*(1), 189–196. doi:10.1111/j.1523-1739.2009.01292.x PMID:19604294

Seavey, J. R., Gilmer, B., & McGarigal, K. M. (2011). Effect of sea-level rise on piping plover (Charadrius melodus) breeding habitat. *Biological Conservation*, *144*(1), 393–401. doi:10.1016/j.biocon.2010.09.017

Senner, N. R. (2012). One species but two patterns: Populations of the Hudsonian Godwit (Limosa haemastica) differ in spring migration timing. *The Auk*, *129*(4), 670–682. doi:10.1525/auk.2012.12029

Simmons, R. E., Kolberg, H., Braby, R., & Erni, B. (2015). Declines in migrant shorebird populations from a winter-quarter perspective. *Conservation Biology*, *29*(3), 877–887. doi:10.1111/cobi.12493 PMID:25858334

Stroud, D. A., Baker, A., Blanco, D. E., Davidson, N. C., Ganter, B., Gill, R. E., Jr., Gonzalez, P. M., Haanstra, L., Morrison, R. I. G., Piersma, T., Scott, D., & Zöckler, C. (2006). The conservation and population status of the world's waders at the turn of the millennium. In *The Waterbirds Around the World Conference* (pp. 643-648).

Studds, C. E., Kendall, B. E., Murray, N. J., Wilson, H. B., Rogers, D. I., Clemens, R. S., Gosbell, K., Hassell, C. J., Jessop, R., Melville, D. S., Milton, D. A., Minton, C. D. T., Possingham, H. P., Riegen, A. C., Straw, P., Woehler, E. J., & Fuller, R. A. (2017). Rapid population declinc in migratory shorebirds relying on Yellow Sea tidal mudflats as stopover sites. *Nature Communications*, *8*(1), 14895. doi:10.1038/ncomms14895 PMID:28406155

Stutchbury, B. J., Tarof, S. A., Done, T., Gow, E., Kramer, P. M., Tautin, J., Fox, J. W., & Afanasyev, V. (2009). Tracking long-distance songbird migration by using geolocators. *Science*, *323*(5916), 896–896. doi:10.1126cience.1166664 PMID:19213909

Sutherland, W. J., Alves, J. A., Amano, T., Chang, C. H., Davidson, N. C., Max Finlayson, C., Gill, J. A., Gill, R. E. Jr, González, P. M., Gunnarsson, T. G., Kleijn, D., Spray, C. J., Székely, T., & Thompson, D. B. (2012). A horizon scanning assessment of current and potential future threats to migratory shorebirds. *The Ibis*, *154*(4), 663–679. doi:10.1111/j.1474-919X.2012.01261.x

Syroechkovski, E. E., Tomkovich, P. S., Kashiwagi, M., Taldenkov, I. A., Buzin, V. A., Lappo, E. G., & Zöckler, C. (2010). Population decline in the spoon-billed sandpiper (Eurynorhynchus pygmeus) in northern Chukotka based on monitoring on breeding grounds. *Biology Bulletin of the Russian Academy of Sciences*, *37*(9), 941–951. doi:10.1134/S1062359010090074

Thibault, J., & Levisen, M. (2013). Red Knot prey availability project report. *Final Report*. South Carolina Department of Natural Resources, Marine Resources Research Institute, Charleston, SC, USA.

Thomas, P., Boyer, D., Oehler, D., Silver, S., & Perrotti, L. (2019). Headstarting as a conservation strategy for threatened and endangered species. *Scientific Foundations of Zoos and Aqimriums: Their Role in Conservation and Research*, 91-111.

Tomkovich, P., Syroechkovskiy, E., Yakushev, N., Loktionov, E., Digby, R., & Shepelev, I. (2016). Spoon-billed Sandpipers in Meinypilgyno: Update of the 2016 breeding season. *Spoon-Bill. Sandpipers Task Force News Bull*, *16*, 13–15.

Tomkovich, P. S., Syroechkovski Jr, E. E., Lappo, E. G., & Zöckler, C. (2002). First indications of a sharp population decline in the globally threatened Spoon-billed Sandpiper Eurynorhynchus pygmeus. *Bird Conservation International*, *12*(1), 1-18.

Touhami, F., Idrissi, H. R., & Benhoussa, A. (2020). Foraging behaviour of wintering shorebirds at Merja Zerga lagoon (Atlantic coast, Morocco). *The Ostrich*, *91*(3), 244–251. doi:10.2989/00306525.2020.1774439

van Heezik, Y., Maloney, R. F., & Seddon, P. J. (2009). Movements of translocated captive-bred and released Critically Endangered kaki (black stilts) Himantopus novaezelandiae and the value of long-term post-release monitoring. *Oryx*, *43*(4), 639–647. doi:10.1017/S0030605309990081

Wauchope, H. S., Shaw, J. D., Varpe, Ø., Lappo, E. G., Boertmann, D., Lanctot, R. B., & Fuller, R. A. (2017). Rapid climate-driven loss of breeding habitat for Arctic migratory birds. *Global Change Biology*, *23*(3), 1085–1094. doi:10.1111/gcb.13404 PMID:27362976

Weiser, E. L., Brown, S. C., Lanctot, R. B., Gates, H. R., Abraham, K. F., Bentzen, R. L., Bêty, J., Boldenow, M. L., Brook, R. W., Donnelly, T. F., English, W. B., Flemming, S. A., Franks, S. E., Gilchrist, H. G., Giroux, M.-A., Johnson, A., Kendall, S., Kennedy, L. V., Koloski, L., & Sandercock, B. K. (2018a). Effects of environmental conditions on reproductive effort and nest success of Arctic-breeding shorebirds. *The Ibis*, *160*(3), 608–623. doi:10.1111/ibi.12571

Weiser, E. L., Lanctot, R. B., Brown, S. C., Alves, J. A., Battley, P. F., Bentzen, R., Bêty, J., Bishop, M. A., Boldenow, M., Bollache, L., Casler, B., Christie, M., Coleman, J. T., Conklin, J. R., English, W. B., Gates, H. R., Gilg, O., Giroux, M.-A., Gosbell, K., & Sandercock, B. K. (2016). Effects of geolocators on hatching success, return rates, breeding movements, and change in body mass in 16 species of Arctic-breeding shorebirds. *Movement Ecology*, *4*(1), 1–19. doi:10.118640462-016-0077-6 PMID:27134752

Weiser, E. L., Lanctot, R. B., Brown, S. C., Gates, H. R., Bentzen, R. L., Bêty, J., Boldenow, M. L., English, W. B., Franks, S. E., Koloski, L., Kwon, E., Lamarre, J.-F., Lank, D. B., Liebezeit, J. R., McKinnon, L., Nol, E., Rausch, J., Saalfeld, S. T., Senner, N. R., ... Sandercock, B. K. (2018b). Environmental and ecological conditions at Arctic breeding sites have limited effects on true survival rates of adult shorebirds. *The Auk*, *135*(1), 29–43. doi:10.1642/AUK-17-107.1

Zöckler, C., Beresford, A. E., Bunting, G., Chowdhury, S. U., Clark, N. A., Fu, V. W. K., Htin Hla, T., Morozov, V., Syroechkovskiy, E., Kashiwagi, M., Lappo, E. G., Tong, M., Long, T. L., Yu, Y.-T., Huettmann, F., Akasofu, H. K., Tomida, H., & Buchanan, G. M. (2016). The winter distribution of the Spoon-billed Sandpiper Calidris pygmaeus. *Bird Conservation International*, *26*(4), 476–489. doi:10.1017/S0959270915000295

Zöckler, C., Chowdhury, S. U., Sun, L., Qing, C., Aung, P. P., Clements, C., ... Syroechkovskiy, E. E. (2020). The Spoon-billed Sandpiper Calidris pygmaea conservation project in 2019 and 2020: Population trends continue to be negative. *BirdingASIA*, *33*, 51–56.

Zöckler, C., Syroechkovskiy, E. E., & Atkinson, P. W. (2010). Rapid and continued population decline in the Spoon-billed Sandpiper Eurynorhynchus pygmeus indicates imminent extinction unless conservation action is taken. *Bird Conservation International*, *20*(2), 95–111. doi:10.1017/S0959270910000316

KEY TERMS AND DEFINITIONS

Breeding Ground: It refers to a specific location or area where animals, including birds, engage in mating and reproduction activities.

Climate Change: Climate change refers to long-term shifts in global or regional climate patterns, primarily attributed to human activities, resulting in alterations to temperature, precipitation, sea levels, and other climatic variables.

Foraging Ground: It is a specific location or area, where animals, including birds, search for food resources.

Headstarting: Headstarting refers to a conservation technique in which young individuals, typically vulnerable or endangered species, are reared in a protected settings until they reach a stage of increased survival capability before being released into the wild.

Migratory Flyways: Migratory flyways are defined pathways or routes that migratory species, including birds, follow during their seasonal migrations, typically characterized by specific geographical landmarks or habitats.

Population Decline: Population decline refers to a significant and sustained decrease in the number of individuals within a specific group or species over a defined period of time.

Roosting Ground: Specific sites where birds gather and rest, typically during the night or period of inactivity.

Chapter 8
Urban Anthropocene-Socio-Ecology Planning Resilience

José G. Vargas-Hernandez
Budapest Centre for Long-term Sustainability, Hungary

Omar C. Vargas-González
https://orcid.org/0000-0002-6089-956X
Tecnológico Nacional de México, Ciudad Guzmán, Mexico

ABSTRACT

This study aims to analyze some of the existing implications between urban anthropology and Anthropocene urban socio-ecology planning resilience, beginning from the assumption that urban anthropology gives support to create and develop any urban planning based on the Anthropocene urban socio-ecology resilience. The methods employed are the analytical-descriptive based on an ethnographic interpretation and reflection of the theoretical and empirical literature review. The analysis concludes that urban anthropology fundamentals give support to strengthen the Anthropocene orientation of the urban socioecological planning resilience.

INTRODUCTION

Urban anthropology is a scientific discipline epistemologically and intricately linked and interrelated with other scientific theories. Anthropology, like any social science, predicts various phenomena (Álvarez, 2016; Brady, 2019; Grimmer et al., 2021; Hindman, 2015; Hofman et al., 2017; Hofman et al., 2021; Taagepera, 2008). The scientific status of anthropology goes beyond the conception of a soft, interpretative, and humanistic phenomenon to interpret meanings, to use quantitative techniques,

DOI: 10.4018/978-1-6684-9034-1.ch008

to apply formal methods in experimental designs, to postulate scientific theories predicting phenomena and formulating scientific laws linked and interrelated with other scientific theories.

On the other hand, the term resilience was introduced by Holling (1973) in the field of socio-ecological research to describe the ability of the socio-ecosystem to return to its original shape after having been deformed by the impact of external pressures from the external environment. Resilience influences urban planning and innovation, urban infrastructure construction and development leading to urban and regional economic resilience as one of the contents of research on resilience in urban planning and governance.

Likewise, the improvement of urban socio-ecological resilience contributes to the prevention, management, and reduction of catastrophes, which strengthens urban resilience processes (Yang, Lili & Hongchi, 2022). Some approaches to urban anthropology do not consider it a science and those who prefer theories or the particular over the universal, and those who choose to interpret rather than explain.

However, the anthropological approach to urban socio-ecological resilience has shorter interactions in time between social groups that aim at urban socio-ecological resilience planning and green innovations. The development of urban anthropological motivation on creative capacities generates the conditions of the local innovation economy based on urban socio-ecological planning and green innovation areas. Being that, currently, international institutions are dedicated to studying lateral work with groups of experts in urban socio-ecological planning and areas of green innovation with the support of urban anthropology (Hannerz, 2004; Holmes & Marcus, 2006, 2021; Ortner, 2010, 2013).

On the other hand, the changes from traditional urban anthropology to the urban sociology of the Anthropocene and resilience have been analyzed by Kuhn (1962/1996) who has focused on paradigm shifts, focusing on the epistemic cycles of the methods of research from its rise and decline. Thus, this study aims to analyze some of the relationships and implications between urban anthropology and urban socio-ecological planning for the resilience of the Anthropocene. To achieve this goal, this study first starts from the analysis through the study of organizational and urban anthropology cases, followed by urban socio-ecological resilience planning and urban green innovation areas. Finally, some conclusions are offered.

THEORETICAL BACKGROUND

The ancient Greeks developed critical thinking about urban socio-ecological resilience based on mutualistic relationships (Egerton, 2001). Currently it is recognized that the socio-ecological change of societies is caused by exogenous factors, such as globalization, colonialism, capitalism, etc. Some colonial societies are composed of heterogeneous elements with instability due to divergent interests (Radcliffe-Brown, 1940). Similarly, the phenomena of sociology, anthropology, and political science differ from individual behavior, according to Bunge (1999), who argues that predicting the behavior of an unknown individual is different from predicting the behavior of a system. However, the social sciences have translated small group prediction into complex social systems using sophisticated computational methods (Bunge, 1999; Kaplan, 1940).

Likewise, it is recognized that a scientific law is a logical-mathematical representation of the relationships between variables (Pfeifer, 2006). Scientific predictions arise from the verification and application of a theory capable of explaining specific phenomena. Research in urban anthropology and socio-ecology of resilience in the fields of urban planning and green innovation using interdisciplinary methodologies is changing people's societies, which is relevant for research on a global scale. However, anthropomorphic representations in research have not been convincing in favor of the functional hypothesis.

Science resolves testable hypotheses (Hirst, 2019). Research in the socioeconomic and organizational sciences focuses on sympathy for the population of subjects who move from the realm of an observer to become an active participant (Bilharz, 1998; Kolb, 1998). The interdisciplinarity and transdisciplinarity of urban anthropology applied to urban socio-ecological planning and green innovation areas are limited in space to conduct the development of disciplines in multidimensional crisis including ethical values and moral issues that drive to an ontological crisis of homo sapiens (Tarantino-Curseri, 2018).

It is recognized that urban anthropology leading to socio-ecological planning processes is evolving as a dialectical process from centralized, sequenced, and systematic rational integral planning to complete fragmentation and disjointed incrementalism (Lindblom, 1959). Strengthening the consolidation of national science, technology and innovation systems has effects on competitiveness derived from incremental socio-ecological and green innovations in urban planning. Therefore, research in urban anthropology and socio-ecology of resilience requires a comprehensive analysis, returning to a scientific perspective, without neglecting issues of human development in an environment that adapts to current demands, both of globalization and awareness of the environment.

METHOD

The analysis of the relations between the establishment of urban anthropology of action is framed in the reflexive dialectic actor and author with specific disciplinary orientation in urban socio-ecology resilience planning and areas of green innovation. The analysis of the relationships between the objectivity of the intersubjectivity of urban anthropology become methodological problems and dilemmas that need to be addressed specifically. The intersubjectivity of urban anthropology is an objective element of construction of reality, not with a relativistic ideology but with a methodological and practical one based on valid assumptions for urban socio-ecological resilience planning and green innovation.

In accordance with the above, a qualitative, analytical, and descriptive study is proposed on the state of urban anthropology and resilient urban socio-ecological planning of the Anthropocene; the foregoing, through the analysis of case studies related to organizational and urban anthropology, urban socio-ecological resilience planning and urban green innovation areas.

CASE STUDIES

The formal methods must meet the requirements of operationalization of terms, data analysis, logical consistency, justification of affirmations and replication of results, although neither the formal theories were dominant nor were the dominant theories ever (Reynoso, 1995b). The discipline of urban anthropology can formalize the knowledge produced. Formal computational methods used for ethnographic analysis in urban anthropology and archaeological data based on agent models, cellular automata, diffusion and percolation models, network analysis, etc., proposed a new form of research open to theoretical development and experimentation (Diaz et al. 2007). Urban anthropological experiments, predictions and theories can be compared with the natural and physical sciences to determine scientific status (Llanos, 2009), but they are not sufficient scientific criteria for good science if it is carried out without laws or predictions.

The most common experiments in urban anthropology and social sciences are natural experiments (Diamond & Robinson, 2010; Dunning, 2012), field experiments (Baldassarri & Abascal, 2017) and naturalistic experiments (Bernard, 2011). Accordingly, ethnographic writing, interpretivism, and perspectives in urban anthropology must be replaced by experimentation, theorizing, causation, prediction, mathematical modeling, and epistemology. Participant observation is an ethnographic methodology and one of the most popularized and important techniques developed by Malinowski (1961) in anthropological research that can be applied to urban anthropology and its impact on urban socio-ecological resilience planning and development (DeWalt, DeWalt, & Wayland, 1998; Kawulich, 2005; Jorgensen, 2015).

It should be noted that unstructured techniques without a specific direction led to collect deeper and broader answers and the original terms to know the native categories and perspectives of urban anthropology in urban socio-ecological planning (Pizarro, 2014). Interviews through defined questions are recommended as an ethnographic research instrument at the lowest level of structuring to extract a greater amount of information that links urban anthropology and urban socio-ecological resilience planning. The qualitative-quantitative division of methods and techniques in urban anthropology and urban socio-ecological resilience planning and innovation is objectionable (Aldrich, 2014; Reynoso, 1995b).

A recent investigation that addresses quantitative and qualitative approaches to reveal the consequences of the COVID-19 pandemic crisis in the theoretical framework of urban resilience, society and institutional governance, was carried out by Olimid, Georgescu & Gherghe, (2022). Quantitative methods and techniques in urban anthropology and urban socio-ecological resilience planning and innovation aim to collect and analyze numerical data (Roni et al., 2020).

ORGANIZATIONAL AND URBAN ANTHROPOLOGY

In this section, an analysis is carried out from a sociohistorical perspective of the interrelationships between urban anthropology and labor organizations, on anthropological practices in organizations and urban environments that have challenges that affect research and academic training and development. Whereas a multidisciplinary perspective on urban anthropology and green and socio-ecological innovation attempts to understand traditional technology in the production process. It should be noted that work on entrepreneurship and business anthropology dates to the study of ethnology in urban societies in the 1970s (Durrenberger, 2007). On the other hand, Warner's work ratified Durkheim's analysis from a sociohistorical perspective has an influence on anthropology and sociology related to the social structures of complex tribal and industrialized societies, advancing the analysis of the interrelationships between urban anthropology and organizations.

Industrial technology and social research acquire relevance and, consequently, can improve the understanding of technology in the anthropology of organizations and in urban socio-ecological planning and innovation (Roca, 2008). Organizational anthropology has close relationships with urban anthropology, and this includes demands for participatory actions in practical anthropology of urban imaginaries and organizational facilitation (Gravano, 1992). This situation integrates the organizational and urban anthropological perspectives, as well as the ethnological methodological tools to the analysis of organizational studies, transversal to the demands for research funding.

The social sciences can make possible scientific predictions about specific events and problems in urban socio-ecological planning and development, despite the arbitrariness of human behavior and the assumption that it is not useful to predict behavior (Nießen et al., 2020). In this same sense, the debates in the analysis of organizational and urban anthropology are building bridges between the studies of organizations and urban areas from an anthropological perspective, which seeks to analyze the interrelationships between urban anthropology and business organizations. and labor, from a socio-historical orientation to develop reflections that situate the analysis of urban areas and neighborhoods in post-industrial and global societies. Thus, urban anthropology can be analyzed from a sociohistorical orientation through its interrelationships with organizations and companies.

It should be noted that organizational and urban anthropology has conceptual, methodological and theoretical tools aimed at developing the interrelationships between organizations and urban settlements, which lead to creating the meanings that both organization workers and citizens have of their heritage. industrial. Regarding the above, the studies of Mayo (1945) and his collaborators Roethlisberger and Dickson (1939) in the Hawthorne Wester Electric Co. industrial plant stand out. They were pioneers in organizational anthropology research applied to work productivity. The design of jobs in settlements and urban labor organizations is characterized by the development of multiple skills based on the redundancy of a variety of functions rather than the redundancy of parts that make those of the job capable of performing.

The approach to the professional practice of organizational facilitation and participatory processes presents theoretical-methodological challenges in metropolitan urban environmental socio-ecological planning (Gravano, 2007b) supported by the work of the organizational culture of the organizational and economic agents and of the political actors. In this regard, the Manchester and Chicago schools have worked on urban anthropology (Cucó Giner, 2004; Gravano, 2005; Hannerz, 1986; Signorelli, 1999; Velho, 1978). Therefore, organizational performance reflects the problems and dilemmas of organizational practices linked to the state and urban anthropology in the socio-ecological planning of control and weighting of the anthropological ethnography method used in facilitating organization in professional practice.

Urban anthropology in socio-ecological planning is a process of negotiation and mobilization that leads to the application of the definition of incremental action (de Melo, 2002a, b). The urban neighborhood, as organizational anthropology in organizational culture, controls the symbolic culture that is manipulated as an ideology aimed at controlling urban anthropology or organizational life conditions and the interrelationships of social subjects.

In another order of ideas, the socio-technical approach has demonstrated its effectiveness in some studies carried out in coal mines and in a project in which operational units of Royal Dutch Shell in England participated, in a textile factory in India, with the participation of unions, employers and the government involved in improving the quality of working life in Norway and Sweden, and other socio-technical projects in Canada, the Netherlands and Australia. The execution of urban socio-technical projects is carried out in a continuous process of urban renewal to reach fullness because of revisions and modifications.

The urban planning of urban socio-technical and socio-ecological innovation in urban regions known as local industrial systems can be an example of an economy derived from the industrial location of self-reinforcing agglomerations of accumulated socio-technical skills, investments and venture capital, infrastructures, suppliers, spillovers of knowledge, services, etc. associated with informal information flows and proximity to research centers and universities (Saxenian, 1994). Likewise, urban techno-scientific models of relationality, such as radioecology, make it possible to visualize cross-border movements in urban areas. The socio-technical approach is based on semi-autonomous groups with roots in self-regulated projects and redundancy of functions (Trist, 1976; de Melo, 1997).

The socio-technical support or urban systems based on promotions and rewards, promotions, must be consistent with the activities and be aligned with the objectives of the institution. Communication across the urban metropolitan boundaries of the organization's departments links the levels of control and supervision. In accordance with the above, it is considered that radioecology offers something more than limited human relationships and delimits an eco-semiotics of ineffable interrelation of living, decomposing and dying things. The application of eco-semiotics considers that humans and other species co-constitute ecological phenomena (Maran & Kull 2014). Radioecology uses specific radioisotopes to investigate the tracking and mapping connections of radiological contaminants and their transboundary movements between the earth and the atmosphere (Odum, 1959).

The resulting production of cumulative nuclear waste that brings urban socio-ecological resilience to collective recovery. The socio-technical urban criterion fosters the autonomy and initiative of employees. However, the variation must be controlled in the sociotechnical urban criteria as the source for each person responsible for controlling job learning and failures. Urban movements reveal the state's limits to post-disaster urban socio-ecological resilience.

URBAN SOCIO-ECOLOGY RESILIENCE PLANNING AND URBAN GREEN INNOVATION AREAS

The theory of urbanism for green and socio-ecological innovation continued to be marginal during the 1970s until, recently, innovation acquired a multidisciplinary character, encompassing knowledge in fields such as sociology, economics, psychology, administration, geography, engineering etc. (Sundbo, 1998; Schlapfer & Marinova, 2001). Throughout the 1980s and 1990s, the economy experienced a socio-ecological restructuring of technological innovation and a process of economic internationalization and market globalization that led to the substitution of the scientific and technological system to focus on the competitiveness of the industrial sector. oriented to satisfy international markets (González Alvarez, 1997).

The urban planning of areas of socio-ecological and green innovation is the process that begins with the invention of a new element until its development for commercial use. The commercialization of green socio-ecological urban planning innovations generated in public universities and research institutions requires clear regulations to share and transfer intellectual property rights to private companies, new forms of partnership between public-private sectors for development technological. Green innovation must be practical, have commercial value and be socially acceptable (Sundbo, 1998).

The sources of knowledge for the planning of urban green areas are rooted in institutional and human forms subject to financial capital and less to geographic movement (Schlapfer & Marinova 2001). On the other hand, the conceptual framework of the institutional analysis model is proposed by René Lourau (1979) and Lapassade (1986) that can be applied to analyze urban anthropology in the Anthropocene of socio-ecological urban planning and green innovation areas.

The ethnomethodological urban anthropology perspective of urban socio-ecological resilience planning and green innovation areas points to the inherent transformation of the sociohistorical process that affects non-contemplative knowledge production that claims meaning and challenges deciphering without typologies. in an ahistorical way. Regarding the above, urban socio-ecological planning processes occur with institutional and community participation, addressing the circularity of management issues and exchanging institutional power relations based on the anthropological object of cultural otherness (Gravano, 2006).

However, anthropological contributions to urban socio-ecological and environmental planning are based on community participation in institutional and organizational processes. Individual and business benefits prevail over the vocations and values of the local socio-ecological communities, which leads to the sacrifice of the common good. The community participatory process imposed by law delimits the specific possibilities, difficulties and opportunities based on the asymmetric rationalities at stake and state action in the context or urban environmental socio-ecological planning. A specific situation of demand for anthropological contribution by the State facilitates the community participatory process. A well-structured State can develop a local system of green innovation and urban socio-ecological resilience to support the activities of companies, institutions, universities and research centers, the economic and private sector, etc.

Likewise, the process of urban planning and green innovation and green technology transfer begins with an idea that research turns into a technology that will be used and applied, followed by dissemination and evaluation of the potential economic value, as well as discussion and consultation with specialists in the urban areas concerned with the discovery, considering criteria based on scientific and technical relevance, existing patents, and market potential. From there, the registration of the patent and intellectual property follows. Therefore, the evolution of scientific knowledge and socio-ecological and technological green innovation in urban planning is driven by the challenges posed by the design of new goods, services, and processes by carrying out internal research and development activities in the companies and in cooperation with other agents, in efforts to improve the efficiency of regions with deficiencies in their scientific, technological and innovation systems.

However, companies that intend to innovate in socio-ecological urban planning face financial obstacles. Fiscal and market incentives leave the decision on which urban development projects to finance in the hands of companies (de Melo, 2001). Therefore, to stimulate investments in green science, green technology and socio-ecological innovation systems for urban planning, the necessary conditions are required to invest in research and development (de Melo, 2001). Unproven green technological innovation has high risks and difficulties in assessing the market for green high-tech socio-ecological goods and services, and green technology-based companies have difficulties in accessing sources of capital. Investment activities in innovation activities have a high level of risk and uncertainty in investments in credit financing and equity financing in research and development of urban green planning innovation in technologies.

Currently, universities offer technical training and development in urban planning and socio-ecological innovation. In addition, university-business interactions occur between socio-ecological and environmental urban planning research institutes and centers with companies interested in joint research programs sponsored by consortia or individual government agencies and corporations to exploit the results through patents and licenses. Thus, through business incubators, universities and research institutions, technology parks and agents of science, green technology and socio-ecological innovation in local university-based green innovation systems, early-stage start-ups may have the opportunity to access government venture capital.

In accordance with the above, government funding of research in green technological innovation for urban planning carried out by companies has a positive effect on business spending on research and development (Guellec & van Pottelsbergue, 2000, in Melo, 2001). Government authorities facilitate the formation of local green innovation and urban socio-ecological resilience systems from universities and research centers by increasing public investment in science foundations, green technology and socio-ecological innovation. Government intervention in university-oriented local urban green innovation can regulate incentives to promote green technology innovation training and strengthen green technology-based start-ups. The technological training of companies reinforces the convergence of scientific research on green innovation in universities.

Companies can share training in green technological innovation, fostered by public green investment. Technological and social capital determines the level of socio-ecological and urban anthropology in companies' green innovation. The accumulation of social, economic, and symbolic capital is relevant to undertake planning activities for urban socio-ecological resilience and green innovation (Reguera, 2009). On the other hand, local urban innovation systems for green socio-ecological areas integrate the definitions considered in Lundvall (1992) and the one used by MIT (Lester, 2005). Local innovation systems in urban planning have some elements that interact with the creation, diffusion, transfer and use of knowledge for economic value in specific regions (Lundvall, 1992). The context of the transfer of urban green technological innovation from the University to companies is a factor of competitiveness of the productive sectors of society.

Thus, government agencies and institutions of research centers and universities make massive investments in infrastructure and equipment for urban planning and green innovation (Porter, 2001). Creation of technical schools and incentive programs for the application of urban planning policies and innovative green practices in the industry. Therefore, urban socio-ecological resilience planning and green innovation networks constitute a related economy of connections, which can be effective in the future of work by fueling debates of ideas, thoughts, and feelings to discover the point of view. of the natives in the local perspective to obtain, analyze and interpret the data.

Some actors such as research laboratories, the network of students and researchers, incubator companies, angel investors and venture capital, etc., participate in urban environmental planning processes, although those who are not involved in the governance of local urban planning and socio-ecological innovation systems are not necessarily less influential. However, non-financial government institutions are hardly available to complement the green investment of private venture capitalists in green companies. Declining government involvement is a principle that applies to mature local urban green innovation systems (OECD, 1997b; de Melo, 2001).

Anchoring organizations and companies focus on urban planning of university-oriented local innovation systems to cooperate in joint efforts of R&D projects directing the focus of research cooperation towards socio-ecological and green areas with which they can gain a competitive advantage with a competent university. On the other hand, the group of green companies participates in an articulating task to support identification, selection and consolidation mechanisms of urban green innovation projects through articulating instruments (Pimienta-Bueno, 1999). The local urban socio-ecological concentration of green companies allows economies of scale, agglomeration and gains from externalities (Audretsch, 1998).

Green urban innovation technologies in technology-intensive emerging green companies can be compared to mature companies in terms of achievements and results. Green technology-based start-ups have higher costs related to their seeding and start-up phases, which are higher than those of small firms (OECD, 1997b).

DISCUSSION

The intersection between expanded urbanization and multilevel analysis in urban socio-ecology and resilience is a promising line of research (Connolly et al., 2021; Wolf, 2016). More recently, the theory and methods of multi-criteria decision analysis (MCDA) assess urban sustainability and socio-ecological resilience complemented with AHP and Electre III meta-analysis techniques forming a hierarchy of problems to make decisions that satisfy the criteria and indicators of urban communities in cities that suffer frequent natural disasters, for example (Kosova, Qendraj & Xhafaj, 2022).

The impact of specific risk disasters in urban areas, such as earthquakes, hurricanes, floods, public health emergencies, etc., leading to urban socio-ecological planning resilience to prevent the possible impact of these extreme events and provide a valuable diagnosis of the weak links in urban resilience and improve the resistance and resilience of the city of the future sustainable development of human society. Regarding the above, the case analyzed by Mu, Fang, Yang, & Guo, 2022 stands out) based on population mobility, centrality, and nodality between cities in China, which was significantly reduced, affecting the urban resilience within the network during the COVID-19 epidemic. Other natural phenomena are also causing urban ecological and resilience dysfunctions, such as strong storms that inundate urban communities, immobilize them, and disrupt urban transport, energy, water, and waste networks.

Cities and any urban areas are practically saturated with fossil fuels controlled to a large extent by corporations. The interactions of stakeholders and actors contribute in disarray to identify the complexity inherent in urban socio-ecological meta-problems and contribute to intensify the complexity that the intertwining process requires to address.

It should be noted that the objective of social ecosystems in urban areas is to maximize the benefits of human needs in the Anthropocene era. The concept of the Anthropocene is linked to the socio-ecological aspects of climate change and to the analysis of the causes of the collapse of civilization and of societies that can learn from archaeological research when dealing with crises (Tainter, 1988; Tainter, 2008; McIntosh, et al., 2000; Trawick & Alf, 2015). Western science as a mode of existence is a source of prescriptions for climate change complemented by interpretation and translation (Latour, 2013; Finucane, 2009). Communities merge global scientific explanations with regional and local knowledge that blame humans for global warming and climate change (Schnegg et al., 2021).

Likewise, the approach of the global framework for climate services has provoked critical reflections on initiatives related to ethical implications, epistemic politics, colonial legacies, etc. (Haines, 2019b; Krauß, 2020; Nost, 2019; Webber, 2019). International political frameworks underpinned discourses combined with the adverse effects of warming (Baer & Singer, 2014). Consistent with anthropogenic climate change perspectives, humans may be responsible for climate. People make sense of and respond to changing patterns of global warming and climate change (Barnes et al., 2013; Fiske et al., 2014; Peterson & Broad, 2009). The global transformation has at the center of the concerns that Anthropocene in climate change.

A sociological approach refers to unpredictability in Bauman and Beck's work research and social acceleration by Hartmut Rosa (Beck, 2009; Hartmut Resonanz, 2016; W. S. et al., 2007). The evolution of human activities has given rise to temporary urban landscapes perceived as natural but shaped by the evolution of generations (Meeus, 1995; Kuna and Dreslerová, 2007). The transformations of urban landscapes by human groups cause anthropogenic impacts on the local socio-ecological balance. The donut economy model proposes knowledge and practices for the resilience, sustainability and well-being of urban socio-ecology that should be applied everywhere (Strang, 2017).

Now, urban resilience is more tangible than invisible greenhouse gases, with serious implications. The isolation of homes is associated with investment and frugality, while organic food with risk and pastoral ideals. (Press and Arnould, 2014; Wilk and Wilhite, 1985). Furthermore, taking up the ability of human beings to make decisions to choose nutritious and healthy foods that increase the chances of survival (Birch et al., 1998; Cashdan, 1998; Wright, 1991).

At the same time, pleasure, holistic health, and purity are values of purchasing behavior in some European countries (Ditlevsen et al., 2019). Where, consumption is relational embedded in social relationships. Consumers of traditional marketing goods and services are alienated (Lambert, 2019). In accordance with the above, Marx's criticisms of consumer capitalism led to a naturalistic epistemology (Kramer, 2021; Stoner, 2021; Trujillo, 2021). In line with the above, social exchange theory is a dyadic approach to marketing that is extended to incorporate philanthropic actors and multiple stakeholders. Marketing theory and practices replace the consumer fetish with value creation by a collective of stakeholders (Arnould & Cayla, 2015; Tsing, 2015; Atanasova, 2021; Casey et al. 2020), Giesler, 2006, 2008; Figueiredo & Scaraboto, 2016).

Without replacing the concept of consumer, consumers can share, gift, barter, and other semiotic consumption processes (Belk, 2014; Wei et al., 2021). Consumption is socially learned and individually determined during primary socialization. The urban anthropology of consumption is the study, analysis and interpretation of consumption as a social phenomenon that incorporates the acquisition, consumption and above all the appropriation of products and brands that serve as a reinforcer of a person's identity (Páramo & Ramírez, 2017).

CONCLUSION

The analysis sought to find some implications and relationships between urban anthropology and the resilience of urban socio-ecological planning in the Anthropocene, based on the assumption that urban anthropology supports the creation and development of any urban planning based on socio-ecological resilience. urban Anthropocene. This analysis concludes that the foundations of urban anthropology serve as support to reinforce the anthropocenic orientation of the resilience of urban socio-ecological planning.

Urban anthropological research is positioned between the limits of global scope in connections and assemblies with urban anthropological socio-ecology and the areas of resilience and green innovation. Anthropoid dominance has resulted in the scarcity of the planet's resources that may be a more acute problem than climate change. The institutional support of urban anthropology to local urban planning and the socio-ecological innovation system is a relevant factor to achieve successful results. Urban planning and eco-innovation institutions are characterized by their ability to create and develop social capital to facilitate relationships of cooperation and conflict.

There are various levels of urban anthropology applied to urban green planning and socio-ecological innovation. The urban green innovation technology park provides green companies that seek the benefits of agglomeration. The Emergence of Innovative Green and Socio-Ecological Enterprises explores opportunities in urban innovation from the main sources and drivers that use the wealth of green knowledge. The generation of scientific green knowledge and technological socio-ecological innovation, extension and consulting activities should be structured through various mechanisms for planning and the formation of cooperative urban communities structured by interactions of deliberative processes to stimulate entrepreneurship and the innovation.

On the other hand, urban anthropology in the local dimension of the university-based socio-ecological urban planning and green innovation system must be supported by financial subsidies to guide the efforts of the governance structure. Another mechanism of urban socio-ecological planning and innovation is the indirect support of green infrastructures and green asset constitutions used by flow companies and a set of projects. Green infrastructure, socio-ecological physical facilities and green technologies in urban planning and innovation centers are green assets that can be financed. The foregoing is an effort that requires shared objectives and collaboration supported by a governance structure capable of synchronizing the mission and objectives of the agents and actors from the perspective of urban anthropology analysis to urban planning and urban centers and parks. socio-ecological innovation.

Deliberative actions on innovative urban green areas are triggered by normative and deliberative actions conducted by agents and actors. The stakeholders involved focused on the urban green innovation technology park which should have autonomy, flexibility and should avoid erosions caused by less naturally occurring regulatory issues. Similarly, research on the resilience mechanisms of socio-ecological urban planning based on cross-cutting linkages should be strengthened.

Finally, the reductionism of the works that address issues related to urban anthropology and urban socio-ecology of the Anthropocene are recognized as limitations of the present work, which present a diffuse causality about planning and resilience, therefore, For future research, an analysis of a greater number of case studies is established as recommendations, which allow a detailed understanding of urban anthropology as a support to create and develop urban planning based on the socio-ecological resilience of the Anthropocene. Regarding the above, the comprehensive analysis of these constructs is recognized as a contribution of this study.

REFERENCES

Aldrich, H. (2014). *Stand up and be counted: Why social science should stop using the qualitative/quantitative dichotomy*. LSE. https://blogs.lse.ac.uk/impactofsocialsciences/2014/11/28/stand-up-and-becounted-social-science-qualitative-quantitative-dichotom

Alvarez, R. (2016). *Computational social science*. Cambridge University Press. doi:10.1017/CBO9781316257340

Arnould, E. J., & Cayla, J. (2015). Consumer fetish: Commercial ethnography and the sovereign consumer. *Organization Studies*, *36*(October), 1361–1386. doi:10.1177/0170840615580012

Atanasova, A. (2021). Re-examining utopia in contemporary consumption: Conceptualization and implications for marketing. *AMS Review*, *11*(1-2), 23–29. doi:10.100713162-021-00193-0

Audretsch, D. B. (1998). *Agglomeration and the Location of Innovative Activity.*

Baer, H., & Singer, M. (2014). *The anthropology of climate change: An integrated critical perspective*. Routledge. doi:10.4324/9781315818702

Baldassarri, D., & Abascal, M. (2017). Field experiments across the social sciences. *Annual Review of Sociology*, *43*(1), 41–73. doi:10.1146/annurev-soc-073014-112445

Barnes, J., Dove, M., Lahsen, M., Mathews, A., McElwee, P., McIntosh, R., Moore, F., O'Reilly, J., Orlove, B., Puri, R., Weiss, H., & Yager, K. (2013). Contribution of anthropology to the study of climate change. *Nature Climate Change*, *3*(6), 541–544. doi:10.1038/nclimate1775

Barrett, S. (2009). *Anthropology*. University of Toronto Press. doi:10.3138/9781442697362

Beck, U. (2009). *World at Risk*. Polity.

Belk, R. W. (2014). You are what you can access: Sharing and collaborative consumption online. *Journal of Business Research*, *67*(August), 1595–1600. doi:10.1016/j.jbusres.2013.10.001

Bernard, H. (2011). *Research methods in anthropology: Qualitative and quantitative approaches*. AltaMira Press.

Bilharz, J. A. (1998). *The Allegany Senecas and Kinzua Dam: Forced Relocation through Two Generations*. University of Nebraska Press.

Birch, L. L., Gunder, L., Grimm-Thomas, K., & Laing, D. G. (1998). Infants' consumption of a new food enhances acceptance of similar foods. *Appetite*, *30*(3), 283–295. doi:10.1006/appe.1997.0146 PMID:9632459

Brady, H. (2019). The challenge of big data and data science. *Annual Review of Political Science*, *22*(1), 297–323. doi:10.1146/annurev-polisci-090216-023229

Bunge, M. (1999). *Buscar la filosofía en las ciencias sociales*. Siglo XXI.

Casey, K., Lichrou, M., & O'Malley, L. (2020). Prefiguring sustainable living: An ecovillage story. *Journal of Marketing Management*, *36*(17–18), 1658–1679. doi:10.1080/0267257X.2020.1801800

Cashdan, E. (1998). Adaptiveness of food learning and food aversions children. *Social Sciences Information. Information Sur les Sciences Sociales, 37*(4), 613–632. doi:10.1177/053901898037004003

Connolly, C., Keil, R., & Ali, S. H. (2021). Extended urbanisation and the spatialities of infectious disease: Demographic change, infrastructure and governance. *Urban Studies (Edinburgh, Scotland), 58*(2), 245–263. doi:10.1177/0042098020910873

de Melo, A. (2001). *The Innovation Systems of Latin America and the Caribbean.* Inter-American Development Bank. Agosto.

de Melo, M. A. C. (1997). *Processo de Planejamento e as Inovações Tecnológicas e Sociais: uma perspectiva sócio-ecológica.* Anais do 5o. Seminário de Modernização Tecnológica.

de Melo, M. A. C. (2002a). *Enriquecendo a Atuação de Incubadora de Emrpesas. Tecnologia e Inovação: experiências de gestão nas micro e pequenas empresas.* PGT/USP.

de Melo, M.A.C. (2002b). Inovação e Modernização Tecnológica e Organizacional nas MPMEs: o domínio interorganizacional. *Seminário Internacional: Políticas*

DeWalt, K., DeWalt, B., & Wayland, C. (1998). Participant observation. In H. R. Bernard (Ed.), *Handbook of methods in cultural anthropology* (pp. 259–299). Altamira.

Diamond, J., & Robinson, J. (2010). *Natural experiments of history.* The Belknap Press of Harvard University Press.

Díaz, D., Kristoff, J., Castro, M., Miceli, J., Castro, D., Quinteros, R., & Guerrero, S. (2007). *Exploraciones en antropología y complejidad.* Editorial Sb.

Ditlevsen, K., Sandøe, P., & Lassen, J. (2019). Healthy food is nutritious, but organic food is healthy because it is pure: The negotiation of healthy food choices by Danish consumers of organic food. *Food Quality and Preference, 71*(January), 46–53. doi:10.1016/j.foodqual.2018.06.001

Dunning, T. (2012). *Natural experiments in the social sciences.* Cambridge University Press. doi:10.1017/CBO9781139084444

Durrenberger, P. (2007) The Anthropology of Organized Labor in the United States. *Ann Review of Anthropology, 36*, 73-88. https://www.annualreviews.org/doi/pdf/10.1146/annurev.anthro.36.081406.094433

Egerton, F. N. (2001). A history of the ecological sciences: Early Greek origins. *Bulletin of the Ecological Society of America, 82*(1), 93–97.

Figueiredo, B., & Scaraboto, D. (2016). The systemic creation of value through circulation in collaborative consumer networks. *The Journal of Consumer Research, 43*(4), 509–533. doi:10.1093/jcr/ucw038

Finucane, M. L. (2009). Why science alone won't solve the climate crisis: Managing climate risks in the Pacific. *Asia Pacific Issues, 89*, 1–8.

Fiske, S. J., Crate, S. A., Crumley, C., Galvin, K., Lazrus, H., Lucero, L., Oliver-Smith, A., Orlove, B., Strauss, S., & Wilk, R. R. (2014). *Changing the atmosphere. Anthropology and climate change. Final report of the AAA global climate change task force*. American Anthropological Association.

Giesler, M. (2008). Conflict and compromise: Drama in marketplace evolution. *Journal of Consumer Research*, 34(April), Giesler, M. (2006). Consumer gift systems. *The Journal of Consumer Research, 33*(September), 283–290. doi:10.1086/506309

Giner, C. (2004). *J. Antropología urbana*. Barcelona. *Ariel*.

González Álvarez, M. D. (1997). Processos de Planejamento nos Pólos Tecnológicos: um enfoque adaptativo. [Tese de Doutorado, Departamento de Engenharia Industrial, PUC-Rio, Rio de Janeiro].

Gravano, A. (1992). Antropología Práctica, muestra y posibilidades de la Antropología Organizacional. *Publicar en Antropología y Ciencias Sociales, Buenos Aires*, (1), 95–126.

Gravano, A. (2005). *El barrio en la teoría social*. Espacio Editorial.

Gravano, A. (2006). Imaginarios regionales y circularidad en la planificación: El caso del TOAR. *Intersecciones*, (7), 305–323.

Gravano, A. (2007b). Claves para la facilitación organizacional del proceso participativo en la planificación urbano-ambiental metropolitana. *Revista Regional de Trabajo Social*. Montevideo.

Grimmer, J., Roberts, M., & Stewart, B. (2021). Machine learning for social science: An agnostic approach. *Annual Review of Political Science, 24*(1), 395–419. doi:10.1146/annurev-polisci-053119-015921

Haines, S. (2019b). Managing expectations: Articulating expertise in climate services for agriculture in Belize. *Climatic Change, 157*(1), 43–59. doi:10.100710584-018-2357-1

Hannerz, U. (1986). *Exploración de la ciudad*. Fondo de Cultura Económica.

Hannerz, U. (2004). *Foreign New: Exploring the World of Foreign Correspondents*. University of Chicago Press.

Hartmut Resonanz, R. (2016). *Eine Soziologie der Weltbeziehung*. Suhrkamp.

Hindman, M. (2015). Building better models: Prediction, replication, and machine learning in the social sciences. *The Annals of the American Academy of Political and Social Science*, *659*(1), 48–62. doi:10.1177/0002716215570279

Hirst, K. (2019). *Is anthropology a science?* ThoughtCo. https://www.thoughtco.com/isanthropology-a-science-3971060

Hofman, J., Sharma, A., & Watts, D. (2017). Prediction and explanation in social systems. *Science*, *355*(6324), 486–488. doi:10.1126cience.aal3856 PMID:28154051

Hofman, J., Watts, D., Athey, S., Garip, F., Griffiths, T., Kleinberg, J., Margetts, H., Mullainathan, S., Salganik, M., Vazire, S., Vespignani, A., & Yarkoni, T. (2021). Integrating explanation and prediction in computational social science. *Nature*, *595*(7866), 181–188. doi:10.103841586-021-03659-0 PMID:34194044

Holling, C. S. (1973). *Resilience and Stability of Ecological Systems. Annual Review of Ecology and Systematics* (Vol. 4). JSTOR. https://www.jstor.org/stable/2096802.

Holmes, D. R., & Marcus, G. E. (2006). Fast capitalism: para-ethnography and the rise of the symbolic analyst. In M. Fisher & G. Downey (Eds.), *Frontiers of Capital: Ethnographic Reflections on the New Economy*. Duke University Press.

Holmes, D. R., & Marcus, G. E. (2021). How do we collaborate? An updated manifesto. In D. Boyer & G. Marcus (Eds.), *Collaborative Anthropology Today: A Collection of Exceptions*. Cornell University Press. doi:10.7591/cornell/9781501753343.003.0002

Jorgensen, D. (2015). Participant observation. In R. Scott & S. Kosslyn (Eds.), *Emerging trends in the social and behavioral sciences* (pp. 1–15). Wiley. doi:10.1002/9781118900772.etrds0247

Kaplan, O. (1940). Prediction in the social sciences. *Philosophy of Science*, *7*(4), 492–498. doi:10.1086/286658

Kawulich, B. (2005). La observación participante como método de recolección de datos. *Forum Qualitative Social Research*, *68*(2), 43.

Kolb, Ch. (1998). Review of Bilharz, Joy A., *The Allegany Senecas and Kinzua Dam: Forced Relocation through Two Generations*. H-AmIndian, H-Net Reviews. https://www.h-net.org/reviews/showrev.php?id=2366

Kosova, R., Qendraj, D.H., Xhafaj, E. (2022). Meta-Analysis ELECTRE III and AHP in Evaluating and Ranking the Urban Resilience. *Journal of Environmental Management and Tourism*, *3*(59), 756 - 768. doi:10.14505/jemt.v13.3(59).15

Kramer, J. M. (2021). From predator to parasite: On private property and our ecological disaster. *Journal of Economic Issues*, *55*(June), 416–422. doi:10.1080/00213624.2021.1908804

Krauß, W. (2020). Narratives of change and the co-development of climate services for action. *Climate Risk Management*, *28*, 100217. doi:10.1016/j.crm.2020.100217

Kuhn, T. S. (1962/1996). *The Structure of Scientific Revolutions* (3rd ed.). University of Chicago Press.

Kuna, M., & Dreslerová, D. (2007). Landscape archaeology and "community areas" in the archaeology of central Europe. D. Hicks, L. McAtackney, G.J. Fairclough (Eds.), Envisioning landscape: Situations and standpoints in archaeology and heritage. Left Coast Press.

Lambert, A. (2019). Psychotic, acritical and precarious? A Lacanian exploration of the neoliberal consumer subject. *Marketing Theory*, *19*(3), 329–346. doi:10.1177/1470593118796704

Lapassade, G. (1986). *Grupos, organizaciones e instituciones: la transformación de la burocracia*. Gedisa.

Latour, B. (2013). *An inquiry into modes of existence: An anthropology of the moderns* (C. Porter, Trans.). Harvard University Press.

Lester, R. K. (2005). *Universities, Innovation, and the Competitiveness of Local Economies A Summary Report from the Local Innovation Systems Project – Phase I.* Industrial Performance Center Massachusetts Institute of Technology.

Lindblom, C. E. (1959). The Science of Muddling through. *Public Administration Review*, *19*(2), 79–88. doi:10.2307/973677

Llanos, M. (2009). *Epistemología de las ciencias sociales*. UNMSM.

Lourau, R. (1979). *El análisis institucional*. Amorrortu.

Lundvall, B.-Å. (1992). (ed.). National Systems of Innovation: towards a theory of innovation and interactive learning. Printer Publisher. Londres.

Malinowski, B. (1961). *Argonauts of the Western Pacific*. Routledge.

Maran, T., & Kull, K. (2014). Ecosemiotics: Main principles and current developments. *Geografiska Annaler. Series B, Human Geography, 96*(1), 41–50. doi:10.1111/geob.12035

Mayo, E. (1945). *The social problems of an Industrial Civilization.* Harvard University Press.

McIntosh, R. J., Tainter, J. A., & McIntosh, S. K. (Eds.). (2000). *The Way the Wind Blows: Climate Change, History, and Human Activity*. Columbia University Press.

Meeus, J. H. A. (1995). Pan-European landscapes. *Landscape and Urban Planning, 31*(1-3), 57–79. doi:10.1016/0169-2046(94)01036-8

Mu, X., Fang, C., Yang, Z., & Guo, X. (2022). Impact of the COVID-19 Epidemic on Population Mobility Networks in the Beijing–Tianjin–Hebei Urban Agglomeration from a Resilience Perspective. *Land (Basel), 2022*(11), 675. doi:10.3390/land11050675

Nießen, D., Danner, D., Spengler, M., & Lechner, C. (2020). Big Five personality traits predict successful transitions from school to vocational education and training: A large-scale study. *Frontiers in Psychology, 11*, 1827. doi:10.3389/fpsyg.2020.01827 PMID:32903700

Nost, E. (2019). Climate services for whom? The political economics of contextualizing climate data in Louisiana's coastal master plan. *Climatic Change, 157*(1), 27–42. doi:10.100710584-019-02383-z

Odum, E. P. (1957). *Fundamentals of ecology*. W. B. Saunders Company.

OECD. (1997b). Government Venture Capital for Technology-Based Firms. Ducument OCDE/GD (97). OECD.

Olimid, A. P., Georgescu, C. M., & Gherghe, C. L. (2022). Influences of Covid-19 Crisis on Resilience Theories: An analysis of Community, Societal and Governance Resilience. *Revista de Ştiinţe Politice. Revue des Sciences Politiques, 73*, 38–51.

Ortner, S. (2010). Access: Reflections on studying up in Hollywood. *Ethnography, 11*(2), 211–233. doi:10.1177/1466138110362006

Ortner, S. (2013). *Not Hollywood: Independent Film at the Twilight of the American Dream*. Duke University Press.

Páramo, D., & Ramírez, E. (2017). *Etnomarketing, la dimensión cultural del marketing*. Klasse Editorial.

Peterson, N., & Broad, K. (2009). Climate and weather discourse in anthropology: From determinism to uncertain futures. In *Anthropology and climate change: From encounters to actions* (pp. 70–86). Left Coast Press.

Pfeifer, J. (2006). Laws of nature. In S. Sarkar & J. Pfeifer (Eds.), *The philosophy of science: An encyclopedia* (pp. 439–444).

Pimienta-Bueno, J.A. (1999). *Parque de Inovação Tecnológica e Cultural da Gávea: A Visão da PUC-Rio.* Documento Interno, 23 págs, 1999.

Pizarro, C. (2014). La entrevista etnográfica como práctica discursiva: Análisis de caso sobre las pistas meta-discursivas y la emergencia de categorías nativas. *Revista de Antropologia, 57*(1), 461–496. doi:10.11606/2179-0892.ra.2014.87770

Porter, M. (2001). Clusters of Innovation: regional foundations of U.S. competitiveness. Council on Competitiveness, Washington, D.C: Monitor Group.

Press, M., & Arnould, E. J. (2014). Narrative transparency. *Journal of Marketing Management, 30*(13–14), 1353–1376. doi:10.1080/0267257X.2014.925958

Reguera, A. (2009). Empresarios de ayer, de hoy y de siempre. *Un recorrido latinoamericano por sus formas espacio-temporales América Latina en la historia económica.* Scielo. https://www.scielo.org.mx/scielo.php?script=sci_arttext&pid=S1405- 22532009000200003&lng=es&nrm=iso

Reynoso, C. (1995b). Hacia la perfección del consenso: Los lugares comunes de la antropología. *Intersecciones, 1*, 51–72.

Roca, A. (2008). *La comprensión de la tecnología en la antropología.* Actas VII ESOCITE UFRJ, Jornadas Latino-Americanas de Estudos Sociais das Ciências e das Tecnologías.

Roethlisberger, F., & Dickson, W. (1939). *Management and the Worker.* Harvard Univ. Press.

Roni, S., Merga, M., & Morris, J. (2020). *Conducting quantitative research in education.* Springer. doi:10.1007/978-981-13-9132-3

Saxenian, A. (1994). *Regional Advantage: culture and competition in Silicon Valley and Route 128.* Harvard Business Press.

Schlapfer, A., & Marinova, D. (2001). *Local Innovation Systems: nature, importance and role.* Conference Proceedings: International Summer Academy on Technological Studies: user involvement in technological innovation, Deutschlandsberg. Austria.

Schnegg, M., O'Brian, C. I., & Sievert, I. J. (2021). It's our fault: A global comparison of different ways of explaining climate change. *Human Ecology: an Interdisciplinary Journal*, *49*(3), 327–339. doi:10.100710745-021-00229-w

Signorelli, A. (1999). Antropología Urbana. Barcelona: Anthropos: UAM, 1999.

Stoner, A. M. (2021). Things are getting worse on our way to catastrophe: Neoliberal environmentalism, repressive desublimation, and the autonomous ecoconsumer. *Critical Sociology*, *47*(May), 491–506. doi:10.1177/0896920520958099

Strang, V. (2017). The Gaia complex: Ethical Challenges to an Anthropocentric 'Common Future. In B. Marc & L. Jerome (Eds.), *The Anthropology of Sustainability* (pp. 255–283). Palgrave. doi:10.1057/978-1-137-56636-2_12

Sundbo, J. (1998). *The Theory of Innovation: entrepreneurs, technology, and strategy*. Edward Elgar.

Taagepera, R. (2008). *Making social sciences more scientific: The need for predictive models*. Oxford University Press. doi:10.1093/acprof:oso/9780199534661.001.0001

Tainter, J. A. (1988). *The Collapse of Complex Societies*. Cambridge University Press.

Tainter, J. A. (2008). Collapse and Sustainability: Rome, the Maya, and the Modern World. *Archeological Papers of the American Anthropological Association*, *24*(1), 201–214. doi:10.1111/apaa.12038

Tarantino-Curseri, S. (2018). Antropología, etología, sociología, y psicología, 4 disciplinas útiles en el ámbito empresarial. *Management Review*, *3*(3). doi:10.18583/umr.v3i3.126

Trawick, P., & Alf, H. (2015). Revisiting the Image of Limited Good: On Sustainability, Thermodynamics, and the Illusion of Creating Wealth. *Current Anthropology*, *56*(1), 1–27. doi:10.1086/679593

Trist, E. L. (1976). *A Concept of Organizational Ecology: an invited address to the three Melbourne universities.*

Trujillo, M. L. N. (2021). Notes for a critical and ecological view of patriarchal capitalism in the web of life. *Capital and Class*, *45*(March), 21–32. doi:10.1177/0309816820929115

Tsing, A. L. (2015). *The mushroom at the end of the world*. Princeton University Press.

Velho, G. (1978). Observando o familiar. In: Nunez, E. de O. (Org.). An aventura sociológica: objetividade, paixao, improviso e método na pesquisa social. Rio de Janeiro: Zahar.

W. S. (2007). The Anthropocene: Are Human Beings Now Overwhelming the Forces of Nature? *Ambio*, *36*, 614–621. doi:10.1579/0044-7447(2007)36[614:TAAHNO]2.0.CO;2 PMID:18240674

Webber, S. (2019). Putting climate services in contexts: Advancing multi-disciplinary understandings: Introduction to the special issue. *Climatic Change*, *157*(1), 1–8. doi:10.100710584-019-02600-9

Wei, X., Lo, C. K. Y., Jung, S., & Choi, T.-M. (2021). From co-consumption to co-production: A systematic review and research synthesis of collaborative consumption practices. *Journal of Business Research*, *129*(May), 282–294. doi:10.1016/j.jbusres.2021.02.027

Wilk, R. R., & Wilhite, H. L. (1985). Why don't people weatherize their homes? An ethnographic solution. *Energy*, *10*(May), 621–629. doi:10.1016/0360-5442(85)90093-3

Wolf, M. (2016). Rethinking urban epidemiology: Natures, networks and materialities. *International Journal of Urban and Regional Research*, *40*(5), 958–982. doi:10.1111/1468-2427.12381 PMID:32336869

Yang, Y., Lili, W., & Hongchi, Z. (2022). *Bibliometric research on the evolution of resilience theme from the perspective of Geographical Science*. 2022 29th International Conference on Geoinformatics, Beijing, China. 10.1109/Geoinformatics57846.2022.9963870

Compilation of References

Abellán, P., Sánchez-Fernández, D., Velasco, J., & Millán, A. (2007). Effectiveness of protected area networks in representing freshwater biodiversity: The case of a Mediterranean river basin (south-eastern Spain). *Aquatic Conservation, 17*(4), 361–374. doi:10.1002/aqc.778

Abrol, D. P., Gorka, A. K., Ansari, M. J., Al-Ghamdi, A., & Al-Kahtani, S. (2019). Impact of insect pollinators on yield and fruit quality of strawberry. *Saudi Journal of Biological Sciences, 26*(3), 524–530. doi:10.1016/j.sjbs.2017.08.003 PMID:30899167

Agathokleous, E., Feng, Z., Oksanen, E., Sicard, P., Wang, Q., Saitanis, C. J., Araminiene, V., Blande, J. D., Hayes, F., Calatayud, V., Domingos, M., Veresoglou, S. D., Peñuelas, J., Wardle, D. A., De Marco, A., Li, Z., Harmens, H., Yuan, X., Vitale, M., & Paoletti, E. (2020). Ozone affects plant, insect, and soil microbial communities: A threat to terrestrial ecosystems and biodiversity. *Science Advances, 6*(33), eabc1176. doi:10.1126ciadv.abc1176 PMID:32851188

Aharon-Rotman, Y., Soloviev, M., Minton, C., Tomkovich, P., Hassell, C., & Klaassen, M. (2015). Loss of periodicity in breeding success of waders links to changes in lemming cycles in Arctic ecosystems. *Oikos, 124*(7), 861–870. doi:10.1111/oik.01730

Akgün, F., Kayseri, M. S., & Akkiraz, M. S. (2007). Palaeoclimatic evolution and vegetational changes during the Late Oligocene–Miocene period in Western and Central Anatolia (Turkey). *Palaeogeography, Palaeoclimatology, Palaeoecology, 253*(1–2), 56–90. doi:10.1016/j.palaeo.2007.03.034

Akıner, M. M., Beriş, F. Ş., Seyis, F., Öztürk, M., Sevgili, H., & Demir, E. (2019). Annual variation of the *Orosanga japonica* Melichar 1898 (Hemiptera: Ricaniidae) populations in the eastern Black Sea region of Turkey and possible molecular separation with based on 28S rDNA sequences from other Ricaniidae groups. *Bitki Koruma Bulteni, 59*(4), 11–19. doi:10.16955/bitkorb.546120

Akıner, M. M., Demirci, B., Babuadze, G., Robert, V., & Schaffner, F. (2016). Spread of the invasive mosquitoes *Aedes aegypti* and *Aedes albopictus* in the Black Sea Region increases risk of Chikungunya, Dengue, and Zika Outbreaks in Europe. *PLoS Neglected Tropical Diseases, 10*(4), e0004664. doi:10.1371/journal.pntd.0004664 PMID:27115737

Akıner, M. M., Demirci, B., Bedir, H., Doğan, A. F., Gökdemir, A., Topluoğlu, S., Altuğ, Ü., Kurtcebe, Z. Ö., & Irmak, H. (2018). Surveillance and control of invasive *Aedes* species in the eastern black sea area of Turkey. *Turk Hijiyen ve Deneysel Biyoloji Dergisi, 75*(3), 225–238. doi:10.5505/TurkHijyen.2018.68736

Ak, K., Uluca, M., Aydin, Ö., & Gokturk, T. (2019). Important invasive species and its pest status in Turkey: *Halyomorpha halys* (Stål) (Heteroptera: Pentatomidae). *Journal of Plant Diseases and Protection, 126*(5), 401–408. doi:10.100741348-019-00243-7

Akyıldırım, H. (2013). Evaluation of the zoogeographical contents of Turkey aphid (Hemiptera, Aphidoidea) fauna and invasive components. *Biyoloji Bilimleri Araştırma Dergisi, 6*(1), 44–48.

Aldrich, H. (2014). *Stand up and be counted: Why social science should stop using the qualitative/ quantitative dichotomy*. LSE. https://blogs.lse.ac.uk/impactofsocialsciences/2014/11/28/stand-up-and-becounted-social-science-qualitative-quantitative-dichotom

Alvarez, R. (2016). *Computational social science*. Cambridge University Press. doi:10.1017/ CBO9781316257340

Amano, T., Székely, T., Sandel, B., Nagy, S., Mundkur, T., Langendoen, T., Blanco, D., Soykan, C. U., & Sutherland, W. J. (2018). Successful conservation of global waterbird populations depends on effective governance. *Nature, 553*(7687), 199–202. doi:10.1038/nature25139 PMID:29258291

Amarsinghe. (2005). Spatial Variation in Water Supply and Demand across river basins in India, IWMI. *Research Reports (Montgomery), 83*, 2005. http://www.iwmi.cgiar.org/pubs/pub083/ RR83_distribution.pdf

Ambarlı, D., Zeydanlı, U. S., Balkız, Ö., Aslan, S., Karaçetin, E., Sözen, M., Ilgaz, Ç., Gürsoy Ergen, A., Lise, Y., Demirbaş Çağlayan, S., Welch, H. J., Welch, G., Turak, A. S., Bilgin, C. C., Özkil, A., & Vural, M. (2016). An overview of biodiversity and conservation status of steppes of the Anatolian Biogeographical Region. *Biodiversity and Conservation, 25*(12), 2491–2519. doi:10.100710531-016-1172-0

Ambrose, D. P. (2003). Biocontrol potential of assassin bugs (Hemiptera: Reduviidae). *Journal of Experimental Zoology India, 6*(1), 1–44.

Anderson, C. M., Fahrig, L., Rausch, J., Martin, J. L., Daufresne, T., & Smith, P. A. (2023). Climate-related range shifts in Arctic-breeding shorebirds. *Ecology and Evolution, 13*(2), e9797. doi:10.1002/ece3.9797 PMID:36778838

Anlaş, Sinan, & Kočárek, P. (2012). Current status of Dermaptera (Insecta) fauna of Turkey and Cyprus. *Turkiye Entomoloji Dergisi, 36*, 43–58.

Anlaş, S. (2020). A new micropterous species of the genus *Sunius* Stephens, 1829 from Central Anatolia (Coleoptera: Staphylinidae). *Zoology in the Middle East, 66*(4), 342–346. doi:10.108 0/09397140.2020.1826676

Anonymous. (2023). *Mağara Turizmi (Webpage of Republic of Turkey Ministry of Culture and Tourism)*. KTB. https://yigm.ktb.gov.tr/TR-10335/magara-turizmi.html

Ansell, S. W., Stenøien, H. K., Grundmann, M., Russell, S. J., Koch, M. A., Schneider, H., & Vogel, J. C. (2011). The importance of Anatolian mountains as the cradle of global diversity in *Arabis alpina*, a key arctic-alpine species. *Annals of Botany*, *108*(2), 241–252. doi:10.1093/aob/mcr134 PMID:21712298

Arı, İ. (2014). A checklist of Turkish Neuropteran species (Insecta: Neuroptera) with remarks on provincial distribution in Turkey. *Acta Entomologica Slovenicaa*, *22*(1), 59–86.

Arnould, E. J., & Cayla, J. (2015). Consumer fetish: Commercial ethnography and the sovereign consumer. *Organization Studies*, *36*(October), 1361–1386. doi:10.1177/0170840615580012

Arslangündoğdu, Z., & Hızal, E. (2010). The Western conifer seed bug, *Leptoglossus occidentalis* (Heidemann, 1910), recorded in Turkey (Heteroptera: Coreidae). *Zoology in the Middle East*, *50*(1), 138–139. doi:10.1080/09397140.2010.10638426

Assing, V., & Schillhammer, H. (2001). On theTurkish species of *Sunius* Curtis 1829 (Coleoptera : Staphylinidae, Paederinae). *Linzer Biologische Beitrage*, *33*, 195–210.

Atakan, E., & Pehlivan, S. (2020). A harmful *Thrips* species on Lemon in the Eastern Mediterranean Region of Turkey: *Thrips hawaiiensis* (Morgan) (Thysanoptera: Thripidae). *Kahramanmaraş Sütçü İmam Üniversitesi Tarım ve Doğa Dergisi*, *23*(1), 19–25. doi:10.18016/ksutarimdoga.vi.578082

Atalay, I. (2006). The effects of mountainous areas on biodiversity: a case study from the northern Anatolian Mountains and the Taurus Mountains. *Proceedings of the 8th International Symposium on High Mountain Remote Sensing Cartography,* (*vol. 41*, pp. 17–26). Uni Graz..

Atalay, I., Efe, R., & Öztürk, M. (2014). Effects of topography and climate on the ecology of taurus mountains in the mediterranean region of Turkey. *Procedia: Social and Behavioral Sciences*, *120*, 142–156. doi:10.1016/j.sbspro.2014.02.091

Atanasova, A. (2021). Re-examining utopia in contemporary consumption: Conceptualization and implications for marketing. *AMS Review*, *11*(1-2), 23–29. doi:10.100713162-021-00193-0

Audretsch, D. B. (1998). *Agglomeration and the Location of Innovative Activity.*

Avcı, M. (2005). Diversity and endemism in Turkey's vegetation. *Journal of Geography*, *13*, 27–55.

Avgin, S. S., & Cavazzuti, P. (2011). The studies made on Turkish Carabinae with checklist and bibliography (Coleoptera: Carabidae). *Turkish Journal of Zoology*, *35*, 403–432. doi:10.3906/zoo-0905-13

Avgin, S. S., & Prunier, D. (2015). Endemic *Carabus* species of Turkey: Some researches with biogeographic and ecological notes (Coleoptera, Carabidae, Carabinae). *Ekoloji Çevre Dergisi*, *24*, 26–40. doi:10.5053/ekoloji.2015.19

Avgın, S., & Özdikmen, H. (2007). Check list of the Tiger Beetles of Turkey with a review of distribution and biogeography Coleoptera Cicindelidae. *Munis Entomology & Zoology*, *2*, 87–102.

Aydin, G., & Şen, İ. (2020). Determination of arthropod biodiversity and some ecological parameters of Erdal Şekeroğlu (Isparta, Turkey) and Kadıini (Antalya, Turkey) cave ecosystems with evaluation of usability of insects in cave mapping. *Turkiye Entomoloji Dergisi*, *44*, 539–557. doi:10.16970/entoted.770018

Aysal, T., & Kıvan, M. (2014). Occurrence of an invasive alien species *Harmonia axyridis* (Pallas) (Coleoptera: Coccinellidae) in Turkey. *Türkiye Entomoloji Bülteni*, *4*(3), 141. doi:10.16969/teb.17261

Aytar, F., Kara, K., & Atay, T. (2020). Tachinid (Diptera: Tachinidae) parasitoids reared from lepidopterous and hymenopterous hosts in southern forests of Turkey. *Turkiye Entomoloji Dergisi*, *45*, 3–7. doi:10.16970/entoted.779465

Azmaz, M., & Katilmiş, Y. (2017). Updated species list of Cynipidae (Hymenoptera) from Turkey. *Zootaxa*, *4303*(3), 361–378. doi:10.11646/zootaxa.4303.3.3

Baer, H., & Singer, M. (2014). *The anthropology of climate change: An integrated critical perspective*. Routledge. doi:10.4324/9781315818702

Baldassarri, D., & Abascal, M. (2017). Field experiments across the social sciences. *Annual Review of Sociology*, *43*(1), 41–73. doi:10.1146/annurev-soc-073014-112445

Barkan, N. P., & Aytekin, A. M. (2013). Systematical studies on the species of the subgenus *Bombus* (*Thoracobombus*) (Hymenoptera: Apidae, *Bombus* Latreille) in Turkey. *Zootaxa*, *3737*(2), 167–183. doi:10.11646/zootaxa.3737.2.5 PMID:25112746

Barnes, J., Dove, M., Lahsen, M., Mathews, A., McElwee, P., McIntosh, R., Moore, F., O'Reilly, J., Orlove, B., Puri, R., Weiss, H., & Yager, K. (2013). Contribution of anthropology to the study of climate change. *Nature Climate Change*, *3*(6), 541–544. doi:10.1038/nclimate1775

Barrett, S. (2009). *Anthropology*. University of Toronto Press. doi:10.3138/9781442697362

Bastmeijer, J. D., & Wongso, S. (2016a). Indian Cryptocoryne. *The Crypts pages*. Www. cryptocoryneworld.org. https://www.cryptocoryneworld.org/india.php

Battiston, R., Amerini, R., Di Pietro, W., Guariento, L. A., Bolognin, L., & Moretto, E. (2020). A new alien mantis in Italy: Is the Indochina mantis *Hierodula patellifera* chasing the train for Europe? *Biodiversity Data Journal*, *8*, e50779. doi:10.3897/BDJ.8.e50779 PMID:32194321

Battley, P. F., Warnock, N., Tibbitts, T. L., Gill, R. E. Jr, Piersma, T., Hassell, C. J., Douglas, D. C., Mulcahy, D. M., Gartrell, B. D., Schuckard, R., Melville, D. S., & Riegen, A. C. (2012). Contrasting extreme long-distance migration patterns in bar-tailed godwits Limosa lapponica. *Journal of Avian Biology*, *43*(1), 21–32. doi:10.1111/j.1600-048X.2011.05473.x

Bayram, A., & Tonğa, A. (2016). First report of *Chilo partellus* in Turkey, a new invasive maize pest for Europe. *Journal of Applied Entomology*, *140*(3), 236–240. doi:10.1111/jen.12232

Beck, U. (2009). *World at Risk*. Polity.

Belgrano, A., Woodward, G., & Jacob, U. (2015). *Aquatic functional biodiversity: an ecological and evolutionary perspective*. Elsevier: Academic Press.

Belk, R. W. (2014). You are what you can access: Sharing and collaborative consumption online. *Journal of Business Research*, *67*(August), 1595–1600. doi:10.1016/j.jbusres.2013.10.001

Bennett, K. D. (1990). Milankovitch cycles and their effects on species in ecological and evolutionary time. *Paleobiology*, *16*(01), 11–21. doi:10.1017/S0094837300009684

Benyamini, D., Bruschini, C., Serra, G., & John, E. (2007). First confirmed records of *Papilio demoleus* (Linnaeus 1758) in Syria, wih comments on the species' appearance in Turkey and the Dominican Republic. *News of the Israeli Lepidopterists Society*, *24*(2), 4–11.

Berlatsky, N. (2014). *Biodiversity*. Greenhaven Press.

Bernard, H. (2011). *Research methods in anthropology: Qualitative and quantitative approaches*. AltaMira Press.

Beyarslan, A. (2014). Checklist of Braconinae species of Turkey (Hymenoptera: Braconidae). *Zootaxa*, *3790*(2), 201–242. doi:10.11646/zootaxa.3790.2.1 PMID:24869868

Bilgin, R. (2011). Back to the suture: The distribution of intraspecific genetic diversity in and around anatolia. *International Journal of Molecular Sciences*, *12*(6), 4080–4103. doi:10.3390/ijms12064080 PMID:21747726

Bilharz, J. A. (1998). *The Allegany Senecas and Kinzua Dam: Forced Relocation through Two Generations*. University of Nebraska Press.

Biodiversity hotspots. (n.d.). *Home - National Geographic Society*. Biodiversity. https://www.nationalgeographic.org/encyclopedia/biodiversity-hotspots/

Birch, L. L., Gunder, L., Grimm-Thomas, K., & Laing, D. G. (1998). Infants' consumption of a new food enhances acceptance of similar foods. *Appetite*, *30*(3), 283–295. doi:10.1006/appe.1997.0146 PMID:9632459

BirdLife International. (2021). *Species factsheets*. BirdLife International. http://www.birdlife.org.

Blaikie, P. M., & Muldavin, J. S. S. (2004). Upstream, downstream, China, India: The politics of environment in the Himalayan region. *Annals of the Association of American Geographers*, *94*(3), 520–548. doi:10.1111/j.1467-8306.2004.00412.x

Blüthgen, N., & Klein, A.-M. (2011). Functional complementarity and specialisation: The role of biodiversity in plant–pollinator interactions. *Basic and Applied Ecology*, *12*(4), 282–291. doi:10.1016/j.baae.2010.11.001

Bolton, B. (2021). *AntCat: An Online Catalog of the Ants of the World.*

Bond, W. J., Woodward, F. I., & Midgley, G. F. (2005). The global distribution of ecosystems in a world without fire. *The New Phytologist*, *165*(2), 525–537. doi:10.1111/j.1469-8137.2004.01252.x PMID:15720663

Boudot, J.-P., Kalkman, V. J., & Azpilicueta AMorin, M. (2009). Atlas of the Odonata of the Mediterranean and North Africa I IUCN Library System. Libellula Supplement, 9, 1–256.

BourgoinT. (2020). Fulgoromorpha Lists On the Web. *The Catalogue of Life Partnership*. doi:10.15468/57cc67

Boyd, H., & Piersma, T. (2001). Changing balance between survival and recruitment explains population trends in Red Knots Calidris canutus islandica wintering in Britain, 1969-1995. *Ardea*, *89*(2), 301–317.

Bozdoğan, H. (2020). Species richness and composition of Neuroptera in the forests fregments of the Taurus Mountains Range, Turkey. *Saudi Journal of Biological Sciences*, *27*(5), 1201–1207. doi:10.1016/j.sjbs.2020.02.018 PMID:32346325

Bozkurt, V. (2018). İstilacı böcek türlerinin mücadelesinin yönetimi: *Anoplophora chinensis* (Forster, 1771) (Coleoptera: Cerambycidae) örneği. *Doğal Afetler ve Çevre Dergisi*, *4*, 25–31. doi:10.21324/dacd.441390

Brady, H. (2019). The challenge of big data and data science. *Annual Review of Political Science*, *22*(1), 297–323. doi:10.1146/annurev-polisci-090216-023229

Briggs, C. J., Knapp, R. A., & Vredenburg, V. T. (2010). Enzootic and epizootic dynamics of the chytrid fungal pathogen of amphibians. *Proceedings of the National Academy of Sciences of the United States of America*, *107*(21), 9695–9700. doi:10.1073/pnas.0912886107 PMID:20457916

Brooks, T. M., Mittermeier, R. A., Mittermeier, C. G., da Fonseca, G. A. B., Rylands, A. B., Konstant, W. R., Flick, P., Pilgrim, J., Oldfield, S., Magin, G., & Hilton-Taylor, C. (2002). Habitat loss and extinction in the hotspots of biodiversity. *Conservation Biology*, *16*(4), 909–923. doi:10.1046/j.1523-1739.2002.00530.x

Brown, J., Scholtz, C. H., Janeau, J.-L., Grellier, S., & Podwojewski, P. (2010). Dung beetles (Coleoptera: Scarabaeidae) can improve soil hydrological properties. *Applied Soil Ecology*, *46*(1), 9–16. doi:10.1016/j.apsoil.2010.05.010

Bruijnzeel, L. A. (2004). Hydrological functions of tropical forests: Not seeing the soil for the trees? *Agriculture, Ecosystems & Environment*, *104*(1), 185–228. doi:10.1016/j.agee.2004.01.015

Bukejs, A., & Telnov, D. (2014). The first record of the invasive lady beetle *Harmonia axyridis* (Pallas, 1773) (Coleoptera: Coccinellidae) in Turkey. *Zoology and Ecology*, *1–4*, 1–4. Advance online publication. doi:10.1080/21658005.2014.984490

Bunge, M. (1999). *Buscar la filosofía en las ciencias sociales*. Siglo XXI.

Burger, J., Niles, L. J., Porter, R. R., Dey, A. D., Koch, S., & Gordon, C. (2012). Migration and over-wintering of red knots (Calidris canutus rufa) along the Atlantic coast of the United States. *The Condor*, *114*(2), 302–313. doi:10.1525/cond.2012.110077

Bush, G. L., & Butlin, R. K. (2004). Sympatric Speciation in Insects. I: U. Dieckmann, M. Doebeli, J. A. J. Metz, & D. Tautz (Eds.), Adaptive Speciation (pp. 229–248). Cambridge University Press.

Cai, R., Cao, L., Griffin, R., Chansai, S., Hardacre, C., & Palmer, R. E. (2020). Scale-up of cluster beam deposition to the gram scale with the matrix assembly cluster source for heterogeneous catalysis (propylene combustion). *AIP Advances*, *10*(2), 025314–025314. doi:10.1063/1.5142836

Çalışkan, H., Şirin, D. Ü., Aslan, A., Fidan, E. C., & Şahin, Y. (2020). Five New Records for Turkish Robber Flies (Diptera: Asilidae) Fauna. *Journal of Entomological Research Society*, *22*(3), 239–254.

Çalışkan, S. S. (2014). Review of the the Genus *Chrysoclista* Stainton, 1854 (Lepidoptera: Agonoxenidae) in Turkey with a New Species. *Journal of Entomological Research Society*, *16*(3), 67–73.

Cameron, S. A., Hines, H. M., & Williams, P. H. (2007). A comprehensive phylogeny of the bumble bees (*Bombus*). *Biological Journal of the Linnean Society. 91*(1), 161–188. doi:10.1111/j.1095-8312.2007.00784.x

Can, F., Ercan, F., & Ulasli, B. (2022). First discovery of the natural egg parasitoid of *Cydalima perspectalis* (Walker, 1859) in Turkey with molecular methods (Lepidoptera: Crambidae). *SHILAP Revista de Lepidopterologia*, *50*(200), 607–616. doi:10.57065hilap.253

Canpolat, E. (2005). *Karacadağ (Diyarbakır) volkanı jeomorfolojisi/Geomorphology of Karacadağ (Diyarbakır) volcano* [Master thesis, İstanbul Üniversitesi].

Cardinale, B. J., Duffy, J. E., Gonzalez, A., Hooper, D. U., Perrings, C., Venail, P., Narwani, A., Mace, G. M., Tilman, D., Wardle, D. A., Kinzig, A. P., Daily, G. C., Loreau, M., Grace, J. B., Larigauderie, A., Srivastava, D. S., & Naeem, S. (2012). Biodiversity loss and its impact on humanity. *Nature*, *486*(7401), 59–67. doi:10.1038/nature11148 PMID:22678280

Cardinale, B. J., Wright, J. P., Cadotte, M. W., Carroll, I. T., Hector, A., Srivastava, D. S., Loreau, M., & Weis, J. J. (2007). Impacts of plant diversity on biomass production increase through time because of species complementarity. *Proceedings of the National Academy of Sciences of the United States of America*, *104*(46), 18123–18128. doi:10.1073/pnas.0709069104 PMID:17991772

Cardoso, P., Barton, P. S., Birkhofer, K., Chichorro, F., Deacon, C., Fartmann, T., Fukushima, C. S., Gaigher, R., Habel, J. C., Hallmann, C. A., Hill, M. J., Hochkirch, A., Kwak, M. L., Mammola, S., Ari Noriega, J., Orfinger, A. B., Pedraza, F., Pryke, J. S., Roque, F. O., & Samways, M. J. (2020). Scientists' warning to humanity on insect extinctions. *Biological Conservation*, *108426*, 108426. Advance online publication. doi:10.1016/j.biocon.2020.108426

Carpaneto, G., M., Piattella, E., & Pittino, R. (2000). The scarab beetles of Turkey: an updated checklist and chorotype analysis (Coleoptera, Scarabaeoidea). *Biogeographia – The Journal of Integrative Biogeography*, *21*. doi:10.21426/B6110170

Casale, A., & Vigna Taglianti, A. (1999). Caraboid beetles (excl. Cicindelidae) of Anatolia, and their biogeographical significance (Coleoptera, Caraboidea). *Biogeographia – The Journal of Integrative Biogeography*, *20*. doi:10.21426/B620110002

Casey, K., Lichrou, M., & O'Malley, L. (2020). Prefiguring sustainable living: An ecovillage story. *Journal of Marketing Management, 36*(17–18), 1658–1679. doi:10.1080/0267257X.2020.1801800

Cashdan, E. (1998). Adaptiveness of food learning and food aversions children. *Social Sciences Information. Information Sur les Sciences Sociales, 37*(4), 613–632. doi:10.1177/053901898037004003

Cebeci, H. H., & Baydemir, M. (2018). Depredadores de escarabajos de la corteza (Coleoptera) en la región de Balikesir de Turquía. *Revista Colombiana de Entomologia, 44*(2), 283–287. doi:10.25100ocolen.v44i2.7326

Cengiz, F. C., Kaya, K., Ulaşlı, B., & Morinière, J. (2016). First record of the egg parasitoids of *Chilo partellus* (Swinhoe, 1885) (Lepidoptera: Crambidae) in Turkey unveiled by DNA Barcoding. *Turkiye Entomoloji Dergisi, 40*(2). doi:10.16970/ted.80849

Central Water Commission. (2014). *Narmada Basin.* CWC.

Central Water Commission. (n.d.). *Home.* CWC. http://cwc.nic.in/main/webpages/rm/activities.html

CGWB. (n.d.). *Mandla Report.* CGWB. http://cgwb.gov.in/District_Profile/MP/Mandla

Chang, Q., Anderson, G. Q., Brides, K., Clark, J. A., Clark, N. A., Hearn, R., & Green, R. (2019). *A high proportion of the world population of the spoon-billed sandpiper occurs at Tiaozini, China, during the post-breeding Moult.* University of Cambridge.

Chitale, V. S., Behera, M. D., & Roy, P. S. (2014). Future of Endemic Flora of Biodiversity Hotspots in India. *PLoS One, 9*(12), e115264. doi:10.1371/journal.pone.0115264 PMID:25501852

Chobanov, D. P., Kaya, S., Grzywacz, B., Warchałowska-Śliwa, E., & Çıplak, B. (2017). The Anatolio-Balkan phylogeographic fault: A snapshot from the genus *Isophya* (Orthoptera, Tettigoniidae). *Zoologica Scripta, 46*(2), 165–179. doi:10.1111/zsc.12194

Chowdhury, S. U., Foysal, M., & Green, R. E. (2022). Accelerating decline of an important wintering population of the critically endangered Spoon-billed Sandpiper Calidris pygmaea at Sonadia Island, Bangladesh. *Journal of Ornithology, 163*(4), 891–901. doi:10.100710336-022-01995-0

Cigliano, M. M., Braun, H., Eades, D. C., & Otte, D. (2023). *Orthoptera Species File. Version 5.0/5.0.* Orthoptera. http://orthoptera.speciesfile.org/HomePage/Orthoptera/HomePage.aspx

Çıplak, B., Demirsoy, A., Yalım, B., & Sevgili, H. (2008). Türkiye Orthoptera (Düzkanatlılar, Çekirgeler) Faunası. I: A. Demirsoy (Ed.), Genel Zoocoğrafya ve Türkiye Zoocoğrafyası "Hayvan Coğrafyası" (8. Basım, pp. 681–707). Meteksan.

Çıplak, B. (2003). Distribution of Tettigoniinae (Orthoptera, Tettigoniidae) bush-crickets in Turkey: The importance of the Anatolian Taurus Mountains in biodiversity and implications for conservation. *Biodiversity and Conservation, 12*(1), 47–64. doi:10.1023/A:1021206732679

Çıplak, B. (2004). Biogeography of Anatolia: The marker group Orthoptera. *Memorie della Societa Entomologica Italiana*, *82*(2), 357–372.

Clark, N. A., Anderson, G. Q., Li, J., Syroechkovskiy, E. E., Tomkovich, P. S., Zöckler, C., Lee, R., & Green, R. E. (2018). First formal estimate of the world population of the Critically Endangered spoon-billed sandpiper Calidris pygmaea. *Oryx*, *52*(1), 137–146. doi:10.1017/S0030605316000806

Clark, N., Pain, D., & Green, R. (2014). Saving the spoon-billed sandpiper: An update on the conservation programme. *British Birds*, *107*(4), 7–475.

Clay, R. P., Lesterhuis, A. J., Schulte, S., Brown, S., Reynolds, D., & Simons, T. R. (2014). A global assessment of the conservation status of the American Oystercatcher Haematopus palliatus. *International Wader Studies*, *20*, 62–82.

Collins, S. A., Sanders, F. J., & Jodice, P. G. (2016). Assessing conservation tools for an at-risk shorebird: Feasibility of headstarting for American Oystercatchers Haematopus palliatus. *Bird Conservation International*, *26*(4), 451–465. doi:10.1017/S0959270916000095

Colwell, M. A. (2010). *Shorebird ecology, conservation, and management*. University of California Press. doi:10.1525/9780520947962

Conklin, J. R., Senner, N. R., Battley, P. F., & Piersma, T. (2017). Extreme migration and the individual quality spectrum. *Journal of Avian Biology*, *48*(1), 19–36. doi:10.1111/jav.01316

Connolly, C., Keil, R., & Ali, S. H. (2021). Extended urbanisation and the spatialities of infectious disease: Demographic change, infrastructure and governance. *Urban Studies (Edinburgh, Scotland)*, *58*(2), 245–263. doi:10.1177/0042098020910873

Coope, G. R. (1994). The response of insect faunas to glacial-interglacial climatic fluctuations. *Philosophical Transactions of the Royal Society of London. Series B, Biological Sciences*, *344*(1307), 19–26. doi:10.1098/rstb.1994.0046

Çoruh, S. (2016). Biogeography and host evaluation of the subfamily Pimplinae (Hymenoptera: Ichneumonidae) in Turkey. *Journal of Entomological Research Society*, *18*(2), 33–60.

Çoruh, S. (2019a). Taxonomic and biogeographic evaluations of the subfamily Cryptinae (Hymenoptera: Ichneumonidae). *Turkiye Entomoloji Dergisi*, 313–337. doi:10.16970/entoted.520717

Çoruh, S. (2019b). Taxonomical and biogeographical evaluation of the subfamily Tryphoninae (Hymenoptera: Ichneumonidae) in Turkey. *Journal of Entomological Research Society*, *21*(3), 301–321.

Çoruh, S., & Özbek, H. (2005). New records of Cryptinae (Hymenoptera: Ichneumonidae) from Turkey with some hosts. *Turkiye Entomoloji Dergisi*, *29*(3), 183–186.

Coyne, J. A., & Orr, H. A. (2004). *Speciation*. Sinauer Associates.

CWC. (n.d.). *Climate change and water scarcity in India*. CWC. https://globalwarming.com/articles/climate-change-and-water-scarcity-in-india/

Dal, N., & Gönençgil, B. (2018). *Türkiye'de dağ ve dağlık alan sınırlandırması için bir yaklaşım/An approach for the limitation of mountain and mountain areas in Turkey.*, International Geography Symposium on the 30th Anniversary of TUCAUM. 907–913.

Daniels, R. J. R. (1992). Geographical Distribution Patterns of Amphibians in the Western Ghats, India. *Journal of Biogeography*, *19*(5), 521. doi:10.2307/2845771

Darilmaz, M. C., Salur, A., Murányi, D., & Vinçon, G. (2016). Contribution to the knowledge of Turkish stoneflies with annotated catalogue (Insecta: Plecoptera). *Zootaxa*, *4074*(1), 1–74. doi:10.11646/zootaxa.4074.1.1 PMID:27395953

Darilmaz, M. C., & İncekara, Ü. (2011). Checklist of hydrophiloidea of Turkey (Coleoptera: Polyphaga). *Journal of Natural History*, *45*(11–12), 685–735. doi:10.1080/00222933.2010.535916

Darılmaz, M. C., Polat, A., & Incekara, U. (2019). Faunistic study on aquatic coleoptera of the eastern mediterranean region of Turkey. *Turkish Journal of Fisheries and Aquatic Sciences*, *19*(5), 409–421. doi:10.4194/1303-2712-v19_5_06

Darılmaz, M. C., & Salur, A. (2015). Annotated Catalogue of the Turkish Caddisflies (Insecta: Trichoptera). *Munis Entomology & Zoology*, *10*(10), 521–734.

de Melo, M. A. C. (1997). *Processo de Planejamento e as Inovações Tecnológicas e Sociais: uma perspectiva sócio-ecológica*. Anais do 5o. Seminário de Modernização Tecnológica.

de Melo, M.A.C. (2002b). Inovação e Modernização Tecnológica e Organizacional nas MPMEs: o domínio interorganizacional. *Seminário Internacional: Políticas*

de Melo, A. (2001). *The Innovation Systems of Latin America and the Caribbean*. Inter-American Development Bank. Agosto.

de Melo, M. A. C. (2002a). *Enriquecendo a Atuação de Incubadora de Emrpesas. Tecnologia e Inovação: experiências de gestão nas micro e pequenas empresas*. PGT/USP.

Delany, S., Scott, D., Dodman, T., & Stroud, D. (2009). An atlas of wader populations in Africa and Western Eurasia. *British Birds*, *102*, 639–642.

Demir, E. (2006). Preliminary report on the Auchenorrhyncha (Hemiptera) fauna of Kazdağı National Park with two new records for Turkey. *Acta Entomologica Slovenica (Ljubljana)*, *14*(1), 89–102.

Demir, E. (2018). The economically important alien invasive planthoppers in Turkey (Hemiptera: Fulgoromorpha). *Acta Entomologica Slovenica (Ljubljana)*, *2*(2), 73–78.

Demirel, E. (2019). Cicadidae (Hemiptera: Auchenorrhyncha) from Mt.Musa, Turkey. *Monographs of the Upper Silesian Museum*, *10*, 29–45.

Demirel, Ö., & Pouya, S. (2020). Sürdürülebilir dağ yönetimi koruma boyutu ve kaynak rezervi olarak Ağrı Dağı/Sustainable Mountain Management, Ağrı Mount example. *International Journal of Mountaineering and Climbing*, *3*(1), 1–14. doi:10.36415/dagcilik.686665

Demirsoy, A. (2008). Genel Zoocoğrafya ve Türkiye Zoocoğrafyası (Hayvan Coğrafyası) (pp. 1007–1007). Palme yayıncılık, Ankara.

Demirsoy, A. (1982). [). TÜBİTAK.]. *Türkiye Faunası Odonata*, *8/14*, 156–156.

Demirsoy, A., Salman, S., & Sevgili, H. (2002). Novadrymadusa, a new genus of bushcricket with a new species and notes on related genera (Orthoptera: Tettigoniidae). *Journal of Orthoptera Research*, *11*(2), 175–183. doi:10.1665/1082-6467(2002)011[0175:NANGOB]2.0.CO;2

Denk, T., Güner, H. T., & Bouchal, J. M. (2019). Early Miocene climate and biomes of Turkey: Evidence from leaf fossils, dispersed pollen, and petrified wood. *Palaeogeography, Palaeoclimatology, Palaeoecology*, *530*, 236–248. doi:10.1016/j.palaeo.2019.05.042

DeWalt, K., DeWalt, B., & Wayland, C. (1998). Participant observation. In H. R. Bernard (Ed.), *Handbook of methods in cultural anthropology* (pp. 259–299). Altamira.

Di Russo, C., Rampini, M., & Landeck, I. (2007). The cave crickets of northeast Turkey and transCaucasian regions, with descriptions of two new species of the genera *Dolichopoda* and *Troglophilus* (Orthoptera, Rhaphidophoridae). *Journal of Orthoptera Research*, *16*(1), 67–76. doi:10.1665/1082-6467(2007)16[67:TCCONT]2.0.CO;2

Diamond, J., & Robinson, J. (2010). *Natural experiments of history*. The Belknap Press of Harvard University Press.

Díaz, D., Kristoff, J., Castro, M., Miceli, J., Castro, D., Quinteros, R., & Guerrero, S. (2007). *Exploraciones en antropología y complejidad*. Editorial Sb.

Dinc, S., Mutun, S., & Melika, G. (2014). A new species of Andricus Hartig oak gall wasp from Turkey (Hymenoptera: Cynipidae, Cynipini). *North-Western Journal of Zoology*, *10*(1), 122–127.

Ditlevsen, K., Sandøe, P., & Lassen, J. (2019). Healthy food is nutritious, but organic food is healthy because it is pure: The negotiation of healthy food choices by Danish consumers of organic food. *Food Quality and Preference*, *71*(January), 46–53. doi:10.1016/j.foodqual.2018.06.001

Dmitriĭ Mikhaĭlovich Bezmaternykh. (2013). *Sustainable Development of Asian Countries, Water Resources and Biodiversity Under Climate Change*. Siberian Branch of the Russian Academy of Sciences.

Dodds, W. K., & Whiles, M. R. (2020). *Freshwater ecology: concepts and environmental applications of limnology*. Academic Press.

Dodson, S. I. (2005). *Introduction to Limnology*. McGraw-Hill Science, Engineering & Mathematics.

Doğan, M., Akaydın, G., & Erdal, J. (2020). Phytogeography, ecology and conservation of the genus *Limonium* (Plumbaginaceae) in Turkey and a taxonomic revision. *Plant Systematics and Evolution, 306*(6), 89. doi:10.100700606-020-01706-3

Doherty, T. S., Glen, A. S., Nimmo, D. G., Ritchie, E. G., & Dickman, C. R. (2016). Invasive predators and global biodiversity loss. *Proceedings of the National Academy of Sciences of the United States of America, 113*(40), 11261–11265. doi:10.1073/pnas.1602480113 PMID:27638204

Dudgeon, D., Arthington, A. H., Gessner, M. O., Kawabata, Z.-I., Knowler, D. J., Lévêque, C., Naiman, R. J., Prieur-Richard, A.-H., Soto, D., Stiassny, M. L. J., & Sullivan, C. A. (2006). Freshwater biodiversity: Importance, threats, status and conservation challenges. *Biological Reviews of the Cambridge Philosophical Society, 81*(02), 163. doi:10.1017/S1464793105006950 PMID:16336747

Dueñas, M.-A., Ruffhead, H. J., Wakefield, N. H., Roberts, P. D., Hemming, D. J., & Diaz-Soltero, H. (2018). The role played by invasive species in interactions with endangered and threatened species in the United States: A systematic review. *Biodiversity and Conservation, 27*(12), 3171–3183. doi:10.100710531-018-1595-x

Dumont, H. J. (1977). A review of the dragonfly fauna of Turkey and adjacent Mediterranean islands (Insecta: Odonata). *Bulletin et Annales de La Société Royale Belge d'entomologie, 113*, 119–171.

Dunning, T. (2012). *Natural experiments in the social sciences*. Cambridge University Press. doi:10.1017/CBO9781139084444

Durrenberger, P. (2007) The Anthropology of Organized Labor in the United States. *Ann Review of Anthropology, 36*, 73-88. https://www.annualreviews.org/doi/pdf/10.1146/annurev.anthro.36.081406.094433

Dursun, A., & Salur, A. (2013). Presence of Sphedanolestes sanguineus (Fabricius, 1794) in Turkey, followed by an annotated checklist of Reduviidae (Hemiptera: Heteroptera). *Turkish Journal of Zoology, 37*, 610–620. doi:10.3906/zoo-1203-17

Egan, S. P., Hood, G. R., Feder, J. L., & Ott, J. R. (2012). Divergent host-plant use promotes reproductive isolation among cynipid gall wasp populations. *Biology Letters, 8*(4), 605–608. doi:10.1098/rsbl.2011.1205 PMID:22337505

Egerton, F. N. (2001). A history of the ecological sciences: Early Greek origins. *Bulletin of the Ecological Society of America, 82*(1), 93–97.

Ehrlich, P. R., & Raven, P. H. (1964). Butterflies and plants: A study in coevolution. *Evolution; International Journal of Organic Evolution, 18*(4), 586–608. doi:10.2307/2406212

Ehrmann, R. (2011). Mantodea from Turkey and Cyprus (Dictyoptera: Mantodea). *Articulata, 26*(1), 1–42.

Eid, E. M., & Shaltout, K. H. (2017). Growth dynamics of water hyacinth (Eichhornia crassipes): A modelling approach. *Rendiconti Lincei. Scienze Fisiche e Naturali*, *28*(1), 169–181. doi:10.100712210-016-0589-4

Ekim, T., & Güner, A. (1986). The Anatolian Diagonal: Fact or fiction? *Proceedings of the Royal Society of Edinburgh. Section B: Biological Sciences*, *89*, 69–77. doi:10.1017/S0269727000008915

Eldredge, N. (2002). Life On Earth. Vol. 1: A-G: An Encyclopedia of Biodiversity, Ecology, and Evolution. Abc-Clio.

Elekcioglu, N. Z., & Olculu, M. (2017). Pest, predator and parasitoid species in persimmon orchards in the eastern Mediterranean region of Turkey, with new records. *Fresenius Environmental Bulletin*, *26*(8), 5170–5176.

Ellis, K. S., Anteau, M. J., Cuthbert, F. J., Gratto-Trevor, C. L., Jorgensen, J. G., Newstead, D. J., Powell, L. A., Ring, M. M., Sherfy, M. H., Swift, R. J., Toy, D. L., & Koons, D. N. (2021). Impacts of extreme environmental disturbances on piping plover survival are partially moderated by migratory connectivity. *Biological Conservation*, *264*, 109371. doi:10.1016/j.biocon.2021.109371

Eriocaulaceae - efloraofindia. (n.d.). Google. https://sites.google.com/site/efloraofindia/species/a---l/e/eriocaulaceae

Erisoz Kasap, O., Linton, Y.-M., Karakus, M., Ozbel, Y., & Alten, B. (2019). Revision of the species composition and distribution of Turkish sand flies using DNA barcodes. *Parasites & Vectors*, *12*(1), 410. doi:10.118613071-019-3669-3 PMID:31439012

Erol, A., & Randhir, T. O. (2012). Climatic change impacts on the ecohydrology of Mediterranean watersheds. *Climatic Change*, *114*(2), 319–341. doi:10.100710584-012-0406-8

Esin, N. V., Yanko-Hombach, V. V., & Esin, N. I. (2018). Evolutionary mechanisms of the Paratethys Sea and its separation into the Black Sea and Caspian Sea. *Quaternary International*, *465*, 46–53. doi:10.1016/j.quaint.2016.06.019

Fahrig, L. (2003). Effects of habitat fragmentation on biodiversity. *Annual Review of Ecology, Evolution, and Systematics*, *34*(1), 487–515. doi:10.1146/annurev.ecolsys.34.011802.132419

Fallahzadeh, M., & Saghaei, N. (2010). Checklist of Braconidae (Insecta: Hymenoptera) from Iran. *Munis Entomology & Zoology*, *5*(1), 170–186.

FAO. (2021). *Global review of forest pests and diseases*. FAO.

Faria, F. A., Dias, R. A., Bencke, G. A., Bugoni, L., Senner, N. R., Almeida, J. B., Nunes, G. T., Gonçalves, M. S., & Lyons, J. E. (2023). Trends and population estimate of the threatened Buff-breasted Sandpiper *Calidris subruficollis* wintering in coastal grasslands of southern Brazil. *Bird Conservation International*, *33*, e61. doi:10.1017/S0959270923000138

Figueiredo, B., & Scaraboto, D. (2016). The systemic creation of value through circulation in collaborative consumer networks. *The Journal of Consumer Research*, *43*(4), 509–533. doi:10.1093/jcr/ucw038

Finucane, M. L. (2009). Why science alone won't solve the climate crisis: Managing climate risks in the Pacific. *Asia Pacific Issues*, *89*, 1–8.

Fisher, M. C., Garner, T. W. J., & Walker, S. F. (2009). Global Emergence ofBatrachochytrium dendrobatidisand Amphibian Chytridiomycosis in Space, Time, and Host. *Annual Review of Microbiology*, *63*(1), 291–310. doi:10.1146/annurev.micro.091208.073435 PMID:19575560

Fiske, S. J., Crate, S. A., Crumley, C., Galvin, K., Lazrus, H., Lucero, L., Oliver-Smith, A., Orlove, B., Strauss, S., & Wilk, R. R. (2014). *Changing the atmosphere. Anthropology and climate change. Final report of the AAA global climate change taskforce*. American Anthropological Association.

Fitzpatrick, B. M., Fordyce, J. A., & Gavrilets, S. (2009). Pattern, process and geographic modes of speciation. *Journal of Evolutionary Biology*, *22*(11), 2342–2347. doi:10.1111/j.1420-9101.2009.01833.x PMID:19732257

Futuyma, D. J., & Agrawal, A. A. (2009). Macroevolution and the biological diversity of plants and herbivores. *Proceedings of the National Academy of Sciences of the United States of America*, *106*(43), 18054–18061. doi:10.1073/pnas.0904106106 PMID:19815508

G., T. (2020). *Ecosystem-Based Management, Ecosystem Services and Aquatic Biodiversity: Theory, Tools, and Applications*. Springer International Publishing, Imprint Springer.

Gadallah, N. S., Ghahari, H., Peris-Felipo, F. J., & Fischer, M. (2016). Updated checklist of Iranian Opiinae (Hymenoptera: Braconidae). *Zootaxa*, *4066*(1), 1–40. doi:10.11646/zootaxa.4066.1.1 PMID:27395527

Galbraith, H., DesRochers, D. W., Brown, S., & Reed, J. M. (2014). Predicting vulnerabilities of North American shorebirds to climate change. *PLoS One*, *9*(9), e108899. doi:10.1371/journal.pone.0108899 PMID:25268907

Galbraith, H., Jones, R., Park, R., Clough, J., Herrod-Julius, S., Harrington, B., & Page, G. (2002). Global climate change and sea level rise: Potential losses of intertidal habitat for shorebirds. *Waterbirds*, *25*(2), 173–183. doi:10.1675/1524-4695(2002)025[0173:GCCASL]2.0.CO;2

Gallo-Cajiao, E., Morrison, T. H., Woodworth, B. K., Lees, A. C., Naves, L. C., Yong, D. L., Choi, C. Y., Mundkur, T., Bird, J., Jain, A., Klokov, K., Syroechkovskiy, E., Chowdhury, S. U., Fu, V. W. K., Watson, J. E. M., & Fuller, R. A. (2020). Extent and potential impact of hunting on migratory shorebirds in the Asia-Pacific. *Biological Conservation*, *246*, 108582. doi:10.1016/j.biocon.2020.108582

Geller-Grimm, F., Dikow, T., & Lavigne, R. J. (2021). Asilidae Homepage. *Information (Basel)*.

Ghahari, H., Hayat, R., Lavigne, R. J., & Ostovan, H. (2014). An annotated checklist of Iranian Asilidae (Insecta: Diptera: Brachycera: Asiloidea). *Linzer Biologische Beitrage*, *46*(2), 1379–1446.

Giesler, M. (2008). Conflict and compromise: Drama in marketplace evolution. *Journal of Consumer Research*, 34(April), Giesler, M. (2006). Consumer gift systems. *The Journal of Consumer Research*, *33*(September), 283–290. doi:10.1086/506309

Gill, R. E. Jr, Tibbitts, T. L., Douglas, D. C., Handel, C. M., Mulcahy, D. M., Gottschalck, J. C., Warnock, N., McCaffery, B. J., Battley, P. F., & Piersma, T. (2009). Extreme endurance flights by landbirds crossing the Pacific Ocean: Ecological corridor rather than barrier? *Proceedings. Biological Sciences*, *276*(1656), 447–457. doi:10.1098/rspb.2008.1142 PMID:18974033

Giner, C. (2004). *J. Antropología urbana*. Barcelona. *Ariel*.

Gnaspini, P. (1997). Climatic fluctuations and tropical troglobitic evolution. *International Journal of Speleology*, *26*(1/2), 33–36. doi:10.5038/1827-806X.26.1.3

Gomez Isaza, D. F., Cramp, R. L., & Franklin, C. E. (2020). Living in polluted waters: A meta-analysis of the effects of nitrate and interactions with other environmental stressors on freshwater taxa. *Environmental Pollution*, *261*, 114091. doi:10.1016/j.envpol.2020.114091 PMID:32062099

González Álvarez, M. D. (1997). Processos de Planejamento nos Pólos Tecnológicos: um enfoque adaptativo. [Tese de Doutorado, Departamento de Engenharia Industrial, PUC-Rio, Rio de Janeiro].

Gorochov, A. V., & Ünal, M. (2012). Review of Turkish crickets from the subfamily Gryllomorphinae (Orthoptera: Gryllidae). *Zoosystematica Rossica*, *21*(1), 28–59. doi:10.31610/zsr/2012.21.1.28

Görür, G., Şenol, Ö., Akyildirim Beğen, H., & Akyürek, B. (2020). Son zamanlarda yapılan çalışmalara göre Türkiye afit faunasıyla ilgili öngörüler. *Atatürk Üniversitesi Ziraat Fakültesi Dergisi*, *51*, 63–68. doi:10.17097/ataunizfd.597925

Government of India. (2011). *Census*. Government of India. https://censusindia.gov.in/2011census/dchb/DCHB.html

Gravano, A. (2007b). Claves para la facilitación organizacional del proceso participativo en la planificación urbano-ambiental metropolitana. *Revista Regional de Trabajo Social*. Montevideo.

Gravano, A. (1992). Antropología Práctica, muestra y posibilidades de la Antropología Organizacional. *Publicar en Antropología y Ciencias Sociales, Buenos Aires*, (1), 95–126.

Gravano, A. (2005). *El barrio en la teoría social*. Espacio Editorial.

Gravano, A. (2006). Imaginarios regionales y circularidad en la planificación: El caso del TOAR. *Intersecciones*, (7), 305–323.

Green, R., Syroechkovskiy, E. E., Anderson, G. Q., Chang, Q., Chowdhury, S., Clark, J. A., & Clark, N. A. (2021). *New estimates of the size and trend of the world population of the Spoon-billed Sandpiper using three independent statistical models*.

Green, J. M., Sripanomyom, S., Giam, X., & Wilcove, D. S. (2015). The ecology and economics of shorebird conservation in a tropical human-modified landscape. *Journal of Applied Ecology*, *52*(6), 1483–1491. doi:10.1111/1365-2664.12508

Grimmer, J., Roberts, M., & Stewart, B. (2021). Machine learning for social science: An agnostic approach. *Annual Review of Political Science*, *24*(1), 395–419. doi:10.1146/annurev-polisci-053119-015921

Grodsky, S. M., Moorman, C. E., & Russel, K. R. (2015). Forest wildlife management. In G. LaRocque (Ed.), *Ecological forest management handbook* (pp. 47–85). CRC Press. doi:10.1201/b19150-5

Gross, M. (2012). Turkey's biodiversity at the crossroads. *Current Biology*, *22*(13), R503–R505. doi:10.1016/j.cub.2012.06.051 PMID:22970449

Guclu, K., & Karahan, F. (2004). A review: The history of conservation programs and development of the national parks concept in Turkey. *Biodiversity and Conservation*, *13*(7), 1373–1390. doi:10.1023/B:BIOC.0000019399.35427.64

Günay, G., & Şimşek, Ş. (2001). *Karst hydrogeology in hydrothermal system*, 1–13.

Güncan, A., & Gümüş, E. (2019). Brown Marmorated Stink Bug, *Halyomorpha halys* (StåL, 1855) (Hemiptera: Heteroptera, Pentatomidae), a new and important pest in Turkey. *Entomological News*, *128*(2), 204. doi:10.3157/021.128.0208

Güner, A., Aslan, S., Ekim, T., Vural, M., & Babaç, M. T. (2012). Türkiye Bitkileri Listesi (Damarlı Bitkiler) (pp. 1290–1290). Nezahat Gökyiğit Botanik bahçesi ve Flora Araştırmaları Derneği Yayını.

Güngöroğlu, C. (2017). Natura 2000: Its status quo, opportunities and challenges for forest biodiversity in Turkey. *Fresenius Environmental Bulletin*, *26*(10), 6217–6224.

Gür, H. (2017). Geç kuvaterner buzul buzullararası döngülerinin anadolu'nun biyolojik çeşitliliği üzerine etkileri. *Türkiye Jeoloji Bülteni / Geological Bulletin of Turkey*, *60*(4), 507–528. doi:10.25288/tjb.363813

Gür, H. (2013). The effects of the Late Quaternary glacial-interglacial cycles on Anatolian ground squirrels: Range expansion during the glacial periods? *Biological Journal of the Linnean Society. Linnean Society of London*, *109*(1), 19–32. doi:10.1111/bij.12026

Gür, H. (2016). The Anatolian diagonal revisited: Testing the ecological basis of a biogeographic boundary. *Zoology in the Middle East*, *62*(3), 189–199. doi:10.1080/09397140.2016.1226544

GWRA MP. (2020). *Assessment.* CGWB. http://cgwb.gov.in/GW-Assessment/GWR-2020-Reports%20State/2022-03-10

Hacet, N. (2017). Updated checklist of Odonata fauna in the Turkish Thrace Region, with additional records of new, rare, and threatened taxa. *Turkish Journal of Zoology*, *41*, 33–42. doi:10.3906/zoo-1601-65

Hacet, N., & Colonnelli, E. (2019). On the Ceutorhynchinae (Coleoptera: Curculionidae) Fauna of Turkish Thrace, with Additional Records for Turkey. *Journal of Entomological Research Society*, *21*(2), 175–183.

Haines, S. (2019b). Managing expectations: Articulating expertise in climate services for agriculture in Belize. *Climatic Change*, *157*(1), 43–59. doi:10.100710584-018-2357-1

Hannerz, U. (1986). *Exploración de la ciudad.* Fondo de Cultura Económica.

Hannerz, U. (2004). *Foreign New: Exploring the World of Foreign Correspondents.* University of Chicago Press.

Hanson, T., Brooks, T. M., Da Fonseca, G. A. B., Hoffmann, M., Lamoreux, J. F., Machlis, G., Mittermeier, C. G., Mittermeier, R. A., & Pilgrim, J. D. (2009). Warfare in biodiversity hotspots. *Conservation Biology, 23*(3), 578–587. doi:10.1111/j.1523-1739.2009.01166.x PMID:19236450

Harrington, L. W., & Fisher, M. J. (2014). *Water Scarcity, Livelihoods and Food Security: Research and Innovation for Development.* Taylor and Francis. doi:10.4324/9781315851662

Harris, J. E., Rodenhouse, N. L., & Holmes, R. T. (2019). Decline in beetle abundance and diversity in an intact temperate forest linked to climate warming. *Biological Conservation, 240,* 108219. doi:10.1016/j.biocon.2019.108219

Hartmut Resonanz, R. (2016). *Eine Soziologie der Weltbeziehung.* Suhrkamp.

Harvey, B. (2013). *Blue Genes.* Taylor & Francis. doi:10.4324/9781849770606

Hasbenli, A., Çiftçi, D., & Çağlar, Ü. (2020). Tanap, a new robber fly genus from Turkey (Diptera: Asilidae: Dioctriinae). *Zootaxa, 4822*(1). doi:10.11646/zootaxa.4822.1.6

Hayat, R. (1997). Prey of some Robber Flies (Diptera: Asilidae) in Turkey. *Zoology in the Middle East, 15*(1), 87–94. doi:10.1080/09397140.1997.10637743

Hayward, A., & Stone, G. N. (2005). Oak gall wasp communities: Evolution and ecology. *Basic and Applied Ecology, 6*(5), 435–443. doi:10.1016/j.baae.2005.07.003

HBW and BirdLife International. (2022). *Handbook of the Birds of the World and BirdLife International digital checklist of the birds of the world. Version 7.* BirdLife International. http://datazone.birdlife.org/userfiles/file/Species/Taxonomy/HBW-BirdLife_Checklist_v7_Dec22.zip

Hedenström, A. (2010). Extreme endurance migration: What is the limit to non-stop flight? *PLoS Biology, 8*(5), e1000362. doi:10.1371/journal.pbio.1000362 PMID:20454558

Heller, K.-G. (1988). *Bioakustik der europäischen Laubheuschrecken* (pp. 358–358). J. Margraf.

Hendrickx, F., Maelfait, J.-P., Van Wingerden, W., Schweiger, O., Speelmans, M., Aviron, S., Augenstein, I., Billeter, R., Bailey, D., Bukacek, R., Burel, F., Diekötter, T. I. M., Dirksen, J., Herzog, F., Liira, J., Roubalova, M., Vandomme, V., & Bugter, R. O. B. (2007). How landscape structure, land-use intensity and habitat diversity affect components of total arthropod diversity in agricultural landscapes. *Journal of Applied Ecology, 44*(2), 340–351. doi:10.1111/j.1365-2664.2006.01270.x

Hesselbarth, G., Wagener, S., & Van Oorschot, H. (1995). *Die Tagfalter der Türkei : unter Berücksichtigung der angrenzenden Länder.* Verlag Goecke & Evers.

Hewitt, G. M. (2011). Mediterranean peninsulas: the evolution of hotspots. I: F. E. Zachos & J. C. Habel (Eds.), Biodiversity Hotspots (pp. 123–147). Springer Berlin Heidelberg. doi:10.1007/978-3-642-20992-5_7

Hewitt, G. (2000). The genetic legacy of the Quaternary ice ages. *Nature*, *405*(6789), 907–913. doi:10.1038/35016000 PMID:10879524

Hewitt, G. M. (1999). Post-glacial re-colonization of European biota. *Biological Journal of the Linnean Society. Linnean Society of London*, *68*(1–2), 87–112. doi:10.1111/j.1095-8312.1999.tb01160.x

Hindman, M. (2015). Building better models: Prediction, replication, and machine learning in the social sciences. *The Annals of the American Academy of Political and Social Science*, *659*(1), 48–62. doi:10.1177/0002716215570279

Hiremath, A. J., & Sundaram, B. (2013). Invasive plant species in India protected Areas: conserving biodiversity in cultural landscape. In L. C. Foxcroft, P. Pyšek, D. M. Richardson, & P. Genovesi (Eds.), *Plant Invasions in Protected Areas: Patterns, Problems and Challenges* (pp. 241–266). Springer. doi:10.1007/978-94-007-7750-7_12

Hirst, K. (2019). *Is anthropology a science?* ThoughtCo. https://www.thoughtco.com/isanthropology-a-science-3971060

Hiscock, N., Brand, E., Calvo-Carrasco, D., & Donaldson, L. (2018). *Headstarting the Black-Tailed Godwit Limosa Limosa Limosa at Key Breeding Sites for the UK Population: 2018 Annual Report.* Wildfowl & Wetlands Trust.

HizalE.OztemizS.GjonovI. (2019). *Ricania shantungensis* Chou & Lu 1977 (Hemiptera: Fulgoromorpha: Ricaniidae) a new invasive insect species in European Turkey. Zenodo. doi:10.5281/zenodo.3547677

Hodkinson, I. D. (2005). Terrestrial insects along elevation gradients: Species and community responses to altitude. *Biological Reviews of the Cambridge Philosophical Society*, *80*(3), 489–513. doi:10.1017/S1464793105006767 PMID:16094810

Hofman, J., Sharma, A., & Watts, D. (2017). Prediction and explanation in social systems. *Science*, *355*(6324), 486–488. doi:10.1126cience.aal3856 PMID:28154051

Hofman, J., Watts, D., Athey, S., Garip, F., Griffiths, T., Kleinberg, J., Margetts, H., Mullainathan, S., Salganik, M., Vazire, S., Vespignani, A., & Yarkoni, T. (2021). Integrating explanation and prediction in computational social science. *Nature*, *595*(7866), 181–188. doi:10.103841586-021-03659-0 PMID:34194044

Holling, C. S. (1973). *Resilience and Stability of Ecological Systems. Annual Review of Ecology and Systematics* (Vol. 4). JSTOR. https://www.jstor.org/stable/2096802.

Holmes, D. R., & Marcus, G. E. (2006). Fast capitalism: para-ethnography and the rise of the symbolic analyst. In M. Fisher & G. Downey (Eds.), *Frontiers of Capital: Ethnographic Reflections on the New Economy*. Duke University Press.

Holmes, D. R., & Marcus, G. E. (2021). How do we collaborate? An updated manifesto. In D. Boyer & G. Marcus (Eds.), *Collaborative Anthropology Today: A Collection of Exceptions*. Cornell University Press. doi:10.7591/cornell/9781501753343.003.0002

Home: ENVIS center on biological diversity. (2023). BSI Envis. https://bsienvis.nic.in/Home.aspx

Hörandl, E., & Emadzade, K. (2011). The evolution and biogeography of alpine species in *Ranunculus* (Ranunculaceae): A global comparison. *Taxon*, *60*(2), 415–426. doi:10.1002/tax.602011

Hornung, L. K., Podschun, S. A., & Pusch, M. (2019). Linking ecosystem services and measures in river and floodplain management. *Ecosystems and People (Abingdon, England)*, *15*(1), 214–231. doi:10.1080/26395916.2019.1656287

Howarth, F. G., & Moldovan, O. T. (2018). The ecological classification of cave animals and their adaptations. I: O. T. Moldovan, Ľ. Kováč, & S. Halse (Eds.), Cave Ecology (Vol. 235, pp. 41–67). Springer International Publishing. doi:10.1007/978-3-319-98852-8_4

Hulme, P. E. (2017). Climate change and biological invasions: Evidence, expectations, and response options. *Biological Reviews of the Cambridge Philosophical Society*, *92*(3), 1297–1313. doi:10.1111/brv.12282 PMID:27241717

Huntington, T. G. (2006). Evidence for intensification of the global water cycle: Review and synthesis. *Journal of Hydrology (Amsterdam)*, *319*(1-4), 83–95. doi:10.1016/j.jhydrol.2005.07.003

Ilgar, A. (2015). MessiniyenTuzluluk Krizi: Akdeniz'in kurumasına ilişkin bir derleme. *Doğal Kaynaklar ve Ekon Bülteni*, *20*, 73–80.

IPCC. (2013). *Climate change 2013: The physical science basis. Contribution of working group I to the fifth assessment report of the intergovernmental panel on climate change*. Cambridge University Press.

Isacch, J. P., & Martínez, M. M. (2003). Habitat use by non-breeding shorebirds in flooding pampas grasslands of Argentina. *Waterbirds*, *26*(4), 494–500. doi:10.1675/1524-4695(2003)026[0494:HUBNSI]2.0.CO;2

IUCN SSC Western Ghats Plant Specialist Group. (n.d.). *IUCN*. IUCN. https://www.iucn.org/our-union/commissions/group/iucn-ssc-western-ghats-plant-specialist-group

IUCN. (2023). *The IUCN Red List of Threatened Species. Version 2022-2*. IUCN. https://www.iucnredlist.org.

Jackson, M. V., Fuller, R. A., Gan, X., Li, J., Mao, D., Melville, D. S., Murray, N. J., Wang, Z., & Choi, C. Y. (2021). Dual threat of tidal flat loss and invasive Spartina alterniflora endanger important shorebird habitat in coastal mainland China. *Journal of Environmental Management*, *278*, 111549. doi:10.1016/j.jenvman.2020.111549 PMID:33260073

Jactel, H., Imler, J.-L., Lambrechts, L., Failloux, A.-B., Lebreton, J. D., Maho, Y. L., Duplessy, J.-C., Cossart, P., & Grandcolas, P. (2021). Insect decline: Immediate action is needed. *Comptes Rendus Biologies*, *343*(3), 267–293. doi:10.5802/crbiol.37 PMID:33621456

Jelínek, J., Audisio, P., Hájek, J., Baviera, C., Moncoutier, B., Barnouin, T., Brustel, H., Genç, H., & Leschen, R. A. B. (2016). *Epuraea Imperialis* (Reitter, 1877). New invasive species of Nitidulidae (Coleoptera) in Europe, with a checklist of Sap Beetles introduced to Europe and Mediterranean areas. *AAPP Atti Della Accademia Peloritana Dei Pericolanti, Classe Di Scienze Fisiche. Matematiche e Naturali*, *94*(2), 1–24.

Jeppesen, E., Brucet, S., Naselli-Flores, L., Papastergiadou, E., Stefanidis, K., Nõges, T., Nõges, P., Attayde, J. L., Zohary, T., Coppens, J., Bucak, T., Menezes, R. F., Freitas, F. R. S., Kernan, M., Søndergaard, M., & Beklioğlu, M. (2015). Ecological impacts of global warming and water abstraction on lakes and reservoirs due to changes in water level and related changes in salinity. *Hydrobiologia*, *750*(1), 201–227. doi:10.100710750-014-2169-x

Johnson, C. N., Balmford, A., Brook, B. W., Buettel, J. C., Galetti, M., Guangchun, L., & Wilmshurst, J. M. (2017). Biodiversity losses and conservation responses in the Anthropocene. *Science*, *356*(6335), 270–275. doi:10.1126cience.aam9317 PMID:28428393

Jones, O. R., Purvis, A., Baumgart, E., & Quicke, D. L. J. (2009). Using taxonomic revision data to estimate the geographic and taxonomic distribution of undescribed species richness in the Braconidae (Hymenoptera: Ichneumonoidea). *Insect Conservation and Diversity*, *2*(3), 204–212. doi:10.1111/j.1752-4598.2009.00057.x

Jorgensen, D. (2015). Participant observation. In R. Scott & S. Kosslyn (Eds.), *Emerging trends in the social and behavioral sciences* (pp. 1–15). Wiley. doi:10.1002/9781118900772.etrds0247

Joseph, S., Sudhakar Reddy, Ch., Thomas, A. P., Srivastava, S. K., & Srivastava, V. K. (2010). Spatial interpolation of carbon stock: A case study from the Western Ghats biodiversity hotspot, India. *International Journal of Sustainable Development and World Ecology*, *17*(6), 481–486. doi:10.1080/13504509.2010.516107

Joulami, L., El Hamoumi, R., Daief, Z., Bazairi, H., & Lopes, R. J. (2019). Impact of shorebird predation on intertidal macroinvertebrates in a key North African Atlantic wintering site: An experimental approach. *African Journal of Marine Science*, *41*(1), 1–9. doi:10.2989/181423 2X.2018.1552193

Jumani, S., Rao, S., Kelkar, N., Machado, S., Krishnaswamy, J., & Vaidyanathan, S. (2018). Fish community responses to stream flow alterations and habitat modifications by small hydropower projects in the Western Ghats biodiversity hotspot, India. *Aquatic Conservation*, *28*(4), 979–993. doi:10.1002/aqc.2904

Kalaentzis, K., Mpamnaras, A., & Kazilas, C. (2019). First record of the alien exotic sap beetle *Phenolia (Lasiodites) picta* (Coleoptera: Nitidulidae) in Greece. *Entomologia Hellenica*, *28*(2), 11. doi:10.12681/eh.21323

Kalkman, V. J., Clausnitzer, V., Dijkstra, K.-D. B., Orr, A. G., Paulson, D. R., & van Tol, J. (2008). Global diversity of dragonflies (Odonata) in freshwater. *Hydrobiologia*, *595*(1), 351–363. doi:10.100710750-007-9029-x

Kalkman, V. J., Wasscher, M., & Vanpelt, G. J. (2003). An annotated checklist of the Odonata of Turkey. *Odonatologica*, *32*(3), 215–236.

Kannan, R., Shackleton, C. M., & Uma Shaanker, R. (2013). Reconstructing the history of introduction and spread of the invasive species, Lantana, at three spatial scales in India. *Biological Invasions*, *15*(6), 1287–1302. doi:10.100710530-012-0365-z

Kaplan, O. (1940). Prediction in the social sciences. *Philosophy of Science*, *7*(4), 492–498. doi:10.1086/286658

Kapli, P., Botoni, D., Ilgaz, C., Kumlutaş, Y., Avcı, A., Rastegar-Pouyani, N., Fathinia, B., Lymberakis, P., Ahmadzadeh, F., & Poulakakis, N. (2013). Molecular phylogeny and historical biogeography of the Anatolian lizard *Apathya* (Squamata, Lacertidae). *Molecular Phylogenetics and Evolution*, *66*(3), 992–1001. doi:10.1016/j.ympev.2012.12.002 PMID:23261710

Karaca, İ., Kayahan, A., Şimşek, B., & Çelikpençe, Y. (2015). First record of *Glycaspis brimblecombei* Moore (Hemiptera: Aphalaridae), in Turkey. *Phytoparasitica*, *43*(2), 171–175. doi:10.100712600-015-0457-2

Kara, K., & Alaoglu, Ö. (2001). Some new host records of Tachinidae (Diptera) from Turkey. *Studia Dipterologica*, *8*(1), 349–351.

Kara, K., Tschorsnig, H. P., & Atay, T. (2020). Checklist of Turkish tachinidae (insecta, diptera) with new records. *Journal of Entomological Research Society*, *22*(2), 163–190.

Karataş, A., Karataş, A., Yavuz, N., & Genç, M. (2020). Distribution and activity period of the invasive *Orosanga japonica* (Melichar, 1898) (Hemiptera: Ricaniidae) in Turkey. *Zoology in the Middle East*, *66*(3), 246–252. doi:10.1080/09397140.2020.1793502

Kawulich, B. (2005). La observación participante como método de recolección de datos. *Forum Qualitative Social Research*, *68*(2), 43.

Kaya, Z., & Raynal, D. J. (2001). Biodiversity and conservation of Turkish forests. *Biological Conservation*, *97*(2), 131–141. doi:10.1016/S0006-3207(00)00069-0

Kaynaş, B Y. (2017). Long-term changes in surface-active beetle communities in a post-fire successional gradient in *Pinus brutia* forests. *IForest - Biogeosciences and Forestry, 10*(2), 376–382. doi:10.3832/ifor2140-009

Kaynaş, B. Y. Kıran, K., & Karaman, C. (2018). Long-term effects of fire on ant communities (Hymenoptera: Formicidae) inPinus brutia forests of south-western Turkey. Zoology in the Middle East, 64(2), 160–168. doi:10.1080/09397140.2018.1442302

Kazancı, N., & Türkmen, G. (2012). The checklist of Ephemeroptera (Insecta) species of Turkey. *Review of Hydrobiology*, *5*(2), 143–156.

Keddy, P. (2016). *Wetland ecology: Principles and conservation*. Cambridge University Press.

Kelly, L. T., Giljohann, K. M., Duane, A., Aquilué, N., Archibald, S., Batllori, E., Bennett, A. F., Buckland, S. T., Canelles, Q., Clarke, M. F., Fortin, M.-J., Hermoso, V., Herrando, S., Keane, R. E., Lake, F. K., McCarthy, M. A., Morán-Ordóñez, A., Parr, C. L., Pausas, J. G., & Brotons, L. (2020). Fire and biodiversity in the Anthropocene. *Science*, *370*(6519), eabb0355. doi:10.1126cience.abb0355 PMID:33214246

Kemal, M., & Koçak, A. Ö. (2013). Notes on the subgenus Polyommatus (Neolysandra) Koçak in East Turkey, with descriptions of new taxa (Lycaenidae, Lepidoptera). *Miscellaneous Papers*, *160*, 5–6.

Kenis, M., Auger-Rozenberg, M.-A., Roques, A., Timms, L., Péré, C., Cock, M. J. W., Settele, J., Augustin, S., & Lopez-Vaamonde, C. (2009). Ecological effects of invasive alien insects. *Biological Invasions*, *11*(1), 21–45. doi:10.100710530-008-9318-y

Kerey, I. E., Meriç, E., Tunoğlu, C., Kelling, G., Brenner, R. L., & Doğan, A. U. (2004). Black Sea–Marmara Sea Quaternary connections: New data from the Bosphorus, İstanbul, Turkey. *Palaeogeography, Palaeoclimatology, Palaeoecology*, *204*(3–4), 277–295. doi:10.1016/S0031-0182(03)00731-4

King, J. M., & Brown, C. (2003). Environmental flows: Case studies. In R. Davis & R. Hirji (Eds.), *Water Resources and Environment: Technical Note C.2*. World Bank.

Kiran, K., & Karaman, C. (2012). First annotated checklist of the ant fauna of Turkey (Hymenoptera: Formicidae). *Zootaxa*, *3548*(1), 1. doi:10.11646/zootaxa.3548.1.1 PMID:23847408

Kirca, S., Kreutz, K., & Çolak, A. H. (2020). A biogeographical and ecological classification of orchids in Turkey. *Phytocoenologia*, *50*(1), 65–77. doi:10.1127/phyto/2019/0292

Kıyak, S. (2019). Dünya ve Türkiye'de Heteroptera Taksonu Biyo-çeşitliliği ve Çalışmalarına Kısa Bir bakış. *Journal of the Heteroptera of Turkey, Special Is*, 4–5.

Kıyak, S. (2020). Analysis of the insect fauna of Turkey and suggestions for future studies. *Analysis of the Insect Fauna of Turkey and Suggestions for Future Studies*, *15*(2), 690–710.

Klein, A. M., Vaissière, B. E., Cane, J. H., Steffan-Dewenter, I., Cunningham, S. A., Kremen, C., & Tscharntke, T. (2007). Importance of pollinators in changing landscapes for world crops. *Proceedings. Biological Sciences*, *274*(1608), 303–313. doi:10.1098/rspb.2006.3721 PMID:17164193

Koçak, A. (2006). A new genus and species to the Lepidoptera Fauna of Turkey and Syria (Papilionidae). *Miscellaneous Papers*, *97*(97), 39–45.

Koçak, A. Ö., & Kemal, M. (2018). A synonymous and distributional list of the species of the Lepidoptera of Turkey. *Memoirs*, *8*, 1–487.

Kolb, Ch. (1998). Review of Bilharz, Joy A., *The Allegany Senecas and Kinzua Dam: Forced Relocation through Two Generations*. H-AmIndian, H-Net Reviews. https://www.h-net.org/reviews/showrev.php?id=2366

Koleček, J., Reif, J., Šálek, M., Hanzelka, J., Sottas, C., & Kubelka, V. (2021). Global population trends in shorebirds: Migratory behaviour makes species at risk. *Naturwissenschaften*, *108*(2), 9. doi:10.100700114-021-01717-1 PMID:33580336

Kolics, B., Ács, Z., Chobanov, D. P., Orci, K. M., Qiang, L. S., Kovács, B., Kondorosy, E., Decsi, K., Taller, J., Specziár, A., Orbán, L., & Müller, T. (2012). Re-visiting phylogenetic and taxonomic relationships in the genus *Saga* (Insecta: Orthoptera). *PLoS One*, *7*(8), e42229. doi:10.1371/journal.pone.0042229 PMID:22912691

Kosova, R., Qendraj, D.H., Xhafaj, E. (2022). Meta-Analysis ELECTRE III and AHP in Evaluating and Ranking the Urban Resilience. *Journal of Environmental Management and Tourism*, *3*(59), 756 - 768. doi:10.14505/jemt.v13.3(59).15

Kovanci, O. B., & Kovanci, B. (2015). Contribution to the knowledge of Mantispoidea, Osmyloidea, and Myrmeleontoidea with new records for the Neuroptera fauna of northwestern Turkey. *Turkish Journal of Zoology*, *39*, 110–118. doi:10.3906/zoo-1401-45

Kramer, J. M. (2021). From predator to parasite: On private property and our ecological disaster. *Journal of Economic Issues*, *55*(June), 416–422. doi:10.1080/00213624.2021.1908804

Krauß, W. (2020). Narratives of change and the co-development of climate services for action. *Climate Risk Management*, *28*, 100217. doi:10.1016/j.crm.2020.100217

Krijgsman, W., Hilgen, F. J., Raffi, I., Sierro, F. J., & Wilson, D. S. (1999). Chronology, causes and progression of the Messinian salinity crisis. *Nature*, *400*(6745), 652–655. doi:10.1038/23231

Kuçlu, Ö., & Dik, B. (2018). Mosquito (Diptera: Culicidae) fauna of western Black Sea region of Turkey. *Turkiye Parazitolojii Dergisi*, *42*(2), 138–143. doi:10.5152/tpd.2018.5339

Küçükbasmacı, İ. (2014). Two new invasive species recorded in Kastamonu (Turkey): Oak lace bug [*Corythucha arcuata* (Say, 1832)] and sycamore lace bug [*Corythucha ciliata* (Say, 1832)] (Heteroptera: Tingidae). *Journal of Entomology and Nematology*, *6*(8), 104–111. doi:10.5897/JEN2014.0102

Küçük, M., & Ertürk, E. (2013). Biodiversity and protected areas in Turkey. *Sains Malaysiana*, *42*(10), 1455–1460.

Kuhn, T. S. (1962/1996). *The Structure of Scientific Revolutions* (3rd ed.). University of Chicago Press.

Kuna, M., & Dreslerová, D. (2007). Landscape archaeology and "community areas" in the archaeology of central Europe. D. Hicks, L. McAtackney, G.J. Fairclough (Eds.), Envisioning landscape: Situations and standpoints in archaeology and heritage. Left Coast Press.

Kunt, B. K., Yağmur, A. E., Özkütük, S., & Anlaş, S. (2010). Checklist of the cave Dwelling Invertebrates (Animalia) of Turkey. *Biological Diversity and Conservation, 3*(2), 26–41.

Kuzucuoğlu, C. (2019). The Physical Geography of Turkey: An Outline. I: C. Kuzucuoğlu, A. Çiner, & N. Kazancı (Eds.), Landscapes and Landforms of Turkey (pp. 7–15). Springer.

Lall, U. (2018). Water. In *Impacts, risks, and adaptation in the United States: Fourth national climate assessment* (Vol. II, p. 147). U.S. Global Change Research Program.

Lambert, A. (2019). Psychotic, acritical and precarious? A Lacanian exploration of the neoliberal consumer subject. *Marketing Theory, 19*(3), 329–346. doi:10.1177/1470593118796704

Lapassade, G. (1986). *Grupos, organizaciones e instituciones: la transformación de la burocracia.* Gedisa.

Lappo, E. G., Tomkovich, P. S., & Syroechkovskiy, E. E. (2012). Atlas of Breeding Waders in the Russian Arctic. Publishing House "UF Ofsetnaya Pechat": Moscow, Russia.

Latour, B. (2013). *An inquiry into modes of existence: An anthropology of the moderns* (C. Porter, Trans.). Harvard University Press.

le Roux, E., van Veenhuisen, L. S., Kerley, G. I. H., & Cromsigt, J. P. G. M. (2020). Animal body size distribution influences the ratios of nutrients supplied to plants. *Proceedings of the National Academy of Sciences of the United States of America, 117*(36), 22256–22263. doi:10.1073/pnas.2003269117 PMID:32839336

Legendre, P., & Legendre, L. (1988). *Numerical Ecology*. Elsevier.

Leopold, A. (1933). *Game management.* Scribner.

Lester, R. K. (2005). *Universities, Innovation, and the Competitiveness of Local Economies A Summary Report from the Local Innovation Systems Project – Phase I.* Industrial Performance Center Massachusetts Institute of Technology.

Levin, S. A. (2001). Encyclopedia of biodiversity. Academic Press and Elsevier Inc.

Lindblom, C. E. (1959). The Science of Muddling through. *Public Administration Review, 19*(2), 79–88. doi:10.2307/973677

Lips, K. R., Brem, F., Brenes, R., Reeve, J. D., Alford, R. A., Voyles, J., Carey, C., Livo, L., Pessier, A. P., & Collins, J. P. (2006). From The Cover: Emerging infectious disease and the loss of biodiversity in a Neotropical amphibian community. *Proceedings of the National Academy of Sciences of the United States of America, 103*(9), 3165–3170. doi:10.1073/pnas.0506889103 PMID:16481617

Lishman, C., & Nol, E. (2012). Ecology and habitat selection of the Magellanic Plover (Pluvianellus socialis): A little-known Patagonian shorebird. *The Wilson Journal of Ornithology, 124*(3), 487–496. doi:10.1676/11-035.1

Llanos, M. (2009). *Epistemología de las ciencias sociales.* UNMSM.

Longcore, J. E., Pessier, A. P., & Nichols, D. K. (1999). Batrachochytrium Dendrobatidis gen. et sp. nov., a Chytrid Pathogenic to Amphibians. *Mycologia*, *91*(2), 219–227. doi:10.1080/00275514.1999.12061011

Lourau, R. (1979). *El análisis institucional.* Amorrortu.

Lovejoy, T. (1994). The quantification of biodiversity: An esoteric quest or a vital component of sustainable development? *Philosophical Transactions of the Royal Society of London. Series B, Biological Sciences*, *345*(1311), 81–87. doi:10.1098/rstb.1994.0089 PMID:7972359

Lundvall, B.-Å. (1992). (ed.). National Systems of Innovation: towards a theory of innovation and interactive learning. Printer Publisher. Londres.

Lutovinovas, E., Tschorsnig, H.-P., Barták, M., Kubík, Š., Dursun, O., Civelek, H.-S., & Kara, K. (2018). Contribution to the tachinid fauna of southwestern Turkey (Diptera: Tachinidae). *Annales de La Société Entomologique de France (N.S.)*, *54*(4), 335–366. doi:10.1080/00379271.2018.1468724

Maheswaran, U., Rajamani, S. K. S. K., Deivasigamani, S., Joshi, K., Verma, D., & Kumar Rajamani, S., & ROSS, D. S. (n.d.). (2022 *Computational Intelligence Approach to Improve The Classification Accuracy of Brain Tumor Detection 1192–1196*https://doi.org/ doi:10.1109/SMART55829.2022.10047792

Mainka, S. A., & Howard, G. W. (2010). Climate change and invasive species: Double jeopardy. *Integrative Zoology*, *5*(2), 102–111. doi:10.1111/j.1749-4877.2010.00193.x PMID:21392328

Malinowski, B. (1961). *Argonauts of the Western Pacific*. Routledge.

Maran, T., & Kull, K. (2014). Ecosemiotics: Main principles and current developments. *Geografiska Annaler. Series B, Human Geography*, *96*(1), 41–50. doi:10.1111/geob.12035

Matthiessen, P. (1973). *The Wind Birds: Shorebirds of North America*. Avery. doi:10.1604/9780670770960

Mayo, E. (1945). *The social problems of an Industrial Civilization.* Harvard University Press.

McIntosh, R. J., Tainter, J. A., & McIntosh, S. K. (Eds.). (2000). *The Way the Wind Blows: Climate Change, History, and Human Activity*. Columbia University Press.

MEA. (2005). *Millennium Ecosystem Assessment. Ecosystems and human well-being: biodiversity synthesis*. World Resources Institute.

MEA. (2005). *Millennium Ecosystem Assessment. Fresh water*. World Resources Institute.

MEA. (2005). *Millennium Ecosystem Assessment.* Freshwater Ecosystem Services, World Resources Institute.

Meeus, J. H. A. (1995). Pan-European landscapes. *Landscape and Urban Planning*, *31*(1-3), 57–79. doi:10.1016/0169-2046(94)01036-8

Meijers, M. J. M., Brocard, G. Y., Whitney, D. L., & Mulch, A. (2020). Paleoenvironmental conditions and drainage evolution of the central Anatolian lake system (Turkey) during late Miocene to Pliocene surface uplift. *Geosphere*, *16*(2), 490–509. doi:10.1130/GES02135.1

Meijers, M. J. M., Peynircioğlu, A. A., Cosca, M. A., Brocard, G. Y., Whitney, D. L., Langereis, C. G., & Mulch, A. (2018). Climate stability in central Anatolia during the Messinian Salinity Crisis. *Palaeogeography, Palaeoclimatology, Palaeoecology*, *498*, 53–67. doi:10.1016/j.palaeo.2018.03.001

Meinzen-Dick, R. S., & Pradhan, R. (2002). *Legal Pluralism and Dynamic Property Rights, Consultative Group on International Agricultural Research system- wide program on collective action and property rights no. 22*. International Food Policy Research Institute.

Menxiu, T., Lin, Z., Li, J., Zöckler, C., & Clark, N. A. (2012). The critical importance of the Rudong mudflats, Jiangsu Province, China in the annual cycle of the Spoon-billed Sandpiper Calidris pygmeus. *Bulletin - Wader Study Group*, *119*(3), 74.

Miroğlu, A., Kartal, V., & Salur, A. (2011). Odonata of the eastern Black Sea region of Turkey, with some taxonomic notes | Request PDF. *Odonatologica*, *40*(2), 105–122.

Mittermeier, R. A., Turner, W. R., Larsen, F. W., Brooks, T. M., & Gascon, C. (2011). Global Biodiversity Conservation: The Critical Role of Hotspots. I: F. E. Zachos & J. C. Habel (Eds.), Biodiversity Hotspots (pp. 3–22). Springer Berlin Heidelberg. doi:10.1007/978-3-642-20992-5_1

Mitton, J. B., & Ferrenberg, S. M. (2012). Mountain pine beetle develops an unprecedented summer generation in response to climate warming. *American Naturalist*, *179*(5), E163–E171. doi:10.1086/665007 PMID:22504550

Mohammadi, R., Talebi, A. A., Sakhvon, V. V., Khaghaninia, S., Fathipour, Y., & Kazerani, F. (2022). Study of the robber flies (Diptera: Asilidae) in northwestern Iran: new records, description of a new species, and an updated checklist of Iranian fauna. *Biologia*, *77*(9), 2487–2496. doi:10.100711756-022-01116-5

Moog, O., Christian, E., & Eis, R. (2021). Increased cave use by butterflies and moths: A response to climate warming? *International Journal of Speleology*, *50*(1), 15–24. doi:10.5038/1827-806X.50.1.2361

Moreira, F., Ascoli, D., Safford, H., Adams, M. A., Moreno, J. M., Pereira, J. M. C., Catry, F. X., Armesto, J., Bond, W., González, M. E., Curt, T., Koutsias, N., McCaw, L., Price, O., Pausas, J. G., Rigolot, E., Stephens, S., Tavsanoglu, C., Vallejo, V. R., & Fernandes, P. M. (2020). Wildfire management in Mediterranean-type regions: Paradigm change needed. *Environmental Research Letters*, *15*(1), 011001. doi:10.1088/1748-9326/ab541e

Moritz, R. F. A., Härtel, S., & Neumann, P. (2005). Global invasions of the western honeybee (*Apis mellifera*) and the consequences for biodiversity. *Ecoscience*, *12*(3), 289–301. doi:10.2980/i1195-6860-12-3-289.1

Mutun, Serap. (2016). Türkiye'deki mazı arılarının filocoğrafi örüntülerine dair inceleme, Anadolu çaprazının temel rolü. *Turkish Journal of Forestry | Türkiye Ormancılık Dergisi, 17*(0), 1. doi:10.18182/tjf.65861

Mutun, S. (2010). Intraspecific genetic variation and phylogeography of the oak gallwasp andricus caputmedusae (hymenoptera: cynipidae): Effects of the anatolian diagonal. *Acta Zoologica Academiae Scientiarum Hungaricae, 56*(2), 153–172.

Mutun, S., & Dínç, S. (2015). Twelve oak gall wasp species (Hymenoptera, Cynipidae) new to the Turkish fauna. *Turkish Journal of Zoology, 39*, 962–964. doi:10.3906/zoo-1409-7

Mutun, S., & Dinç, S. (2019). The Anatolian Diagonal and Paleoclimatic Changes Shaped the Phylogeography of Cynips quercus (Hymenoptera, Cynipidae). *Annales Zoologici Fennici, 56*(1–6), 65. doi:10.5735/086.056.0107

Mu, X., Fang, C., Yang, Z., & Guo, X. (2022). Impact of the COVID-19 Epidemic on Population Mobility Networks in the Beijing–Tianjin–Hebei Urban Agglomeration from a Resilience Perspective. *Land (Basel), 2022*(11), 675. doi:10.3390/land11050675

Myers, N., Mittermeier, R. A., Mittermeier, C. G., da Fonseca, G. A., & Kent, J. (2000). Biodiversity hotspots for conservation priorities. *Nature, 403*(6772), 853–858. doi:10.1038/35002501 PMID:10706275

Narmada Valley Development Authority. (2014) *Government of Madhya Pradesh, Narmada Basin, Narmada Water Dispute.* Nvda.nic.in..

Neuman, K. K., Stenzel, L. E., Warriner, J. C., Page, G. W., Erbes, J. L., Eyster, C. R., Miller, E., & Henkel, L. A. (2013). Success of captive-rearing for a threatened shorebird. *Endangered Species Research, 22*(1), 85–94. doi:10.3354/esr00531

Nicholls, R. J., & Cazenave, A. (2010). Sea-level rise and its impact on coastal zones. *Science, 328*(5985), 1517-1520.

Nießen, D., Danner, D., Spengler, M., & Lechner, C. (2020). Big Five personality traits predict successful transitions from school to vocational education and training: A large-scale study. *Frontiers in Psychology, 11*, 1827. doi:10.3389/fpsyg.2020.01827 PMID:32903700

Noroozi, J., Talebi, A., Doostmohammadi, M., Rumpf, S. B., Linder, H. P., & Schneeweiss, G. M. (2018). Hotspots within a global biodiversity hotspot - areas of endemism are associated with high mountain ranges. *Scientific Reports, 8*(1), 10345. doi:10.103841598-018-28504-9 PMID:29985437

Noroozi, J., Zare, G., Sherafati, M., Mahmoodi, M., Moser, D., Asgarpour, Z., & Schneeweiss, G. M. (2019). Patterns of endemism in Turkey, the meeting point of three global biodiversity hotspots, based on three diverse families of vascular plants. *Frontiers in Ecology and Evolution, 7*, 159. Advance online publication. doi:10.3389/fevo.2019.00159

Nost, E. (2019). Climate services for whom? The political economics of contextualizing climate data in Louisiana's coastal master plan. *Climatic Change*, *157*(1), 27–42. doi:10.100710584-019-02383-z

O'Hara, J. E., Henderson, S. J., & Wood, D. M. (2020). *Preliminary checklist of the Tachinidae (Diptera) of the world. Version 2.1.* World Checklist of Teh Tachinidae. http://www.nadsdiptera.org/Tach/WorldTachs/Checklist/Worldchecklist.html

O'Sullivan, P., & Reynolds, C. S. (2008). *The Lakes Handbook.* John Wiley & Sons.

Odum, E. P. (1957). *Fundamentals of ecology.* W. B. Saunders Company.

OECD. (1997b). Government Venture Capital for Technology-Based Firms. Ducument OCDE/GD (97). OECD.

Oğuzoğlu, Ş., Avci, M., Şenal, D., & Karaca, İ. (2017). *Anatis ocellata* (Linnaeus, 1758) (Coleoptera: Coccinellidae)'nın Türkiye'deki ilk kaydı. *Türkiye Entomoloji Bülteni*, *7*(2), 197–202. doi:10.16969/entoteb.331402

Okay, A. I. (2008). Geology of Turkey : A Synopsis. *Anschnitt*, *21*, 19–42.

Okay, A. I., Zattin, M., Özcan, E., & Sunal, G. (2020). Uplift of Anatolia. *Turkish Journal of Earth Sciences*, *29*(5), 696–713. doi:10.3906/yer-2003-10

Oki, T., & Kanae, S. (2006). Global hydrological cycles and world water resources. *Science*, *313*(5790), 1068–1072. doi:10.1126cience.1128845 PMID:16931749

Olimid, A. P., Georgescu, C. M., & Gherghe, C. L. (2022). Influences of Covid-19 Crisis on Resilience Theories: An analysis of Community, Societal and Governance Resilience. *Revista de Ştiinţe Politice. Revue des Sciences Politiques*, *73*, 38–51.

Ortner, S. (2010). Access: Reflections on studying up in Hollywood. *Ethnography*, *11*(2), 211–233. doi:10.1177/1466138110362006

Ortner, S. (2013). *Not Hollywood: Independent Film at the Twilight of the American Dream.* Duke University Press.

Otte, D., Spearman, L., & Stiewe, M. B. D. (2022). *Mantodea Species File Online. Version 5.0/5.0 (Online Database).* Mantodea Home Page. http://mantodea.speciesfile.org

Ouda, S., Zohry, A. E.-H., & Noreldin, T. (2020). *Deficit irrigation: A remedy for water scarcity.* Springer. doi:10.1007/978-3-030-35586-9

Özbek, H. (2016). Türkiye'de arı (Hymneoptera: Apoidea: Apiformes) avcısı olan Asilidae (Diptera) türleri. *Uludağ Arıcılık Dergisi*, *15*(1), 10–15. doi:10.31467/uluaricilik.377559

Özbek, H., & Terzo, M. (2016). Distribution data for the Tribes Ceratinini and Allodapini (Hymenoptera: Apidae) with a checklist of the Subfamily Xylocopinae of Turkey. *Acta Entomologica Serbica*, *21*, 93–112.

Özdan, A., & Gürbüz, M. F. (2019). Ichneumonidae (Hymenoptera) fauna of Kovada Lake National Park, Isparta, Turkey. *Turkiye Entomoloji Dergisi*, 301–312. doi:10.16970/entoted.537395

Özdikmen, H. (2017). A complete list of invasive alien longhorned beetles species for Turkey (Coleoptera: Cerambycidae). *Munis Entomology & Zoology*, *12*(2), 508–515.

Oztas, T., Koc, A., & Comakli, B. (2003). Changes in vegetation and soil properties along a slope on overgrazed and eroded rangelands. *Journal of Arid Environments*, *55*(1), 93–100. doi:10.1016/S0140-1963(02)00267-7

Öztemiz, S., & Doğanlar, M. (2015). Invasive plant pests (Insecta and Acarina) of Turkey. *Munis Entomology & Zoology*, *10*(101), 144–159.

Páramo, D., & Ramírez, E. (2017). *Etnomarketing, la dimensión cultural del marketing*. Klasse Editorial.

Parolly, G. (2020). The Anatolian high-mountain ranges – Plant diversity between two seas. I: J. Noroozi (Ed.), Plant Biogeography and Vegetation of High Mountains of Central and South-West Asia (Vol. 17, pp. 215–286). Springer International Publishing. doi:10.1007/978-3-030-45212-4_7

Parratt, S. R., Walsh, B. S., Metelmann, S., White, N., Manser, A., Bretman, A. J., Hoffmann, A. A., Snook, R. R., & Price, T. A. R. (2021). Temperatures that sterilize males better match global species distributions than lethal temperatures. *Nature Climate Change*, *11*(6), 481–484. doi:10.103841558-021-01047-0

Patil, T., Jamdade, R. A., Patil, S., Govindwar, S. P., & Muley, D. V. (2018). DNA barcode based delineation of freshwater fishes from northern Western Ghats of India, one of the world's biodiversity hotspots. *Biodiversity and Conservation*, *27*(13), 3349–3371. doi:10.100710531-018-1604-0

Pausas, J. G., & Keeley, J. E. (2019). Wildfires as an ecosystem service. *Frontiers in Ecology and the Environment*, *17*(5), 289–295. doi:10.1002/fee.2044

Pearce-Higgins, J. W., Brown, D. J., Douglas, D. J., Alves, J. A., Bellio, M., Bocher, P., Buchanan, G. M., Clay, R. P., Conklin, J., Crockford, N., Dann, P., Elts, J., Friis, C., Fuller, R. A., Gill, J. A., Gosbell, K., Johnson, J. A., Marquez-Ferrando, R., Masero, J. A., & Verkuil, Y. I. (2017). A global threats overview for Numeniini populations: Synthesising expert knowledge for a group of declining migratory birds. *Bird Conservation International*, *27*(1), 6–34. doi:10.1017/S0959270916000678

Peng, H. B., Anderson, G. Q., Chang, Q., Choi, C. Y., Chowdhury, S. U., Clark, N. A., Gan, X., Hearn, R. D., Li, J., Lappo, E. G., Liu, W., Ma, Z., Melville, D. S., Phillips, J. F., Syroechkovskiy, E., Tong, M., Wang, S., Zhang, L., & Zöckler, C. (2017). The intertidal wetlands of southern Jiangsu Province, China–globally important for Spoon-billed Sandpipers and other threatened waterbirds, but facing multiple serious threats. *Bird Conservation International*, *27*(3), 305–322. doi:10.1017/S0959270917000223

Peterson, N., & Broad, K. (2009). Climate and weather discourse in anthropology: From determinism to uncertain futures. In *Anthropology and climate change: From encounters to actions* (pp. 70–86). Left Coast Press.

Pfeifer, J. (2006). Laws of nature. In S. Sarkar & J. Pfeifer (Eds.), *The philosophy of science: An encyclopedia* (pp. 439–444).

Pfister, C., Harrington, B. A., & Lavine, M. (1992). The impact of human disturbance on shorebirds at a migration staging area. *Biological Conservation*, *60*(2), 115–126. doi:10.1016/0006-3207(92)91162-L

Piersma, T., & Lindström, Å. (2004). Migrating shorebirds as integrative sentinels of global environmental change. *The Ibis*, *146*(s1), 61–69. doi:10.1111/j.1474-919X.2004.00329.x

Pimienta-Bueno, J.A. (1999). *Parque de Inovação Tecnológica e Cultural da Gávea: A Visão da PUC-Rio*. Documento Interno, 23 págs, 1999.

Pintilioaie, A.-M., Spaseni, P., Jurjescu, A., & Rădac, I. A. (2021). First record of the alien mantid *Hierodula tenuidentata* (Insecta: Mantodea) in Romania. *Travaux Du Muséum National d'Histoire Naturelle. Grigore Antipa*, *64*(1), 37–49. doi:10.3897/travaux.64.e65489

Pizarro, C. (2014). La entrevista etnográfica como práctica discursiva: Análisis de caso sobre las pistas meta-discursivas y la emergencia de categorías nativas. *Revista de Antropologia*, *57*(1), 461–496. doi:10.11606/2179-0892.ra.2014.87770

Poff, N. L., Allan, J. D., Bain, M. B., Karr, J. R., Prestegaard, K. L., Richter, B. D., Sparks, R. E., & Stromberg, J. C. (1997). The natural flow regime. *Bioscience*, *47*(11), 769–784. doi:10.2307/1313099

Porter, M. (2001). Clusters of Innovation: regional foundations of U.S. competitiveness. Council on Competitiveness, Washington, D.C: Monitor Group.

Powell, A. N., Cuthbert, F. J., Wemmer, L. C., Doolittle, A. W., & Feirer, S. T. (1997). Captive-rearing piping plovers: developing techniques to augment wild populations. *Zoo Biology: Published in affiliation with the American Zoo and Aquarium Association, 16*(6), 461-477.

Prather, C. M., Pelini, S. L., Laws, A., Rivest, E., Woltz, M., Bloch, C. P., Del Toro, I., Ho, C.-K., Kominoski, J., Newbold, T. A. S., Parsons, S., & Joern, A. (2013). Invertebrates, ecosystem services and climate change. *Biological Reviews of the Cambridge Philosophical Society*, *88*(2), 327–348. doi:10.1111/brv.12002 PMID:23217156

Press, M., & Arnould, E. J. (2014). Narrative transparency. *Journal of Marketing Management*, *30*(13–14), 1353–1376. doi:10.1080/0267257X.2014.925958

Pujolar, J. M., Vincenz, I. S., Zane, L., Jesensek, D., De Leo, G. A., & Crivelli, A. J. (2011). The effect of recurrent floods on genetic composition of marble trout populations. *PLoS One*, *6*(9), e23822. doi:10.1371/journal.pone.0023822 PMID:21931617

QGIS Development Team. (2023). *QGIS Geographic Information System*. Open Source Geospatial Foundation Project.

Raghavan, R., Prasad, G., Ali, P. H. A., & Pereira, B. (2007). Fish fauna of Chalakudy River, part of Western Ghats biodiversity hotspot, Kerala, India: Patterns of distribution, threats and conservation needs. *Biodiversity and Conservation*, *17*(13), 3119–3131. doi:10.100710531-007-9293-0

Rajamani, S. K., & Iyer, R. S. (2022). Development Of an Android Mobile Phone Application for Finding Closed-Loop, Analytical Solutions to Dense Linear, Algebraic Equations for The Purpose of Mathematical Modelling in Healthcare and Neuroscience Research. *NeuroQuantology : An Interdisciplinary Journal of Neuroscience and Quantum Physics*, *20*, 4959–4973. doi:10.6084/m9.figshare.c.6156024.v1

Rajamani, S. K., & Iyer, R. S. (2023a). A Scoping Review of Current Developments in the Field of Machine Learning and Artificial Intelligence. In D. Samantha (Ed.), *Designing and Developing Innovative Mobile Applications* (pp. 138–164). IGI Global. doi:10.4018/978-1-6684-8582-8.ch009

Rajamani, S. K., & Iyer, R. S. (2023b). Machine Learning Based Mobile Applications Using Python and ScikitLearn. In D. Samanta (Ed.), *Designing and Developing Innovative Mobile Applications* (pp. 282–306). IGI Global. doi:10.4018/978-1-6684-8582-8.ch016

Raju, K., & Taron, A. (2007). River Basins Organizations in India: An Overview. Conference. *IWMI-TATA*, *2007*, 613–646.

Rampini, M., Russo, C. D., Taylan, M. S., Gelosa, A., & Cobolli, M. (2012). Four new species of *Dolichopoda* Bolivar, 1880 from Southern Sporades and Western Turkey (Orthoptera, Rhaphidophoridae, Dolichopodainae). *ZooKeys201*, 43–58. doi:10.3897/zookeys.201.2609

Rashiba, A. P., Jishnu, K., Byju, H., Shifa, C. T., Anand, J., Vichithra, K., Xu, Y., Nefla, A., Muzaffar, S. B., Aarif, K. M., & Rubeena, K. A. (2022). The Paradox of Shorebird Diversity and Abundance in the West Coast and East Coast of India: A Comparative Analysis. *Diversity (Basel)*, *14*(10), 885. doi:10.3390/d14100885

Ratner, B. D. (2003). The politics of regional overnance in the Mekong River Basin. *Global Change*, *15*, 59–76.

Raven, P. H., & Wagner, D. L. (2021). Agricultural intensification and climate change are rapidly decreasing insect biodiversity. *Proceedings of the National Academy of Sciences of the United States of America*, *118*(2), e2002548117. doi:10.1073/pnas.2002548117 PMID:33431564

Rawat, M. Sumit Dookia, & Chandrakasan Sivaperuman. (2015). Aquatic Ecosystem: Biodiversity, Ecology and Conservation. Springer.

Reddy, C. S., Jha, C. S., & Dadhwal, V. K. (2016). Assessment and monitoring of long-term forest cover changes (1920–2013) in Western Ghats biodiversity hotspot. *Journal of Earth System Science*, *125*(1), 103–114. doi:10.100712040-015-0645-y

Regional Remote Sensing Center West. (2012). *NRSC, ISRO*. River Basin Atlas of India.

Reguera, A. (2009). Empresarios de ayer, de hoy y de siempre. *Un recorrido latinoamericano por sus formas espacio-temporales América Latina en la historia económica.* Scielo. https://www.scielo.org.mx/scielo.php?script=sci_arttext&pid=S1405- 22532009000200003&lng=es &nrm=iso

Reynoso, C. (1995b). Hacia la perfección del consenso: Los lugares comunes de la antropología. *Intersecciones, 1*, 51–72.

Richman, N. I., Böhm, M., Adams, S. B., Alvarez, F., Bergey, E. A., Bunn, J. J. S., Burnham, Q., Cordeiro, J., Coughran, J., Crandall, K. A., Dawkins, K. L., DiStefano, R. J., Doran, N. E., Edsman, L., Eversole, A. G., Füreder, L., Furse, J. M., Gherardi, F., Hamr, P., & Collen, B. (2015). Multiple drivers of decline in the global status of freshwater crayfish (Decapoda: Astacidea). *Philosophical Transactions of the Royal Society B: Biological Sciences, 370*(1662), 20140060. doi:10.1098/rstb.2014.0060

Ricklefs, R. E. (2007). History and diversity: Explorations at the intersection of ecology and evolution. *American Naturalist, 170*(S2, Suppl 2), S56–S70. doi:10.1086/519402 PMID:17874385

Riddiford, J. (2021). *Current integrated catchment management policy and management settings in the Murray–Darling Basin.* Murray-Darling Basin. doi:10.1016/B978-0-12-818152-2.00009-7

Riedel, M., Diller, E., & Çoruh, S. (2018). New contributions to the Ichneumoninae (Hymenoptera, Ichneumonidae) from Turkey. *Journal of Entomological Research Society, 20*(1), 59–72.

Riedel, M., Kolarov, J., Çoruh, S., & Özbek, H. (2014). A contribution to the Mesochorinae (Hymenoptera: Ichneumonidae) of Turkey. *Zoology in the Middle East, 60*(3), 217–221. doi:10 .1080/09397140.2014.939815

River Basins Organizations in India – Institutional Frameworks and Management Options. (1999). A Case for Fundamental Review, published by South Asian Network of Rivers, Dams and people. *Thematic Reviews: Institutional and Governance Issues.* http://www.zef.de/module/register/media/ 0f22_Saravanan-RiverBasin_wcd.pdf

Roberts, D. L., Elphick, C. S., & Reed, J. M. (2010). Identifying anomalous reports of putatively extinct species and why it matters. *Conservation Biology, 24*(1), 189–196. doi:10.1111/j.1523-1739.2009.01292.x PMID:19604294

Roca, A. (2008). *La comprensión de la tecnología en la antropología.* Actas VII ESOCITE UFRJ, Jornadas Latino-Americanas de Estudos Sociais das Ciências e das Tecnologías.

Roethlisberger, F., & Dickson, W. (1939). *Management and the Worker.* Harvard Univ. Press.

Rögl, F., & Steininger, F. F. (1983). Vom Zerfall der Tethys zu Mediterran und Paratethys. *Annalen Des Naturhistorischen Museums Wien Serie A, 85*(April), 135–163.

Roni, S., Merga, M., & Morris, J. (2020). *Conducting quantitative research in education.* Springer. doi:10.1007/978-981-13-9132-3

Roy, H. E., Bacher, S., Essl, F., Adriaens, T., Aldridge, D. C., Bishop, J. D. D., Blackburn, T. M., Branquart, E., Brodie, J., Carboneras, C., Cottier-Cook, E. J., Copp, G. H., Dean, H. J., Eilenberg, J., Gallardo, B., Garcia, M., García-Berthou, E., Genovesi, P., Hulme, P. E., & Rabitsch, W. (2019). Developing a list of invasive alien species likely to threaten biodiversity and ecosystems in the European Union. *Global Change Biology*, *25*(3), 1032–1048. doi:10.1111/gcb.14527 PMID:30548757

Saghaei, N., Hayat, R., Ostovan, H., & Fallahzadeh, M. (2010). Prey of some robber flies (Diptera: Asilidae) in fars province, Iran. *Journal of Entomological Research Society*, *12*(1), 17–26.

Saleth, R. M. (2004). *Strategic analysis of water institutions in India: Application of a new research paradigm*. IWMI.

Salur, A., Darilmaz, M. C., & Bauernfeind, E. (2016). An annotated catalogue of the mayfly fauna of Turkey (Insecta, Ephemeroptera). *ZooKeys*, *620*, 67–118. doi:10.3897/zookeys.620.9405

Sands, D. P. A. (2018). Important issues facing insect conservation in Australia: Now and into the future. *Austral Entomology*, *57*(2), 150–172. doi:10.1111/aen.12342

Sarikaya, O., & Ibis, H. M. (2016). Predatory species of bark beetles in the pine forests of izmir region in Turkey with new records for Turkish Fauna. *Egyptian Journal of Biological Pest Control*, *26*(3), 651–656.

Saxenian, A. (1994). *Regional Advantage: culture and competition in Silicon Valley and Route 128*. Harvard Business Press.

Scheele, B. C., Pasmans, F., Skerratt, L. F., Berger, L., Martel, A., Beukema, W., Acevedo, A. A., Burrowes, P. A., Carvalho, T., Catenazzi, A., De la Riva, I., Fisher, M. C., Flechas, S. V., Foster, C. N., Frías-Álvarez, P., Garner, T. W. J., Gratwicke, B., Guayasamin, J. M., Hirschfeld, M., & Canessa, S. (2019). Amphibian fungal panzootic causes catastrophic and ongoing loss of biodiversity. *Science*, *363*(6434), 1459–1463. doi:10.1126cience.aav0379 PMID:30923224

Schindler, S., Sebesvari, Z., Damm, C., Euller, K., Mauerhofer, V., Schneidergruber, A., Biró, M., & Essl, F. (2014). Multifunctionality of floodplain landscapes: Relating management options to ecosystem services. *Landscape Ecology*, *29*, 229–244. doi:10.100710980-014-9989-y

Schlapfer, A., & Marinova, D. (2001). *Local Innovation Systems: nature, importance and role*. Conference Proceedings: International Summer Academy on Technological Studies: user involvement in technological innovation, Deutschlandsberg. Austria.

Schnegg, M., O'Brian, C. I., & Sievert, I. J. (2021). It's our fault: A global comparison of different ways of explaining climate change. *Human Ecology: an Interdisciplinary Journal*, *49*(3), 327–339. doi:10.100710745-021-00229-w

Seavey, J. R., Gilmer, B., & McGarigal, K. M. (2011). Effect of sea-level rise on piping plover (Charadrius melodus) breeding habitat. *Biological Conservation*, *144*(1), 393–401. doi:10.1016/j.biocon.2010.09.017

Sehgal, J., Mandal, D. K., Mandal, C., & Vadivelu, S. (1992). Agro-Ecological Regions of India. *Tech. Bull. No. 24.*

Şengör, A. M. C., Özeren, M. S., Keskin, M., Sakınç, M., Özbakır, A. D., & Kayan, İ. (2008). Eastern Turkish high plateau as a small Turkic-type orogen: Implications for post-collisional crust-forming processes in Turkic-type orogens. *Earth-Science Reviews, 90*(1–2), 1–48. doi:10.1016/j.earscirev.2008.05.002

Şenkul, Ç., & Kaya, S. (2017). Türkiye endemik bitkilerinin coğrafi dağılışları. *Türkiye Coğrafya Dergisi, 69*, 109–120. doi:10.17211/tcd.322515

Senner, N. R. (2012). One species but two patterns: Populations of the Hudsonian Godwit (Limosa haemastica) differ in spring migration timing. *The Auk, 129*(4), 670–682. doi:10.1525/auk.2012.12029

Şenol, Ö., Akyildirim Beğen, H., Görür, G., & Demirtaş, E. (2015). New additions and invasive aphids for Turkey's aphidofauna (Hemiptera: Aphidoidea). *Turkish Journal of Zoology, 39*, 39–45. doi:10.3906/zoo-1308-12

Şenol, Ö., Görür, G., & Beğen, H. A. (2017). Recent findings on aphid fauna from east and south eastern parts of Turkey. *Biological Diversity and Conservation, 10*(3), 76–78.

Seven, S. (2014). A new species of blue from Turkey, "*Neolycaena soezen*" Seven, sp. n. (Lepidoptera: Lycaenidae). *SHILAP Revista de Lepidopterologia, 42*(166), 311–317.

Sevgili, H. (2004). *A revision of Turkish species of Isophya Brunner von Wattenwyl (Orthoptera: Tettigoniidae: Phaneropterinae)*, 397–397.

Sevgili, H. (2016). Spermatophore weight and sperm number in the southernmost species of the genus *Poecilimon* Fischer, 1853 (Orthoptera: Phaneropterinae). *Acta Biologica Turcica, 29*(3), 83–88.

Sevgili, H., Caglar, S. S., & Saglam, I. K. (2010). Re-evaluation of the genus *Phonochorion* (Orthoptera: Tettigoniidae: Phaneropterinae). *European Journal of Entomology, 107*(4), 631–645. doi:10.14411/eje.2010.072

Sevgili, H., & Çıplak, B. (2000). The Orthoptera of Şanliurfa province from the Mesopotamian part of Turkey. *The Italian Journal of Zoology, 67*(2), 229–240. doi:10.1080/11250000009356316

Sevgili, H., Çiplak, B., Heller, K. G., & Demirsoy, A. (2006). Morphology, bioacoustics and phylogeography of the *Isophya major* group (Orthoptera: Tettigoniidae: Phaneropterinae): A species complex occurring in Anatolia and Cyprus. *European Journal of Entomology, 103*(3), 657–671. doi:10.14411/eje.2006.086

Sevgili, H., Heller, K.-G., & Reinhold, K. (2008). A re-assessment of the Poecilimon syriacus group (Orthoptera Tettigonioidea, Phaneropteridae) based on bioacoustics, morphology and molecular data. *Insect Systematics & Evolution, 39*(4), 361–379. doi:10.1163/187631208788784309

Sevgili, H., Önal, H., & Yiğit, A. (2015). Mating behavior and spermatophore characteristics in two closely related bushcricket species of the genus *Phonochorion* (Orthoptera: Phaneropterinae). *Journal of Insect Behavior*, *28*(4), 369–386. doi:10.100710905-015-9509-1

Sevgili, H., Şirin, D., Heller, K.-G., & Lemonnier-Darcemont, M. (2018). Review of the *Poecilimon (Poecilimon) zonatus* species group and description of new species from Turkey with data on bioacoustics and morphology (Orthoptera: Phaneropterinae). *Zootaxa*, *4417*(1), 1–62. doi:10.11646/zootaxa.4417.1.1 PMID:30313607

Sevgili, H., & Yılmaz, K. (2022). Contributions of citizen scientists to monitoring alien species: the case study on Giant Asian Mantes, *Hierodula tenuidentata* and *H. patellifera* (Mantodea: Mantidae). *Zoology in the Middle East*, *68*(4), 1–9. doi:10.1080/09397140.2022.2145802

Shahidi, F., & Fereidoon Naczk. (1992). *Ecological Effects of Wastewater*. CRC Press.

Shrivastava, R. N., Nema, R. K., & Awasthi, M. K. (2018). Strategy of Water Resource Utilization in Upper Narmada Basin. *International Journal of Current Microbiology and Applied Sciences*, *7*(05), 591–602. doi:10.20546/ijcmas.2018.705.073

Shuter, B. J., & Meisner, J. D. (1992). Tools for assessing the impact of climate change on freshwater fish populations. *GeoJournal*, *28*(1), 7–20. doi:10.1007/BF00216402

Sigee, D. C. (2005). *Freshwater Microbiology*. John Wiley & Sons.

Signorelli, A. (1999). Antropología Urbana. Barcelona: Anthropos: UAM, 1999.

Sileshi, G., & Mafongoya, P. L. (2006). The short-term impact of forest fire on soil invertebrates in the Miombo. *Biodiversity and Conservation*, *15*(10), 3153–3160. doi:10.100710531-005-5411-z

Simmons, R. E., Kolberg, H., Braby, R., & Erni, B. (2015). Declines in migrant shorebird populations from a winter-quarter perspective. *Conservation Biology*, *29*(3), 877–887. doi:10.1111/cobi.12493 PMID:25858334

Sipahiler, F. (1996). Studies on the Trichoptera fauna of southern Anatolia. *Part 1. Biological Sciences and Living Resources*, *16*(16), 293–312.

Sipahiler, F. (2005). In K. Tanida (Ed.), *A checklist of the caddisflies of Turkey (Trichoptera)* (pp. 393–405).

Sipahiler, F. (2008). Zoogeographical characteristics of the Trichoptera fauna of Turkey. *Ferrantia*, *55*, 93–109.

Şirin, D., Taylan, M. S., Sevgili, H., & Mol, A. (2019). Bioacoustics review of Anatolian species of the predatory bush-cricket genus *Saga* (Orthoptera: Tettigoniidae: Saginae) with the description of a new species. *Zootaxa*, *4664*(1), zootaxa.4664.1.3. doi:10.11646/zootaxa.4664.1.3

Şirin, D., Sevgili, H., & Taylan, M. S. (2018). *Distribution of endemic Ensifera species in Anatolia*. II. *European Congress on Orthoptera Conservation*.

Sivaramakrishna, P., & Yugandhar, P. (2020). A new species of the genus Portulaca L. (Portulacaceae) from the Eastern Ghats, India. *Journal of Asia-Pacific Biodiversity*, *13*(4), 755–761. doi:10.1016/j.japb.2020.08.010

Skuhravá, M. (2016). Checklist of gall midges (Diptera: Cecidomyiidae) of Turkey. *Acta Societatis Zoologicae Bohemicae*, *80*(2–3), 197–199.

Smith, C. I., & Farrell, B. D. (2005). Phylogeography of the longhorn cactus beetle Moneilema appressum LeConte (Coleoptera: Cerambycidae): was the differentiation of the Madrean sky islands driven by Pleistocene climate changes? *Molecular Ecology*, *14*(10), 3049–3065. doi:10.1111/j.1365-294X.2005.02647.x PMID:16101773

Snyder, W. E., & Evans, E. W. (2006). Ecological Effects of Invasive Arthropod Generalist Predators. *Annual Review of Ecology, Evolution, and Systematics*, *37*(1), 95–122. doi:10.1146/annurev.ecolsys.37.091305.110107

Soria, M., Leigh, C., Datry, T., Bini, L. M., & Bonada, N. (2017). Biodiversity in perennial and intermittent rivers: A meta-analysis. *Oikos*, *126*(8), 1078–1089. doi:10.1111/oik.04118

Soyhan, T., Başer, S., & Nazari, V. (2013). First record of *Cacyreus marshalli* Butler, 1898 (Lycaenidae) from Turkey. *Nota Lepidopterologica*, *36*(2), 189–190.

Stewart, J. R., Lister, A. M., Barnes, I., & Dalén, L. (2010). Refugia revisited: individualistic responses of species in space and time. *Proceedings. Biological Sciences / the Royal Society*, *277*(1682), 661–671. doi:10.1098/rspb.2009.1272

Stoner, A. M. (2021). Things are getting worse on our way to catastrophe: Neoliberal environmentalism, repressive desublimation, and the autonomous ecoconsumer. *Critical Sociology*, *47*(May), 491–506. doi:10.1177/0896920520958099

Stork, N. E. (2018). How many species of insects and other terrestrial arthropods are there on earth? *Annual Review of Entomology*, *63*(1), 31–45. doi:10.1146/annurev-ento-020117-043348 PMID:28938083

Strang, V. (2017). The Gaia complex: Ethical Challenges to an Anthropocentric 'Common Future. In B. Marc & L. Jerome (Eds.), *The Anthropology of Sustainability* (pp. 255–283). Palgrave. doi:10.1057/978-1-137-56636-2_12

Stroud, D. A., Baker, A., Blanco, D. E., Davidson, N. C., Ganter, B., Gill, R. E., Jr., Gonzalez, P. M., Haanstra, L., Morrison, R. I. G., Piersma, T., Scott, D., & Zöckler, C. (2006). The conservation and population status of the world's waders at the turn of the millennium. In *The Waterbirds Around the World Conference* (pp. 643-648).

Studds, C. E., Kendall, B. E., Murray, N. J., Wilson, H. B., Rogers, D. I., Clemens, R. S., Gosbell, K., Hassell, C. J., Jessop, R., Melville, D. S., Milton, D. A., Minton, C. D. T., Possingham, H. P., Riegen, A. C., Straw, P., Woehler, E. J., & Fuller, R. A. (2017). Rapid population decline in migratory shorebirds relying on Yellow Sea tidal mudflats as stopover sites. *Nature Communications*, *8*(1), 14895. doi:10.1038/ncomms14895 PMID:28406155

Stutchbury, B. J., Tarof, S. A., Done, T., Gow, E., Kramer, P. M., Tautin, J., Fox, J. W., & Afanasyev, V. (2009). Tracking long-distance songbird migration by using geolocators. *Science, 323*(5916), 896–896. doi:10.1126cience.1166664 PMID:19213909

Sundbo, J. (1998). *The Theory of Innovation: entrepreneurs, technology, and strategy*. Edward Elgar.

Sutherland, W. J., Alves, J. A., Amano, T., Chang, C. H., Davidson, N. C., Max Finlayson, C., Gill, J. A., Gill, R. E. Jr, González, P. M., Gunnarsson, T. G., Kleijn, D., Spray, C. J., Székely, T., & Thompson, D. B. (2012). A horizon scanning assessment of current and potential future threats to migratory shorebirds. *The Ibis, 154*(4), 663–679. doi:10.1111/j.1474-919X.2012.01261.x

Swallow, B. M., Garrity, D. P., & van Noordwijk, M. (2001). *The Effects of Scales,Flows and Filters on Property Rights and Collective Action in Watershed Management*. International Food Policy Research Institute.

Syroechkovski, E. E., Tomkovich, P. S., Kashiwagi, M., Taldenkov, I. A., Buzin, V. A., Lappo, E. G., & Zöckler, C. (2010). Population decline in the spoon-billed sandpiper (Eurynorhynchus pygmeus) in northern Chukotka based on monitoring on breeding grounds. *Biology Bulletin of the Russian Academy of Sciences, 37*(9), 941–951. doi:10.1134/S1062359010090074

Taagepera, R. (2008). *Making social sciences more scientific: The need for predictive models*. Oxford University Press. doi:10.1093/acprof:oso/9780199534661.001.0001

Tainter, J. A. (1988). *The Collapse of Complex Societies*. Cambridge University Press.

Tainter, J. A. (2008). Collapse and Sustainability: Rome, the Maya, and the Modern World. *Archeological Papers of the American Anthropological Association, 24*(1), 201–214. doi:10.1111/apaa.12038

Tarantino-Curseri, S. (2018). Antropología, etología, sociología, y psicología, 4 disciplinas útiles en el ámbito empresarial. *Management Review, 3*(3). doi:10.18583/umr.v3i3.126

Taşdemir, A., & Ustaoğlu, M. R. (2005). Göller Bölgesi İçsularının Chironomidae ve Chaoboridae (Diptera) Faunasının Taksonomik Yönden İncelenmesi. *Su Ürünleri Dergisi, 22*(3), 377–384.

Tavşanoğlu, Ç., & Serter Çatav, Ş. (2012). Seed size explains within-population variability in post-fire germination of *Cistus salviifolius*. *Annales Botanici Fennici, 49*(5–6), 331–340. doi:10.5735/085.049.0604

Taylan, Mehmet Sait, Russo, C. D., Rampini, M., & Ketmaier, V. (2013). Molecular systematics of the genus *Troglophilus* (Rhaphidophoridae, Orthoptera) in Turkey: mitochondrial 16S rDNA evidences. *ZooKeys257*, 33–46. doi:10.3897/zookeys.257.4133

Taylan, M. S., di Russo, C., & Cobolli, M. (2011). The Dolichopodainae and Troglophilinae cave crickets of Turkey: An update of taxonomy and geographic distribution (Orthoptera, Rhaphidophoridae). *Zootaxa, 2829*(1), 59. doi:10.11646/zootaxa.2829.1.3

Taylan, M. S., Russo, C. D., Cobolli, M., & Rampini, M. (2012). New species of the genus *Troglophilus* Krauss, 1879 (Orthoptera: Rhaphidophoridae) from Western and Southern Anatolian caves, Turkey. *Zootaxa, 3597*(1), 33. doi:10.11646/zootaxa.3597.1.4

Taylan, M. S., & Şirin, D. (2016). Speciation of the genus *Dolichopoda* in Anatolia with reference to the role of ancient central lake system. *Insect Systematics & Evolution, 47*(3), 267–283. doi:10.1163/1876312X-47032143

Taylan, Mehmet Sait, Mol, A., & Şirin, D. (2015). Description of a new species of the genus *Troglophilus* krauss, 1879 (Orthoptera: Rhaphidophoridae) from Northern Anatolia, Turkey. *Journal of Entomological Research Society, 17*(2), 51–57.

Tekinoğlu, V., & Çalıskan, H. (2019). The Crane Flies (Tipulidae: Diptera) fauna of the Uludağ Mountain (Turkey). *Biological Diversity and Conservation, 12*(2), 57–64. doi:10.5505/biodicon.2019.32042

Tezcan, S. (2020). Analysis of the insect fauna of Turkey and suggestions for future studies. *Munis Entomology & Zoology Journal, 15*(2), 690–710.

The IUCN Red List of Threatened Species. (n.d.). *IUCN Red List of Threatened Species*. IUCN. https://www.iucnredlist.org/resources/molur2011

Thibault, J., & Levisen, M. (2013). Red Knot prey availability project report. *Final Report*. South Carolina Department of Natural Resources, Marine Resources Research Institute, Charleston, SC, USA.

Thomas, P., Boyer, D., Oehler, D., Silver, S., & Perrotti, L. (2019). Headstarting as a conservation strategy for threatened and endangered species. *Scientific Foundations of Zoos and Aqimriums: Their Role in Conservation and Research*, 91-111.

Tilman, D., Clark, M., Williams, D. R., Kimmel, K., Polasky, S., & Packer, C. (2017). Future threats to biodiversity and pathways to their prevention. *Nature, 546*(7656), 73–81. doi:10.1038/nature22900 PMID:28569796

Tomkovich, P. S., Syroechkovski Jr, E. E., Lappo, E. G., & Zöckler, C. (2002). First indications of a sharp population decline in the globally threatened Spoon-billed Sandpiper Eurynorhynchus pygmeus. *Bird Conservation International, 12*(1), 1-18.

Tomkovich, P., Syroechkovskiy, E., Yakushev, N., Loktionov, E., Digby, R., & Shepelev, I. (2016). Spoon-billed Sandpipers in Meinypilgyno: Update of the 2016 breeding season. *Spoon-Bill. Sandpipers Task Force News Bull, 16*, 13–15.

Tougeron, K., Brodeur, J., Le Lann, C., & Baaren, J. (2020). How climate change affects the seasonal ecology of insect parasitoids. *Ecological Entomology, 45*(2), 167–181. doi:10.1111/een.12792

Touhami, F., Idrissi, H. R., & Benhoussa, A. (2020). Foraging behaviour of wintering shorebirds at Merja Zerga lagoon (Atlantic coast, Morocco). *The Ostrich, 91*(3), 244–251. doi:10.2989/00306525.2020.1774439

Toussaint, E. F. A., Sagata, K., Surbakti, S., Hendrich, L., & Balke, M. (2013). Australasian sky islands act as a diversity pump facilitating peripheral speciation and complex reversal from narrow endemic to widespread ecological supertramp. *Ecology and Evolution*, *3*(4), 1031–1049. doi:10.1002/ece3.517 PMID:23610642

Trakel Web Page. (2021). *TRAKEL kelebek fotoğrafları, kelebek türleri, kelebekler hakkında detaylı Türkçe bilgi*. Trakel.

Trawick, P., & Alf, H. (2015). Revisiting the Image of Limited Good: On Sustainability, Thermodynamics, and the Illusion of Creating Wealth. *Current Anthropology*, *56*(1), 1–27. doi:10.1086/679593

Trist, E. L. (1976). *A Concept of Organizational Ecology: an invited address to the three Melbourne universities.*

Trujillo, M. L. N. (2021). Notes for a critical and ecological view of patriarchal capitalism in the web of life. *Capital and Class*, *45*(March), 21–32. doi:10.1177/0309816820929115

Tschopp, A., Riedel, M., Kropf, C., Nentwig, W., & Klopfstein, S. (2013). The evolution of host associations in the parasitic wasp genus *Ichneumon* (Hymenoptera: Ichneumonidae): convergent adaptations to host pupation sites. *BMC Evolutionary Biology*, *13*(1), 74. doi:10.1186/1471-2148-13-74 PMID:23537515

Tsing, A. L. (2015). *The mushroom at the end of the world*. Princeton University Press.

Tuncer, C., Özdemir, İ. O., & Kushiyev, R. (2020). Türkiye fındık bahçelerinde yeni zararlı türler: *Xylosandrus germanus* Blandford (Col.: Curculionidae: Scolytınae), *Metcalfa pruinosa* Say (Hem.: Flatidae), *Croesus septentrionalis* Linnaeus (Hym.: Tenthredinidae) ve *Anoplophora chinensis* Forster (Col.: Cerambycidae). *Black Sea Journal of Agriculture*, *3*(1), 74–81.

Tunç, İ., & Hastenpflug-Vesmanis, A. (2016). Records and checklist of Thysanoptera in Turkey. *Turkish Journal of Zoology*, *40*, 769–778. doi:10.3906/zoo-1512-37

Tunçyürek-Soydanbay, M. (1978). Türkiye'de bitki zararlısı bazı böceklerin doğal düşman listesi, Kısım II. *Turkiye Entomoloji Dergisi*, *2*(2), 61–92.

Tundisi, J. G., & Tundisi, T. M. (2016). *Limnologia*. Oficina de Textos.

Uitto, J. (2001). World Forests, Markets and Policies. VL - 3 Publisher: Kluwer Academic Publishers pp.65-76. DO - doi:10.1007/978-94-010-0664-4_3

Ülgentürk, S. (2016). A new introduced wax scale to Turkey: *Ceroplastes ceriferus* (Fabricius) (Hemiptera: Coccidae). *Mun. Ent. Zool.*, *11*(2), 359–362.

Uma, R., & Sevgili, H. (2015). Spermatophore allocation strategy over successive matings in the bushcricket *Isophya sikorai* (Orthoptera, Phaneropterinae). *Ethology Ecology and Evolution*, *27*(2), 129–147. doi:10.1080/03949370.2014.896830

Ünal, M. (2016). Pamphagidae (Orthoptera: Acridoidea) from the Palaearctic Region: taxonomy, classification, keys to genera and a review of the tribe Nocarodeini I.Bolívar. *Zootaxa, 4206*(1), zootaxa.4206.1.1. doi:10.11646/zootaxa.4206.1.1

Ünal, M. (2021). *Turkish Orthoptera Site.*

Underwood, E. C., & Quinn, J. F. (2010). Response of ants and spiders to prescribed fire in oak woodlands of California. *Journal of Insect Conservation, 14*(4), 359–366. doi:10.100710841-010-9265-7

US EPA. (n.d.). *Ecoregions.* EPA. https://www.epa.gov/eco-research/ecoregions

USGCRP. (2014). Water Resources. Climate Change Impacts in the United States: The Third National Climate Assessment. U.S. Global Change Research Program, 69-112.

Uslu, E., & Bakış, Y. (2012). Geographic distribution of Turkish Oaks. *Dendrobiology (Poznan), 67*, 41–48.

Uzun, I., & Bayır, A. (2009). Horticultural Biodiversity in Turkey. *Bulletin UASVM Horticulture, 66*, 536–543.

van Heezik, Y., Maloney, R. F., & Seddon, P. J. (2009). Movements of translocated captive-bred and released Critically Endangered kaki (black stilts) Himantopus novaezelandiae and the value of long-term post-release monitoring. *Oryx, 43*(4), 639–647. doi:10.1017/S0030605309990081

Velho, G. (1978). Observando o familiar. In: Nunez, E. de O. (Org.). An aventura sociológica: objetividade, paixao, improviso e método na pesquisa social. Rio de Janeiro: Zahar.

Verves, Y., Barták, M., & Kubík, Š. (2018). Checklist of flesh flies of Turkey (Diptera, Sarcophagidae). *ZooKeys, 743*, 95–136. doi:10.3897/zookeys.743.22022

Verves, Y., Barták, M., Kubík, Š., & Civelek, H. S. (2017). New records of Sarcophagidae from Turkey (Diptera). *ZooKeys, 703*, 129–158. doi:10.3897/zookeys.703.12377

Vijayakumar, S. P., Menezes, R. C., Jayarajan, A., & Shanker, K. (2016). Glaciations, gradients, and geography: multiple drivers of diversification of bush frogs in the Western Ghats Escarpment. *Proceedings of the Royal Society B: Biological Sciences, 283*(1836), 20161011. 10.1098/rspb.2016.1011

Vorosmarty, C. J. (2002). Global change, the water cycle, and our search for Mauna Loa. *Hydrological Processes, 16*(1), 1335–1339. doi:10.1002/hyp.527

Vorosmarty, C. J., Green, P., Salisbury, J., & Lammers, R. (2000). Global water resources: Vulnerability from climate change and population growth. *Science, 289*(5477), 284–288. doi:10.1126cience.289.5477.284 PMID:10894773

Vorosmarty, C. J., & Sahagian, D. (2000). Anthropogenic disturbance of the terrestrial water cycle. *Bioscience, 50*(9), 753–765. doi:10.1641/0006-3568(2000)050[0753:ADOTTW]2.0.CO;2

W. S. (2007). The Anthropocene: Are Human Beings Now Overwhelming the Forces of Nature? *Ambio, 36*, 614–621. doi:10.1579/0044-7447(2007)36[614:TAAHNO]2.0.CO;2 PMID:18240674

Wagener, S. (2006). Butterfly diversity and protection in Turkey. *Bonner Zoologische Beitrage, 54*(1), 3–23.

Wainright, C. A., Muhlfeld, C. C., Elser, J. J., Bourret, S. L., & Devlin, S. P. (2021). Species invasion progressively disrupts the trophic structure of native food webs. *Proceedings of the National Academy of Sciences of the United States of America, 118*(45), e2102179118. doi:10.1073/pnas.2102179118 PMID:34725150

Wauchope, H. S., Shaw, J. D., Varpe, Ø., Lappo, E. G., Boertmann, D., Lanctot, R. B., & Fuller, R. A. (2017). Rapid climate-driven loss of breeding habitat for Arctic migratory birds. *Global Change Biology, 23*(3), 1085–1094. doi:10.1111/gcb.13404 PMID:27362976

Webber, S. (2019). Putting climate services in contexts: Advancing multi-disciplinary understandings: Introduction to the special issue. *Climatic Change, 157*(1), 1–8. doi:10.100710584-019-02600-9

Weiser, E. L., Brown, S. C., Lanctot, R. B., Gates, H. R., Abraham, K. F., Bentzen, R. L., Bêty, J., Boldenow, M. L., Brook, R. W., Donnelly, T. F., English, W. B., Flemming, S. A., Franks, S. E., Gilchrist, H. G., Giroux, M.-A., Johnson, A., Kendall, S., Kennedy, L. V., Koloski, L., & Sandercock, B. K. (2018a). Effects of environmental conditions on reproductive effort and nest success of Arctic-breeding shorebirds. *The Ibis, 160*(3), 608–623. doi:10.1111/ibi.12571

Weiser, E. L., Lanctot, R. B., Brown, S. C., Alves, J. A., Battley, P. F., Bentzen, R., Bêty, J., Bishop, M. A., Boldenow, M., Bollache, L., Casler, B., Christie, M., Coleman, J. T., Conklin, J. R., English, W. B., Gates, H. R., Gilg, O., Giroux, M.-A., Gosbell, K., & Sandercock, B. K. (2016). Effects of geolocators on hatching success, return rates, breeding movements, and change in body mass in 16 species of Arctic-breeding shorebirds. *Movement Ecology, 4*(1), 1–19. doi:10.118640462-016-0077-6 PMID:27134752

Weiser, E. L., Lanctot, R. B., Brown, S. C., Gates, H. R., Bentzen, R. L., Bêty, J., Boldenow, M. L., English, W. B., Franks, S. E., Koloski, L., Kwon, E., Lamarre, J.-F., Lank, D. B., Liebezeit, J. R., McKinnon, L., Nol, E., Rausch, J., Saalfeld, S. T., Senner, N. R., ... Sandercock, B. K. (2018b). Environmental and ecological conditions at Arctic breeding sites have limited effects on true survival rates of adult shorebirds. *The Auk, 135*(1), 29–43. doi:10.1642/AUK-17-107.1

Wei, X., Lo, C. K. Y., Jung, S., & Choi, T.-M. (2021). From co-consumption to co-production: A systematic review and research synthesis of collaborative consumption practices. *Journal of Business Research, 129*(May), 282–294. doi:10.1016/j.jbusres.2021.02.027

Welch, E. B., & Lindell, T. (1992). *Ecological effects of wastewater: applied limnology and pollution effects*. E & Fn Spon. doi:10.4324/9780203038499

Westphal, M. I., Browne, M., MacKinnon, K., & Noble, I. (2008). The link between international trade and the global distribution of invasive alien species. *Biological Invasions, 10*(4), 391–398. doi:10.100710530-007-9138-5

Whiting, P. J. (2002). Streamflow necessary for environmental maintenance. *Annual Review of Earth and Planetary Sciences*, *30*(1), 181–206. doi:10.1146/annurev.earth.30.083001.161748

Wiemers, M., Balletto, E., Dincă, V., Faltynek Fric, Z., Gerardo Lamas, Lukhtanov, V., Munguira, M. L., Chris A M van Swaay, Vila, R., Vliegenthart, A., Wahlberg, N., & Verovnik, R. (2018). An updated checklist of the European Butterflies (Lepidoptera, Papilionoidea). *ZooKeys811*, 9–45. doi:10.3897/zookeys.811.28712

Wietzke, A., Westphal, C., Gras, P., Kraft, M., Pfohl, K., Karlovsky, P., Pawelzik, E., Tscharntke, T., & Smit, I. (2018). Insect pollination as a key factor for strawberry physiology and marketable fruit quality. *Agriculture, Ecosystems & Environment*, *258*, 197–204. doi:10.1016/j.agee.2018.01.036

Wilk, R. R., & Wilhite, H. L. (1985). Why don't people weatherize their homes? An ethnographic solution. *Energy*, *10*(May), 621–629. doi:10.1016/0360-5442(85)90093-3

Wilson, R. J., & Maclean, I. M. D. (2011). Recent evidence for the climate change threat to Lepidoptera and other insects. *Journal of Insect Conservation*, *15*(1–2), 259–268. doi:10.100710841-010-9342-y

Winn, M. I., & Pogutz, S. (2013). Business, ecosystems, and biodiversity. *Organization & Environment*, *26*(2), 203–229. doi:10.1177/1086026613490173

Wolf, M. (2016). Rethinking urban epidemiology: Natures, networks and materialities. *International Journal of Urban and Regional Research*, *40*(5), 958–982. doi:10.1111/1468-2427.12381 PMID:32336869

WRI (World Resources Institute), UNEP, UNDP, & World Bank. (1998). World Resources 1998–99: Environmental Change and Human Health. Oxford University Press, New York, NY.

Yalçın, M., Akçay, Ç., Taşçıoğlu, C., & Yüksel, B. (2019). Insect species damaging industrial wood in western black sea region of turkey. *Drvna Industrija*, *70*(2), 183–191. doi:10.5552/drvind.2019.1818

Yamaç, A., Gilli, E., Tok, E., & Törk, K. (2021). *History, Archaeology and Caves* (Vol. 1). Natural sciences in the caves of turkey. I: Caves and Karst of Turkey. Springer International Publishing., doi:10.1007/978-3-030-65501-3_3

Yang, Y., Lili, W., & Hongchi, Z. (2022). *Bibliometric research on the evolution of resilience theme from the perspective of Geographical Science*. 2022 29th International Conference on Geoinformatics, Beijing, China. 10.1109/Geoinformatics57846.2022.9963870

Yilmaz, K., & Sevgili, H. (2023). The genus *Holaptilon* Beier, 1964: Discussion on the poorly known boxer mantis genus, with a new species (Mantodea, Gonypetidae). *Zootaxa*, *5231*(4), 427–444. doi:10.11646/zootaxa.5231.4.5 PMID:37045135

Yu, D. S., van Achterberg, C., & Horstmann, K. (2016). *Taxapad 2016, Ichneumonoidea 2015.*

Zachos, J., Pagani, M., Sloan, L., Thomas, E., & Billups, K. (2001). Trends, rhythms, and aberrations in global climate 65 Ma to present. *Science*, *292*(5517), 686–693. doi:10.1126cience.1059412 PMID:11326091

Zaitsev, A. S., Gongalsky, K. B., Malmström, A., Persson, T., & Bengtsson, J. (2016). Why are forest fires generally neglected in soil fauna research? A mini-review. *Applied Soil Ecology*, *98*, 261–271. doi:10.1016/j.apsoil.2015.10.012

Zöckler, C., Beresford, A. E., Bunting, G., Chowdhury, S. U., Clark, N. A., Fu, V. W. K., Htin Hla, T., Morozov, V., Syroechkovskiy, E., Kashiwagi, M., Lappo, E. G., Tong, M., Long, T. L., Yu, Y.-T., Huettmann, F., Akasofu, H. K., Tomida, H., & Buchanan, G. M. (2016). The winter distribution of the Spoon-billed Sandpiper Calidris pygmaeus. *Bird Conservation International*, *26*(4), 476–489. doi:10.1017/S0959270915000295

Zöckler, C., Chowdhury, S. U., Sun, L., Qing, C., Aung, P. P., Clements, C., ... Syroechkovskiy, E. E. (2020). The Spoon-billed Sandpiper Calidris pygmaea conservation project in 2019 and 2020: Population trends continue to be negative. *BirdingASIA*, *33*, 51–56.

Zöckler, C., Syroechkovskiy, E. E., & Atkinson, P. W. (2010). Rapid and continued population decline in the Spoon-billed Sandpiper Eurynorhynchus pygmeus indicates imminent extinction unless conservation action is taken. *Bird Conservation International*, *20*(2), 95–111. doi:10.1017/S0959270910000316

About the Contributors

Kavita Chahal is an Assistant Professor in the Department of Botany, Government College Bichhua, Chhindwara, M.P. She obtained M.Sc & Ph.D in Microbiology; and has cleared CSIR-NET in Life Sciences and ASRB-NET in Agricultural Microbiology. She is an Editorial Board Member of American Journal of Life Sciences; Editorial Board Member of International Journal of Current Science Research & Review; Active RMS member (ID: 116866) of International Journal of Creative Research Thoughts; and Member of Indian Academy of Sciences, Delhi. She has to her credit 27 Research papers, 9 Book chapters, 6 Books as an Editor, and 3 books as co-author.

Pravinsang Dodia is an Associate Professor of Zoology at Maharaja Krishnakumarsinhji Bhavnagar University. He has been in the field of research from past 3 decades. His research areas includes Ornithology, Entomology and Wildlife Conservation.

Radha Srinivasan Iyer is a Physiotherapist with Masters in Neurosciences from the prestigious D.Y Patil University, Pimpri, Pune. She works in a residential school for special children, with neurological problems. She is actively involved in academia; she has authored several books and research papers.

Vishal Makwana is currently pursuing his doctoral research in Avian Ecology at Maharaja Krishnakumarsinhji Bhavnagar University. His research areas includes Ornithology, Biodiversity monitoring and Conservation.

Santhosh Kumar Rajamani is a surgeon by profession and a naturalist by passion. He is enthusiastic about diverse subjects like artificial intelligence, computer programming, network sciences, applied mathematics, conservation, wetland ecology, and aquatic botany. He has authored research papers and book chapters on all these subjects. He is a supporter of the conservation of aquatic ecosystems and has written numerous petitions and letters to the Indian government to step in and stop habitat pollution and the destruction of aquatic ecosystems in India. This chapter represents his ongoing efforts to save freshwater bodies, conserve biodiversity, and raise public awareness about this burning problem.

Omar Vargas-González is Professor and Head of Systems and Computing Department at Tecnologico Nacional de Mexico Campus Ciudad Guzman, professor at Telematic Engineering at Centro Universitario del Sur Universidad de Guadalajara with a master degree in Computer Systems. Has been trained in Innovation and Multidisciplinary Entrepreneurship at Arizona State University (2018) and a Generation of Ecosystems of Innovation, Entrepreneurship and Sustainability for Jalisco course by Harvard University T.H. Chan School of Health. At present conduct research on diverse fields such as Entrepreneurship, Economy, Statistics, Mathematics and Information and Computer Sciences. Has colaborated in the publication of over 20 scientific articles and conducted diverse Innovation and Technological Development projects

Index

G

H

I

L

M

P

R

S

T

U

V

W

www.ingramcontent.com/pod-product-compliance
Lightning Source LLC
LaVergne TN
LVHW061541290926
849162LV00004B/865